Particulate Discrete Modelling

T0227698

This is the first dedicated work on the use of particulate DEM in geomechanics and provides key information needed for engineers and scientists who want to start using this powerful numerical modelling approach. The book is a concise point of reference for users of DEM, allowing them to maximize the insight they can gain their material response using DEM covering:

- The background theory

- Details of the numerical method

- Advice on running simulations

- Approaches for interpreting results of simulations

- Issues related to available particle types, contact modelling and boundary conditions.

Particulate Discrete Element Modelling is suitable both for first time DEM analysts as well as more experienced users. It will be of use to professionals, researchers and higher level students, as it presents a theoretical overview of DEM as well as practical guidance on how to set up and run DEM simulations and how to interpret DEM simulation results.

Catherine O'Sullivan is a Senior Lecturer in the Department of Civil and Environmental Engineering at Imperial College, UK. She obtained her undergraduate and master's degrees at University College Cork, Ireland. Dr. O'Sullivan's interest in DEM was sparked during her doctoral studies in Civil Engineering at the University of California at Berkeley, USA. Following graduation from UC Berkeley in 2002, she spent two years working as a lecturer at University College Dublin, prior to moving to Imperial College in 2004.

Applied Geotechnics

Titles currently in this series:

Geotechnical Modelling
David Muir Wood
Hardback ISBN 978-0-415-34304-6
Paperback ISBN 978-0-419-23730-3

Sprayed Concrete Lined Tunnels
Alun Thomas
Hardback ISBN 978-0-415-36864-3

Introduction to Tunnel Construction
David Chapman et al.
Hardback ISBN 978-0-415-46841-1
Paperback ISBN 978-0-415-46842-8

Forthcoming:

Practical Engineering Geology
Steve Hencher
Hardback ISBN 978-0-415-46908-1
Paperback ISBN 978-0-415-46909-8

Landfill Engineering
Geoff Card
Hardback ISBN 978-0-415-37006-6

Advanced Soil Mechanics Laboratory Testing
Richard Jardine et al.
Hardback ISBN 978-0-415-46483-3

Particulate Discrete Element Modelling
a Geomechanics Perspective

Catherine O'Sullivan

Routledge
Taylor & Francis Group

LONDON AND NEW YORK

First published 2011 by Spon Press

2 Park Square, Milton Park, Abingdon, Oxon OX14 4RN
711 Third Avenue, New York, NY 10017, USA

Routledge is an imprint of the Taylor & Francis Group, an informa business

First issued in paperback 2017

British Library Cataloguing in Publication Data
A catalogue record for this book is available from the British Library
Library of Congress Cataloging-in-Publication Data
A catalog record has been requested for this book
ISBN13: 978-0-415-49036-8 (hbk)
ISBN13: 978-1-138-07489-7 (pbk)

Acknowledgements

My objective in preparing this text is to introduce potential users to DEM, and to point them in the correct direction by collating in a single volume what I believe to be the important basic background information needed to develop the understanding needed to successfully complete DEM analyses and interpret the results from the analyses. While I have been carrying out research in this area for about the past decade, I am very aware that there are many other researchers who have more experience than me. However, following my own experience, I know it can be difficult and time consuming to identify the key elements of information necessary to have a handle on this field. I hope that this book will fill the current gap caused by the absence of a introductory text and so smooth the way for future DEM analysts.

My own knowledge of DEM has evolved over the past decade and many of my ideas have developed through interactions with colleagues and other researchers. My initial research at UC Berkeley was completed under the supervision of Prof. Jonathan Bray and Prof. Michael Riemer, and I also gained much from my interactions with other faculty and students at Berkeley, most notably Dr. David Doolin and Prof. Nick Sitar. All of my colleagues at Imperial College have been very supportive over the past six years. In particular my discussions with Prof. Matthew Coop have advanced my understanding of soil mechanics and helped form many of the ideas presented in this text and conversations with Dr. Berend van Wachem have advanced my understanding of DEM. Outside of my own institutions I would like to acknowledge Prof. Malcolm Bolton, Dr. Colin Thornton, Dr. Dave Potyondy and Prof. Stefan Luding who have been willing to engage in discussions on DEM and granular materials and who have been particularly generous in sharing their knowledge, opinions and ideas. I am lucky to have had the opportunity to work with a number of talented Ph.D. and Master's students over the past 8 years and I have learned a lot through discussions with these students. I would particularly like to thank my family who have been very encouraging of my academic career and who helped in the proof reading of this text, I could not have done this without their support.

Contents

Chapter 1

Introduction

1.1 Overview

Particulate DEM in geomechanics

Discrete element modelling (DEM) is a numerical modelling or computer simulation approach that can simulate soil and other granular materials. The unique feature of this approach is that it explicitly considers the individual particles in a granular material and their interactions. DEM presents an alternative to the typical approach adopted when simulating the mechanical behaviour of granular materials (soils in particular), which uses a continuum mechanics framework. In a continuum model soil is assumed to behave as a continuous material and the relative movements and rotations of the particles inside the material are not considered. Sophisticated constitutive models (i.e. equations relating the stress and strain in the soil) are then needed to capture the complexity of the material behaviour that arises owing to the particulate nature of the material. In DEM, even if simple numerical models are used to simulate the inter-particle contacts, and ideal, approximate, particle geometries are used, many of the mechanical response features associated with soil can be captured. Simplifying the particle shapes (e.g. using spheres) and adopting very basic models of the contact response reduces the computational cost of the simulation and thus allows systems involving relatively large numbers of particles to be analysed while still capturing the salient response characteristics of soil behaviour.

There are a range of established and emerging numerical methods that can be used to simulate granular material response and so it is worth clarifying what the term "discrete element method" means in the context of this text. In a discrete element simulation a numerical model made up of a large number of discrete particles or bodies is created. A discrete element method is a simulation method where the finite displacements and rotations of discrete bodies are simulated (e.g. Cundall and Hart (1993)). Within the system it is possible for the particles to come into contact with each other and lose contact, and these changes in contact status are automatically determined. This definition excludes from consideration the meshless or meshfree continuum methods including smoothed particle hydrodynamics (SPH). In these methods the "particles" are interpolation points, rather than being physical particles, and so they are very similar to the nodes in a finite element model.

Particulate DEM is used across a variety of disciplines, ranging from food technology to mining engineering, however the seminal publication in this area by Cundall and Strack (1979a), was published in a soil mechanics journal (*Géotechnique*). Interest in the method amongst geotechnical engineers has grown since this original publication, with a marked increase in interest in recent years as a result of the increase in computing power.

$\gamma=0\%$ $\gamma=15.3\%$

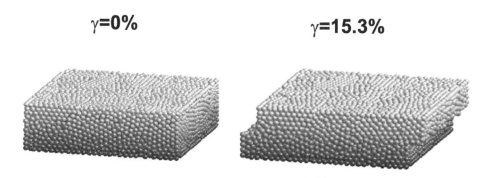

Figure 1.1: Simulation of a direct shear test using DEM

There are two main motivations to use DEM amongst both researchers and practitioners in the area of geomechanics. In the first case, in a DEM model, loads and deformations can be applied to virtual samples to simulate physical laboratory tests, and the particle scale mechanisms that underlie the complex overall material response can be monitored and analysed. In a DEM model the evolution of the contact forces, the particle and contact orientations, the particle rotations, etc., can all easily be measured. It is incredibly difficult (and arguably impossible) to access all this information in a physical laboratory test. Figure 1.1 illustrates a simulation of a direct shear test using particulate DEM. The DEM model allows us to look inside the material and understand the fundamental particle interactions underlying the complex, macro-scale response. To date knowledge of soil response has relied largely on empirical observation of the overall material response in laboratory and field tests. DEM simulations thus present geotechnical engineers with a valuable set of tools to complement existing techniques as they seek to develop a scientifically rigorous understanding of soil behaviour with likely improvements in our ability to predict response in the field. DEM therefore is now established as an essential tool in basic research in geomechanics.

A second, more applied, motivation for the use of DEM is that it allows analysis of the mechanisms involved in large-displacement problems in geomechanics. These problems cannot easily be modelled using more widespread continuum approaches such as the finite element method. Figure 1.2 illustrates a two-dimensional DEM simulation of the insertion of a cone penetrometer into a container of 117,828 disks (for details refer to Kinlock and O'Sullivan (2007)). The particles are shaded according to the amount of rotation they experience, with the particles distant from the penetrometer coloured white as they experience little disturbance, and those closest to the cone penetrometer (coloured black) being rotated and displaced during the penetration. This figure indicates that DEM can effectively accommodate the large displacements involved in the penetration mechanism. Failures in geomechanics often involve very large displacements or deformations, DEM models can therefore inform our understanding of important failure mechanisms. Examples of mechanisms that cannot be simulated using a continuum approach include internal erosion, scour and sand production in oil reservoirs. Figure 1.3 shows a bridge that collapsed in Ireland in 2009 following scour of its foundations, highlighting the importance of being able to simulate this class of problem.

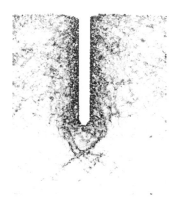

Figure 1.2: Two-dimensional DEM simulation of cone penetrometer penetrating a granular material (disk shading indicates magnitude of rotation)

Outline of book

The objective of this book is to serve as an introduction to the use of discrete element modelling to analyse the response of granular materials, focussing on applications in soil mechanics and geotechnical engineering. The intended audience is people who are thinking about using DEM, or people who are just starting to use DEM, rather than those with years of experience. However, hopefully users with some experience and DEM code developers will also find aspects of the text interesting and useful. In any case, it is assumed that someone interested in DEM is likely to be a graduate or post graduate engineer or scientist with some idea

Figure 1.3: Collapse of railway bridge in Malahide, Dublin, Ireland in August 2009, *Photo Courtesy Sarah McAllister*

of the basic principles of numerical modelling and a knowledge of mechanics.

The overall aim is to provide answers to a few key questions:

1. What is the theoretical basis of DEM ? What is the fundamental modelling approach used? (Chapters 2, 3, 4 and 6).

2. How does someone run a DEM simulation and what information can they get from it? (Chapters 5, 7, 8 and 11).

3. How do you interpret data from a DEM simulation? (Chapters 9 and 10).

4. What has already been achieved using DEM? (Chapter 12).

There is an emphasis on soil mechanics-related applications; however much of the content of this book has a broader application and should prove useful to those working in the fields of in powder technology, chemical engineering, geology, mining engineering, physics, and other disciplines where there is interest in analysing material response at a particulate scale. There are many particulate discrete element codes in use at present, some of which have been developed by individuals solely for research applications, while others are commercially available. This book is not written with any particular code in mind, rather the material and discussions presented here should be of interest to users (and possibly developers) of many different codes.

This initial Chapter aims to introduce the general principles of DEM and presents some of the mathematical concepts used in later Chapters.

1.2 Particulate Scale Modelling of Granular Materials

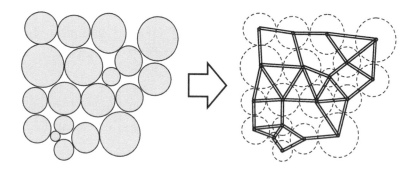

Figure 1.4: Analogy between a granular material and a highly complex, statically redundant structural frame

In discussing the need for computer simulation to facilitate analysis of particulate systems at the micro-scale, Rapaport (2004) points out the similarity between the interaction of a large system of particles and the classical "n-body" problem that has attracted the attention of physicists for hundreds of years. The n-body problem considers the evolution of a system of n "bodies" subject to Newtonian gravitational forces. The initial motivation to analyse this problem was a desire to understand the dynamics of the solar system. There is no general closed-form solution to this problem for systems with more than 3 bodies, consequently numerical methods and computer models are required to analyse these systems.

The need to adopt a computer-based model to analyse granular materials at the particle scale can be appreciated by looking at the system from the perspective of a structural engineer. As illustrated in Figure 1.4, an analogy can be drawn between an assembly of contacting particles and a structure with many elements connecting the nodes of the structure. Engineers, in particular civil engineers, understand that a structure with a large number of connections is statically indeterminate. In a statically indeterminate structure, the forces in each structural member cannot be calculated by considering the static equilibrium of the system alone. More more sophisticated (and nowadays) computer-based models that include consideration of the deformations and hence the stiffness of the structural elements are required to determine the forces within the structure.

Both Duran (2000) and Zhu et al. (2007) divide the numerical techniques used in DEM into two categories called *soft sphere* models and *hard sphere* models. A major differentiation between the methods in each category is whether the particles are approximated to be "soft", in which case penetration is allowed at the particle contacts or "hard", when no deformation or penetration is considered. Figure 1.5 illustrates schematically both approaches. Both types of simulation are transient, or time dependent. This means that the evolution of the system over a period of time is considered by examining the state of the assembly of particles at distinct time intervals.

The hard particle, or hard sphere, approximation is at the basis of the so-called "collisional" or "event driven" (ED) models. The word *hard* refers to the absense of interpenetration or deformation during impact of particles. The collision itself is not necessarily of interest and may be assumed to be instantaneous. The ED models start from the equations governing momentum exchange and the particle contact force is often not explicitly considered (Zhu et al., 2007). This type of model recognizes that when particles collide

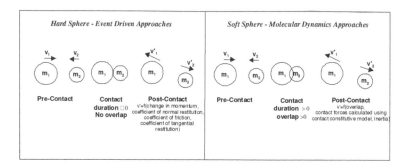

Figure 1.5: "Hard Sphere" and "Soft Sphere" approaches to DEM

energy is dissipated by plastic deformation and heat. The resultant loss of momentum when a collision occurs is characterized solely by means of the coefficients of elastic restitution. Different values for the normal and tangential coefficients of restitution are specified.

Event driven algorithms analyse events sequentially in the order in which they occur. This means that at any time during the simulation at most only one collision can occur at a given time in the analysis. The time increment used in the simulations varies, and equals the time between one collision and the next. Between collisions the particles move along a uniform trajectory.

Applications suited to the use of the event driven modelling approach are generally those involving rapid granular flow, where the granular material has been partially or completely fluidized, e.g. avalanches, or rapid flow through conduits in manufacturing processes. For example, Hoomans et al. (1996) used this approach to simulate fluidized beds for process engineering applications, and Campbell and Brennan (1985) used a hard sphere approach to simulate granular material flow. Delaney et al. (2007) correctly argue that, while it is computationally cheaper than other methods, the hard sphere approach fails to capture the fine details of the response of dense materials involving multiple simultaneous contacts. Delaney et al. also highlight the limitation in the ability to accurately model the tangential or frictional forces between interacting particles. Campbell (2006) considered this method to be inappropriate for considering dense systems as it is non-physical:

the real mechanism of force transfer in a granular material involves deformation of the contacting particles. For further information on event driven approaches refer to Brilliantov et al. (2996) or Rapaport (2009). Pöschel and Schwager (2005) describe two alternative algorithms for implementation of an event driven computer code. As hard sphere approaches are not commonly considered in current geotechnical engineering research or practice, they are not considered further here.

The principle behind the soft sphere approach is to solve, in increments of discrete time, the equations governing the linear and angular dynamic equilibrium of the colliding or contacting particles. This contrasts with the strategy used in ED models, which start from the equations governing momentum exchange. The word "soft" is a misnomer; the particles in the "soft sphere" simulations are rigid, however they can overlap at the contact points. (As discussed above, no overlap is allowed in the "event driven" methods.) In this approach, friction and elastic restitution come into effect only when spheres penetrate each other. In the soft sphere models, the normal component of the inter-particle force is calculated considering either the particle overlap at the contact point (for compressive forces) or the particle separation at the contact point (where tensile force transmission occurs). In geomechanics applications in particular, a key assumption is that the compressive overlap or tensile separation will be small. The shear or tangential forces are calculated from the cumulative relative displacement at the contact points in a direction orthogonal to the contact normal orientation. In contrast to the hard sphere approach where only one collision is considered at each time increment, the soft sphere models can handle systems with multiple simultaneous contacts, as typically arise in static or quasi-static problems. As outlined by O'Sullivan (2002), various algorithms that fall within this "soft sphere" category exist however, the most commonly used approach is the *distinct element method*, as originally described by Cundall and Strack (1979a). Given the prevalence of Cundall and Strack's approach the terms "discrete element method" and "distinct element method" are essentially used interchangeably. Strictly speaking the distinct element method really is a type of discrete element method. The distinct element method is the method given the most consideration in this text. Other soft sphere approaches that are algorithmically similar to DEM include the discontinuous deformation analysis method (DDA) (Shi (1988), adapted for particle systems by Ke and Bray (1995)), and the implicit methods proposed by Kishino (1989) and Holtzman et al. (2008).

There are a few documented geomechanics research studies that have adopted a method called contact dynamics (e.g. Lanier and Jean (2000)). This method does not strictly fall within either the event driven or soft sphere frameworks and is sometimes referred to as rigid body dynamics (Pöschel and Schwager, 2005). The general idea is that the contact forces between the particles are determined so that there is no particle deformation (i.e. "hard spheres", but with finite contact durations). The tangential forces are determined by considering the forces required to keep the particles from sliding. Pöschel and Schwager (2005) state that while the algorithm associated with this method is more complex than DEM or molecular dynamics, and there are more calculations involved in each time increment, there is not a corresponding increase in computational cost, as the time increments in the analysis are larger.

Another particle-scale approach that is used to analyse granular materials is the Monte Carlo method. As in the event driven approach, penetration of particles is not allowed; however the contacts are finite in duration. As outlined by Sutmann (2002), amongst others, in this simulation approach at each iteration each particle is subject to a number of trial moves. The change in energy generated by each of these moves is calculated and the movement leading to the lowest energy is that selected for progressing to the next configuration. This approach is applicable only to the study of systems in static equilibrium, i.e. it cannot be applied to consider flow of granular materials. A less well-established statistically based approach involving the application of the Markov stochastic process was described by Kitamura(1981a,b).

Molecular dynamics

It is important to be aware of the similarities between particulate DEM and molecular dynamics. Molecular dynamics is an analysis tool used in chemistry, biochemistry and materials science. Using this method, materials are studied at the most fundamental level by simulating the interactions between individual molecules or atoms. The objective of these simulations is to relate the bulk properties of a material (be it liquid, solid or gas) and fundamental atomistic interactions. These particles are modelled as point-like centres that interact via pair or multi-particle interaction potentials (e.g. the Leonard-Jones potential). The time scales of interest in molecular dynamics are of the order of 1 μs, and the trajectory lengths are between 10 and 100 Ångstroms (Sutmann, 2002).

Liquids tend to be the materials most commonly considered in molecular dynamics simulations, with consideration often being given to analysis of phase transformation, for example. In fact the method was initially proposed by Alder and Wainwright (1957) who described the phase transformation of a system of rigid spheres, these authors later outlined the general methodology of molecular dynamics in Alder and Wainwright (1959). Sutmann (2002) outlines the history of molecular dynamics, while Rapaport (2004) provides an overview of molecular dynamics, including details of the implementation of a molecular dynamics code. Pöschel and Schwager (2005) suggest that typical molecular dynamics simulations are less computationally intensive than particulate DEM simulation as in DEM the particles exert forces on each other only when they are in contact. The numerical stability requirements necessitate a smaller time step for particulate DEM as the contact response is relatively stiff (the influence of contact stiffness on the simulation time increment is considered in some detail in Chapter 2). However, some molecular dynamics methods (ab initio molecular dynamics) consider explicitly the interaction of the particles at the electron scale and are significantly more complex than granular DEM (e.g. the ONETEP algorithm proposed by Skylaris et al. (2005)).

As noted above meshless methods, including SPH, are another

type of particle-based model used in geomechanics. The basic idea in meshless methods is that the "particles" are used as interpolation points where the material displacement is tracked, and the material is continuous between these points. These methods differ significantly from the particulate DEM methods considered in this text and they are not given further consideration here. Readers seeking additional information on the meshless methods may wish to refer to Belytschko et al. (1996).

1.3 Use of Block DEM Codes in Geomechanics

Two types of discrete element model are used in geomechanics, referred to here as block DEM and particulate DEM. Both types of model considers systems made up of numerous individual bodies, either blocks or particles. These discrete bodies can move relative to each other and they can rotate. Contacts can form between the bodies, and as the system deforms, these contacts can break and new contacts can form. Typically a small amount of overlap is allowed at the contact between the bodies, and this overlap is analogous to the deformation that occurs at the contacts between the real bodies. Simple "contact constitutive models" are used to relate the contact forces between the bodies to the contact overlap. The shear components of the contact force impart a moment to the bodies. Knowing the contact forces and the inertia of the body, by considering the dynamic equilibrium of each body, its acceleration can be calculated. From these accelerations, displacements of the particles over small time increments can be determined. By advancing forward using these small time steps the evolution of the system can be simulated.

While the focus of this book is on particulate DEM, it is important to be aware of the use of block DEM simulations in geomechanics. This type of analysis is used to model systems of polygonal rock blocks or masonry structures; for example Powrie et al. (2002) analysed dry stone retaining walls, while Basarir et al. (2008) simulated excavation of rock. Examples of block discrete element codes include the commercial code UDEC (Itasca (1998)) and the Discontinuous Deformation Analysis code (DDA) (e.g. Shi (1988), MacLaughlin (1997), Doolin (2002)). In these codes a system of orthogonal, stiff ("penalty") springs are used to calculate the contact forces, while minimizing the overlap between the blocks. The blocks are typically simply deformable (linear elastic). The ability of the blocks to deform is the principal difference between the block codes and the particle codes. As a consequence of the block deformability, for two equivalent simulations using the same number of particles and same particle geometries, the calculations are more time-consuming in comparison with a simulation using a particle code with rigid particles.

Figure 1.6 illustrates the application of the DDA block code to analyse the Vaiont landslide that took place in Italy in 1963. As described by Sitar et al. (2005), when compared with limit equilibrium analyses, the DDA simulations yielded reasonable results and facilitated parametric studies considering the influence of the number of discontinuities on the deformation mode. This approach to discrete element modelling is not considered in detail in this text, however many of the basic principles underlying the particulate discrete element modelling codes described here also apply to block discrete element codes.

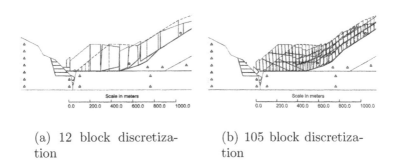

(a) 12 block discretization

(b) 105 block discretization

Figure 1.6: Back analysis of Vaiont Landslide using the Block Discrete Element Method, DDA. *Solid lines indicate deformed block configuration, dashed lines indicate original slope geometry.* Sitar et al. (2005)

13

1.4   Overview of Particulate DEM

As noted above, the distinct element method is the type of discrete element method that is currently most popular in geomechanics. The basic formulation for the distinct element method for granular materials was proposed and described by Peter Cundall and Otto Strack in two reports to the US National Science Foundation, Cundall and Strack(1978 and 1979b) and a subsequent paper in the journal *Géotechnique* (Cundall and Strack, 1979a).

An overview of the sequence of calculations involved in a DEM simulation is given in Figure 1.7. To carry out a DEM simulation initially the user inputs the geometry of the system to be analysed, including the particle coordinates and boundary conditions. The material properties are usually input by specifying the contact model parameters, including stiffness and friction coefficient. The user specifies a schedule for loading or deforming the system. Then the simulation progresses as a transient, or dynamic, analysis, typically for a specified number of time increments. At each time step the contacting particles are identified. The magnitude of the inter-particle forces relate to the distance between contacting particles. Having calculated these inter-particle forces, the resultant force and moment or torque acting on each particle can be determined. Except when particle rotation is inhibited, at each time increment two sets of equations for the dynamic equilibrium of the particles are solved. The translational movement of each particle is determined from the resultant applied force, and the resultant applied moment is used to calculate the rotational motion. Knowing the particle inertia, the translational and rotational accelerations of the particles can be calculated. The displacement and rotation of the particles over the current time-step is then found through a simple central-difference-type integration through time. The resultant forces and moments that impart these translational and rotational accelerations on the particles are sometimes called "out-of-balance" forces (e.g. Thornton and Antony (2000), Itasca (2004)). Using these incremental displacements and rotations, the particle positions and orientations are updated, in the next time step the contact forces are then calculated using this updated geometry, and the series of calculations are repeated. A discrete element analysis is therefore a transient or dynamic, analysis, even if the system of interest is responding in an almost static manner.

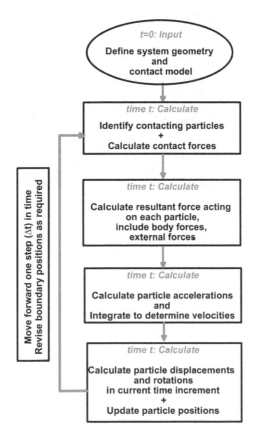

Figure 1.7: Schematic diagram of sequence of calculations in a DEM simulation

As illustrated in Figure 1.8 within each time increment there are two main series of calculations. In the first instance the particle velocities and incremental displacements are calculated by considering the equilibrium of each particle in sequence. Then having updated the system geometry the forces at each contact in the system are calculated. The tangential component of the contact force will always impart a rotational moment to the particles, and in many cases the normal contact force component will also generate a moment. These forces and moments are distributed to the particles and then used to adjust the particle positions in the next time increment.

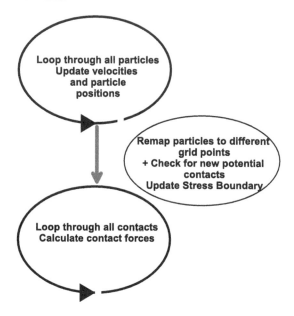

Figure 1.8: Indication of calculation sequence within a DEM time step

A clear statement of the assumptions inherent in DEM is important from the outset, although it must be acknowledged that not every implementation of DEM may adhere exactly to these assumptions, particularly as the complexity of DEM codes increases. However, using the lists proposed by Kishino (1999) and Potyondy and Cundall (2004) as a basis, the following key assumptions typically made in particle-based DEM simulations can be stated:

1. The basic particles are rigid, they possess a finite inertia (mass and rotational inertia) and they can be analytically described.

2. The particles can move independently of each other and can translate and rotate.

3. The program automatically identifies new contacts between particles.

4. The contact between particles occurs over an infinitesimal area and each contact involves only two particles.

5. The particles are allowed to overlap slightly at the contact points and this overlap is analagous to the deformation that occurs between real particles. The magnitude of the deformation of the each particle at the contact point is assumed to be small.

6. The compressive inter-particle forces can be calculated from the magnitude of the overlap.

7. At the contact points, it is possible for particles to transmit tensile and compressive forces in the contact normal direction as well as a tangential force orthogonal to the normal contact force.

8. Tensile inter-particle forces can be calculated by considering the separation distance between two particles. Once the tensile force exceeds the maximum tensile force for that contact (which may be 0), the particles can move away from each other and the contact is deleted and no longer considered when calculating the contact forces.

9. The time increment chosen in a DEM simulation should be small enough that the motion of a particle over a given time step is sufficiently small to only influence its immediate neighbouring particles.

10. Agglomerates of the rigid base particles can be used to represent a single physical particle, and the relative motion of these base particles within the agglomerate may cause a measurable deformation of the composite particles. Alternatively these agglomerates may themselves be rigid.

From the analyst's point of view there are many similarities between the overall process involved in a DEM simulation and the

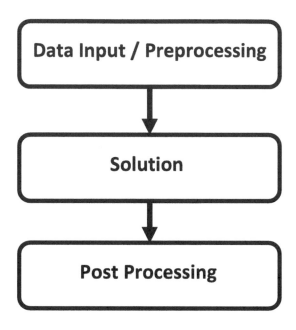

Figure 1.9: Generic Flow Chart for Numerical Analysis in Mechanics

process involved in a continuum-based analysis, e.g. using finite element analysis. A generic flowchart for numerical analysis in mechanics is given in Figure 1.9. There are some key differences between the effort associated with a DEM analysis and a conventional continuum analysis. Undoubtedly mesh generation for finite element analysis of bodies with highly complex geometries is non-trivial. However, t generating the initial positions of the particles in the problem domain to be analysed is probably more difficult and typically involves DEM calculation cycles. In fact, it is possible for this model creation phase to be at least as computationally expensive as the main simulation. As calculation cycles are involved in the specimen generation stage, the discussion of this phase of the analysis is given in Chapter 7, after the details on the method have been considered in Chapters 2–5.

The non-linearity of the systems considered and the explicit approach to time integration used mean that a small time increment must be adopted in DEM simulations. It is these considerations, combined with the need to include large numbers of particles, that make DEM simulations so computationally intensive.

A DEM simulation generates basic results in terms of individual particle positions and inter particle contact forces, rather than in terms of stress and strain. A postprocessing procedure is required to interpret these results in a useful or meaningful manner and relate them to our continuum-mechanics based understanding of soil behaviour. A wide variety of interpretation techniques have been proposed in the literature, some of which are not easy to implement and typically involve greater effort and more abstract concepts (including statistical mechanics) than the methods used to interpret continuum analyses. Chapters 8–10 provide overviews of various interpretation approaches.

For readers accustomed to continuum-based geomechanics analyses it may be useful to consider how DEM meets the theoretical requirements for a valid analysis. In conventional continuum mechanics a method of analysis is typically required to satisfy four theoretical requirements, namely equilibrium, compatibility, constitutive behaviour and boundary conditions. In a DEM simulation equilibrium is accounted for by considering the dynamic equilibrium of each particle at each time increment during the analysis. As discussed further in Chapter 11, for quasi-static analyses the user must also consider the overall equilibrium of the system as a test to establish the validity of a particular simulation. In a continuum analysis the compatibility requirement is satisfied, meaning that as the system deforms holes should not appear and the material does not develop overlaps. As outlined by Potts (2003), amongst others, from a mathematical perspective this requirement implies that components of strain exist and are continuous and the derivatives of strain exist to at least second-order. This requirement is effectively violated in a particulate DEM simulation. No strain occurs within the rigid bodies, they are allowed to overlap and the displacement field is highly non-uniform. A discussion on interpreting DEM analysis by calculating strains from the particle displacements is given in Chapter 9.

In continuum-based materials modelling, the constitutive matrix relates the stresses and strains within the material and this relationship can be linear or non-linear. No constitutive model is required in a DEM model; rather, as discussed in many DEM related papers, the constitutive model "emerges" from the DEM simulation results. A model describing the response at the particle contacts is required and this is somewhat analogous to the constitutive model. A direct mapping of the contact model to a continuum constitutive model would be inappropriate. The macro-scale or continuum response will depend on the response at the contacts, the geometry of the granular material and the ability of the particles to crush, fail or deform. Even if a linear contact model is adopted, the overall response will be non-linear as a consequence of the evolution of the inter-particle contacts.

Finally a statement of the boundary conditions is required; these boundary conditions play a large role in defining the problem to be analysed. The concepts of boundary conditions are similar in both continuum and DEM analyses; however, the details differ and a discussion on the various boundary conditions used in DEM simulations is given in Chapter 5.

1.5 Use of DEM Outside of Geomechanics

Granular materials are encountered in a variety of disciplines outside of soil mechanics and geotechnical engineering. Most notably, chemical and process engineers also regularly adopt DEM in their research. The complexity of granular material response has attracted interest from mathematicians and physicists who use DEM simulations to generate data for subsequent detailed analysis of the fundamentals of granular material response. As in the case of geomechanics applications, there is potential, with increasing computational power, to apply DEM to solve industrial problems. Recent conference proceedings, e.g. Nakagawa and Luding (2009), illustrate the range of applications of DEM across these disciplines. Much information on the applicability of DEM to advance understanding of granular materials for geomechanics applications can therefore be gained by reference to journals in these other disciplines. Two particularly useful publications are Zhu et al. (2007 and 2008) which provide reviews of the development of DEM algorithms and the application of DEM respectively from a chemical engineering perspective. The recent special editions of the journals *Powder Technology*, Thornton (2009), and *Particuology*, Zhu and Yu (2008), also contain papers of interest to the geomechanics community.

1.6 Introduction to Tensorial Notation

Tensorial notation (sometimes called index notation) is adopted throughout this book. Most publications referred to in the book also use tensorial notation and, while some authors (e.g. Potyondy and Cundall (2004)) provide clarification, familiarity with this notation tends to be assumed. This section is included to give the reader a very brief overview of tensorial notation both to facilitate understanding of the material in this book as well as the broader set of publications associated with the topic. For more detailed explanation reference to a continuum mechanics textbook (e.g. Shames and Cozzarelli (1997)) is recommended.

Tensorial notation is attractive as it is allows vectors and operations on vectors to be described concisely. It has particular advantages when developing computer programs where data are stored in arrays that are accessed using integer indices. In particulate DEM there are calculations and operations involving force vectors, position vectors, displacement vectors, etc. In this book the intrinsic form of the vector is denoted in bold typeface; thus the particle displacements are given by $\mathbf{u}$, the resultant force acting on a particle is given by $\mathbf{f}$, and the particle position is given by $\mathbf{x}$. These terms are then used to refer to the vectors in a general sense as entities with a specific magnitude ($|\mathbf{u}|$ or $|\mathbf{f}|$) and whose directions can be described relative to a specified coordinate system.

Every vector will have components parallel to each of the co-ordinate axes. Tensorial notation provides a convenient means of describing operations on each of these components. When these vectors are expressed in tensorial, or indicial, form they are denoted u_i, f_i and x_i with the subscript i indicating that the vector component parallel to a specific coordinate axis, i, is under consideration. For example, if the term u_i is used to describe the displacement of a particle, this vector may have either 2 or 3 components, depending on whether we are considering a two-dimensional or three-dimensional analysis. In the Cartesian coordinate system the displacement denoted u_i is given by $u_i = (u_x, u_y)$ and $u_i = (u_x, u_y, u_z)$ in two-dimensional and three-dimensional analyses respectively. As there is only one index (i) the vector u_i is a first-order tensor.

Extending consideration to two-dimensional tensors, the stress tensor is given by $\boldsymbol{\sigma}$ or σ_{ij} and this tensor can represent either a two-dimensional or three-dimensional state of stress. In this case, even if the stress state is fully three-dimensional, there are two indices $(i$ and $j)$ and this tensor is then a second-order tensor. The indices i and j are considered "free indices" as they are both "free" to adopt independently any of the values x, y (and z in 3D). The stress tensor for two-dimensional analysis is represented in matrix form as

$$\sigma_{ij} = \begin{pmatrix} \sigma_{xx} & \sigma_{xy} \\ \sigma_{yx} & \sigma_{yy} \end{pmatrix} \qquad (1.1)$$

while in 3D the stress tensor is given by

$$\sigma_{ij} = \begin{pmatrix} \sigma_{xx} & \sigma_{xy} & \sigma_{xz} \\ \sigma_{yx} & \sigma_{yy} & \sigma_{yz} \\ \sigma_{zx} & \sigma_{zy} & \sigma_{zz} \end{pmatrix} \qquad (1.2)$$

Here compressive stresses and forces are taken to be positive as this is the convention typically adopted in geomechanics (refer to Figure 1.10(a)). The components along the diagonal (σ_{xx}, σ_{yy}, σ_{zz}) are the *normal* or *direct* stresses, while the off-diagonal terms (σ_{xy}, σ_{yx}, σ_{zx}, etc.) are the shear stresses. When a material is in a state of static equilibrium with equal complementary shear stresses, then the stress tensor is symmetric and we can say $\sigma_{ij} = \sigma_{ji}$. As illustrated in Figure 1.10 for every stress state (two- or three-dimensional) planes oriented at θ and $\theta + \frac{1}{2}\pi$ to the horizontal can be found in the material along which no shear stresses are felt. The direct stresses acting on these planes are called the principal stresses and the normals to the planes give the principal stress orientations. The principal stresses are given by the eigenvalues of the stress tensor, while the eigenvectors give the principal stress orientations. Typically the maximum or *major* principal stress is denoted by σ_1 and the minimum or *minor* principal stress is denoted by σ_3. In 3D there will also be an *intermediate* principal stress σ_2, with $\sigma_1 > \sigma_2 > \sigma_3$.

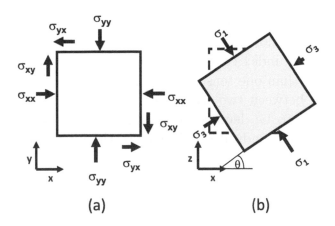

Figure 1.10: Illustration of two-dimensional stress state

The notation for addition and subtraction of vectors using tensors is straightforward. For example consider two contacting (touching) particles a and b. If the centroid of particle a has a position vector x_i^a and the centroid of particle b has a position vector x_i^b, then the vector giving the location of particle b relative to particle a (called the branch vector) is given by $l_i = x_i^b - x_i^a$. In 3D this operation expands as follows:

$$l_i = \begin{pmatrix} l_x \\ l_y \\ l_z \end{pmatrix} = \begin{pmatrix} x_x^b - x_x^a \\ x_y^b - x_y^a \\ x_z^b - x_z^a \end{pmatrix} \tag{1.3}$$

As noted above, the tensorial notation system includes many ways for expressing mathematical operations involving vectors (1D arrays) and matrices concisely. The "dummy index" is used to indicate that we are considering terms along the diagonal, and less intuitively it denotes summation. Using this approach the trace of the stress tensor is given by σ_{ii}, and this is given by

$$\begin{aligned} \sigma_{ii} &= \sigma_{xx} + \sigma_{yy} & \text{(2D)} \\ \sigma_{ii} &= \sigma_{xx} + \sigma_{yy} + \sigma_{zz} & \text{(3D)} \end{aligned} \tag{1.4}$$

The sum σ_{ii} is the first invariant of the stress tensor (I_σ). This parameter is "invariant" (i.e. unchanging) if the tensor is subject to an orthogonal rotation, e.g. if the tensor is rotated to consider the components along the principal axes of stress.

The dummy index concept can be extended to operations involving more than one tensor. For example, consider the contact force vector between two particles to be denoted by f_i and the branch vector to be denoted by l_i. In the expression $f_i l_i$, i is a dummy index and indicates reference to the inner product, i.e.

$$\begin{aligned} f_i l_i &= f_x l_x + f_y l_y & \text{(2D)} \\ \\ f_i l_i &= f_x l_x + f_y l_y + f_z l_z & \text{(3D)} \end{aligned} \tag{1.5}$$

In a similar manner, in three dimensions, the magnitude of a vector $|\mathbf{v}|$ is given by

$$| \mathbf{v} | = \sqrt{v_i v_i} = \sqrt{v_x v_x + v_y v_y + v_z v_z} \tag{1.6}$$

The use of free indices gives the expression $f_i l_j$ and this can be used to represent $f_x l_y$ when $i = x$ and $j = y$ or $f_x l_x$ when $i = x$ and $j = x$. Expressions similar to $\sum_N f_i l_j$ are used throughout this book. The expansions of this expression in two and three dimensions are:

$$\sum_N f_i l_j = \begin{pmatrix} \sum_N f_x l_x & \sum_N f_x l_y \\ \sum_N f_y l_x & \sum_N f_y l_y \end{pmatrix} \quad \text{(2D)}$$

$$\sum_N f_i l_j = \begin{pmatrix} \sum_N f_x l_x & \sum_N f_x l_y & \sum_N f_x l_z \\ \sum_N f_y l_x & \sum_N f_y l_y & \sum_N f_y l_z \\ \sum_N f_z l_x & \sum_N f_z l_y & \sum_N f_z l_z \end{pmatrix} \quad \text{(3D)}$$

$$(1.7)$$

Note that the product $f_i l_j$ is called the *dyadic* product of the two vectors $\mathbf{f}$ and $\mathbf{l}$, and this can also be expressed as $\mathbf{f} \otimes \mathbf{l}$.

In another example involving the use of the dummy index, the stress acting along a direction specified by the normal (unit) vector n_j can be calculated by multiplying the normal vector by the stress tensor. In tensorial notation this operation is expressed as $\sigma_{ij} n_j$. As above, repetition of the index j (the dummy index in this case) in the term $\sigma_{ij} n_j$ indicates that there will be a summation. The expansion (in 3D) is given by

$$\sigma_{ij} n_j = \sigma_{ix} n_x + \sigma_{iy} n_y + \sigma_{iz} n_z = \begin{pmatrix} \sigma_{xx} n_x + \sigma_{xy} n_y + \sigma_{xz} n_z \\ \sigma_{yx} n_x + \sigma_{yy} n_y + \sigma_{yz} n_z \\ \sigma_{zx} n_x + \sigma_{zy} n_y + \sigma_{zz} n_z \end{pmatrix} \quad (1.8)$$

Gradients are often of interest in geomechanics, and in the current context the use of a deformation gradient to calculate strain is important. Tensorial notation provides a concise notation for partial derivatives. In this case a comma, " , ", is used to indicate a partial derivative, i.e. the notation $v_{i,j}$ indicates the spatial partial derivative of the terms in vector v_i with respect to coordinate j. For example, if the vector describing the incremental displacement of a particle is given by u_i the displacement gradient is given by $u_{i,j}$ and in the 3D case this expands to

$$u_{i,j} = \begin{pmatrix} \frac{\partial u_x}{\partial x} & \frac{\partial u_x}{\partial y} & \frac{\partial u_x}{\partial z} \\ \frac{\partial u_y}{\partial x} & \frac{\partial u_y}{\partial y} & \frac{\partial u_y}{\partial z} \\ \frac{\partial u_z}{\partial x} & \frac{\partial u_z}{\partial y} & \frac{\partial u_z}{\partial z} \end{pmatrix} \tag{1.9}$$

In addition to spatial derivatives we also need to consider temporal derivatives, i.e. rates. The notation $\dot{u}_i$ is used to denote the rate of change of the tensor u_i with respect to time, i.e.

$$\dot{u}_i = \begin{pmatrix} \frac{\partial u_x}{\partial t} \\ \frac{\partial u_y}{\partial t} \\ \frac{\partial u_z}{\partial t} \end{pmatrix} \tag{1.10}$$

Finally in relation to tensorial notation it is useful to introduce two specific tensors, the Kronecker delta δ_{ij} and the alternating tensor e_{ijk}. The Kronecker delta is defined to have the property

$$\delta_{ij} = \begin{cases} 1 & \text{when } i = j \\ 0 & \text{when } i \neq j \end{cases} \tag{1.11}$$

The product of the Kronecker delta and a second-order tensor is given by

$$\sigma_{ij}\delta_{jk} = \sigma_{ik} \tag{1.12}$$

In this expression j is a dummy index and the free indices on each side of the equation are the same.

The alternating tensor is given by

- $e_{ijk} = 1$ when the indices are in the order xyz, yzx, zxy, i.e. cyclic order of indices.

- $e_{ijk} = -1$ when the indices are in the order xzy, yzx, zyx, i.e. anticyclic order of indices.

- $e_{ijk} = 0$ when there are repeated indices, e.g. xxy, xxz, xyy, etc.

The alternating tensor can be used to calculate the cross product of two tensors; in three dimensions the cross product is given by $\mathbf{c} = \mathbf{a} \times \mathbf{b}$ where

$$c_i = e_{ijk} a_j b_k \tag{1.13}$$

The vector $\mathbf{c}$ will be orthogonal to both $\mathbf{a}$ and $\mathbf{b}$.

1.7 Orthogonal Rotations

Chapters 2, 5 and 8 all refer to rotation of parameters. For example, when moving from a coordinate system defined by the principal axes of inertia of a given particle to the global coordinate system. To achieve this rotation an orthogonal rotation tensor is required. If a rotation is orthogonal then the product of two successive rotations is given by

$$T_{ij} T_{kj} = \delta_{ik} \tag{1.14}$$

Furthermore the transpose of $\mathbf{T}$ will equal the inverse of $\mathbf{T}$, $\mathbf{T}^T = \mathbf{T}^{-1}$. We can rotate any vector $\mathbf{a}$, with components (a_x, a_y, a_z) using a rotation tensor, using the tensor product $a_i' = T_{ij} a_j$ where the tensor a_i' gives the rotated components of the vector $\mathbf{a}$. The magnitude of the vector will remain unchanged, i.e. $|a_j| = |a_i'|$.

In three dimensions to rotate a vector a_i through an angle θ about the $z-$axis the following expression is used:

$$\begin{pmatrix} a_x' \\ a_y' \\ a_z' \end{pmatrix} = \mathbf{T} \begin{pmatrix} a_x \\ a_y \\ a_z \end{pmatrix} = \begin{pmatrix} \cos\theta & -\sin\theta & 0 \\ \sin\theta & \cos\theta & 0 \\ 0 & 0 & 1 \end{pmatrix} \begin{pmatrix} a_x \\ a_y \\ a_z \end{pmatrix} \tag{1.15}$$

where a_i represents the original vector and a_i' is the rotated vector. When a vector is multiplied by an orthogonal rotation matrix a rigid body rotation is achieved, i.e. the vector length is preserved as the orientation changes.

While the basic DEM calculations are almost exclusively operations on one-dimensional vectors (i.e. particle velocity vector, contact force vector), analysis of the system typically involves the use of second-order tensors (2D matrices), including the stress tensor, the strain tensor and the fabric tensor. To rotate a second-order tensor (σ_{ij}) from one coordinate system to another the operation is given by

$$
\begin{pmatrix} \sigma'_{xx} & \sigma'_{xy} \\ \sigma'_{yx} & \sigma'_{yy} \end{pmatrix} = \mathbf{T} \begin{pmatrix} \sigma_{xx} & \sigma_{xy} \\ \sigma_{yx} & \sigma_{yy} \end{pmatrix} \mathbf{T}^T
\tag{1.16}
$$

1.8 Tessellation

The particulate systems considered in this book comprise discrete particles and their contacts. The creation of triangulations of the system is useful for applications including construction of the initial specimen geometry (Chapter 7), application of boundary stresses (Chapter 5), calculation of strain (Chapter 9), and analysis of the material fabric (Chapter 10). An overview of triangulation is therefore included at this point. More detailed considerations of the application of Delaunay triangulation in granular mechanics are given by Li and Li (2009), Goddard (2001), Ferrez (2001) and Bagi (1999a). Rapaport (2004) describes the implementation of a subroutine to construct a Voronoi polygon to analyse the structure of particulate systems in a molecular dynamics code, while Ferrez (2001) discusses the use of triangulation for contact detection. It may also be possible to use triangulation to couple DEM particle codes with continuum mechanics to represent a fluid phase.

A tessellation is a general term to describe the division of a space into a set of subspaces that do not overlap and that fill the space completely (i.e. with no gaps). These tessellations can exist in two- and three-dimensional space. Amongst the most commonly used tessellations are the Delaunay triangulation and the Voronoi diagram; these geometrical constructs are closely related and each is said to be the "dual" of the other. From a geomechanics perspective it is useful to realize that Delaunay triangulation is often used in mesh generation for finite element analyses of complex geometries.

Referring to Shewchuk (1999), for example, a triangulation of a set of n points or nodes, $\mathbf{P} = \{P_1, P_k, P_n\}$, is a set of m ($m \neq n$) triangles, $\mathbf{T} = \{T_1, T_k, T_m\}$ whose interiors do not intersect each other. A Delaunay triangulation of a nodal set has the property that no node in the nodal set falls in the interior of the circumcircle (circle that passes through all three vertices) of any triangle in the triangulation. The Delaunay triangulation of the vertex set is unique. Higher-dimensional Delaunay triangulations are a generalization of the two-dimensional Delaunay triangulation. In three dimensions, the triangulation of V yields a set, T, of tetrahedra, whose vertices are V, and whose interiors do not intersect each other. In this case no node in the nodal set falls in the interior of the circumsphere (sphere that passes through all four vertices) of any tetrahedron in the triangulation. The Delaunay triangulation of 10 random points (nodes) in two dimensions is illustrated in Figure 1.11(b) and a three-dimensional triangulation is illustrated in Figure 1.12.

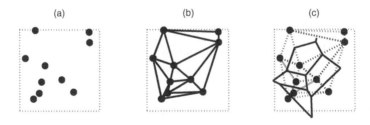

Figure 1.11: (a) 10 random points (b) Delaunay triangulation (c) Voronoi diagram

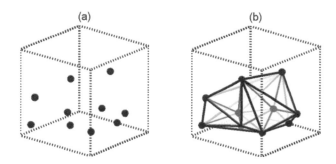

Figure 1.12: (a) 10 random points (b) Tetrahedra generated by 3D Delaunay triangulation

As discussed by Okabe et al. (2000), there are a number of different algorithms available for the implementation of Delaunay triangulation. Most of the triangulations used in this work were calculated using MATLAB, which uses the qhull algorithm (Barber et al., 1996).

As noted above, the Delaunay triangulation is related to (or is the dual of) a second geometrical construct called the Voronoi diagram or the Voronoi tessellation. The Voronoi diagram of a set of n nodes $\mathbf{P} = \{P_1, P_k, P_n\}$, is a set of n polygons, $\mathbf{V} = \{V_1, V_k, V_m\}$. Each polygon V_k is centred around a corresponding node P_k. The polygon V_k encloses an area or volume, such that every point within that polygon that is closer to the node P_k than to any other node in the set $\mathbf{P}$. The Voronoi diagram for the system of points given in Figure 1.11(a) is illustrated in Figure 1.11(c).

1.9 General Comments on Computer Modelling

A DEM model is an idealization of the real physical system and the extent of the idealizations used in creating the model will be discussed at various points in this text. It is important also to always be aware that a DEM simulation is a computer simulation and the calculations are performed using finite, floating point representations of the real numbers, i.e. representations of the real numbers containing only a finite number of digits. An introduction to some of the issues associated with floating point arithmetic is given by Burden and Faires (1997) and a more detailed discussion is given by Goldberg (1991). The error associated with representing a real number in the floating point format used by computers is called a round-off error. The calculations in DEM simulations are therefore carried out on approximate representation of real numbers and the results of the calculations themselves are also subject to a round-off, which will introduce a further error into the system. One way to reduce round-off error is to reduce the number of error generating calculations. Care should also be taken in the choice of algorithm used to accurately resolve the contact geometry or the time integration approach. These issues are considered further in Chapter 4.

Chapter 2

Particle Motion

2.1 Introduction

A discrete element analysis is a dynamic or transient analysis that considers the dynamic interaction of a system of interacting particles. A particulate DEM model creates an ideal system of rigid particles that can move, connected by rigid springs that simulate the contact interactions. (The contact spring formulations are outlined in Chapter 3). As particles move away from each other contacts are broken and some of the springs will be removed; at the same time additional springs will be introduced as new contacts are formed. The continuous removal and introduction of contact springs results in a change in the overall system stiffness. A reduction in stiffness will also occur if a contact starts to slide. Therefore the analysis is non-linear. This non-linearity could be described as a geometrical non-linearity as it arises owing to a change in the local packing geometry of the particles. As will be discussed in Chapter 3, the contact constitutive model used to describe the force displacement response at the contacts is often non-linear, and this adds a material non-linearity to the system. These two particle-scale sources of non-linearity combine to give an overall non-linear macro-scale material response. At larger strains where sliding occurs the geometric non-linearity caused by gross movements at the contacts and "buckling" mechanisms that can develop in local groups of particles will dominate the response, while the influence of the non-linear response at the contacts will be more evident at small strain levels, before the onset of sliding.

The basic principles of DEM are normally introduced by directly considering the dynamic equilibrium of the individual particles. Here DEM is introduced in a slightly different way. Civil engineers are usually familiar with the basic theories of matrix structural analysis and finite element analysis. In these approaches typically a large system of linear equations or stiffness matrix is formed. The displacements of the structural elements are determined by inverting this stiffness matrix. Particulate DEM uses a different solution strategy that introduces a greater risk of numerical instability (it is conditionally rather than unconditionally stable). To understand why the conditionally stable approach is preferred in particulate DEM, it is useful to initially consider DEM from the perspective of matrix structural analysis. As was already shown in Figure 1.4, the particles are analogous to the degrees of freedom in a matrix structural analysis (i.e. the end points of the structural elements) or alternatively the nodes in a finite element mesh. Using this analogy, the overall governing equation for the system can be expressed as the standard governing equation for a dynamic analysis in structures or continuum finite element or finite difference analysis, so that

$$\mathbf{M\ddot{u} + C\dot{u} + K(u) = \Delta F} \tag{2.1}$$

where $\mathbf{M}$ is the mass matrix (or more correctly the inertia matrix , including both mass and rotational inertia), $\mathbf{C}$ is a damping matrix, $\mathbf{u}$ is the incremental displacement vector (including both translational and rotational displacements) and $\mathbf{\Delta F}$ is the incremental force vector (including moments). The global stiffness matrix $\mathbf{K}$ depends upon the system geometry, i.e. which particles are contacting. The incremental displacements are the movements of the particles over the current time step. The objective of the analysis is to solve for the incremental displacements. The velocity and acceleration vectors are given by $\mathbf{\dot{u}}$ and $\mathbf{\ddot{u}}$. The particles in a DEM model are then analagous to the nodes in a finite element analysis. However, as the particles are free to rotate, a particle in a 2D DEM analysis has three degrees of freedom (two translational and one rotational), while a particle in a 3D DEM analysis has six degrees of freedom (three translational and three rotational).

Equation 2.1 is the dynamic equilibrium equation for the system. Broadly speaking, two approaches can be used to solve the dynamic equilibrium equation for a multi-nodal system. These approaches are termed *implicit* and *explicit*. In the implicit approach a single vector **u** can be created to represent the combined incremental displacements for all the particle centroids in the system. This is similar to the use of a single vector to represent the displacements of all the nodes in a finite element analysis, i.e.

$$\mathbf{u} \;=\; \begin{pmatrix} u_x^1 \\ u_y^1 \\ u_z^1 \\ u_x^p \\ u_y^p \\ u_z^p \\ u_x^{N_p} \\ u_y^{N_p} \\ u_z^{N_p} \end{pmatrix} \tag{2.2}$$

where u_x^p, u_y^p, u_z^p, are the incremental translational displacements of particle p in the three coordinate directions respectively and there are N_p particles in the system. (To simplify the discussion rotations are not considered at this point.) The incremental force vector $\Delta \mathbf{F}$ is constructed in a similar manner. The global mass **M**, stiffness **K** and damping **C** matrices are combined as for the finite element method or in structural analysis. The global stiffness matrix construction is not detailed here and interested readers should refer to the finite element or structural analysis texts of Zienkiewicz and Taylor (2000a) or Sack (1989) for guidance on stiffness matrix construction. Ke and Bray (1995) discuss the stiffness matrix formation for the implicitly particulate DEM algorithm DDAD (Discontinuous Deformation Analysis for Disks).

Where algorithms that involve assembly of a stiffness matrix are adopted to solve the dynamic equilibrium equation (Equation 2.33), a large system of simultaneous equations is generated, as in the finite element method, and solution will involve inversion of a highly sparse stiffness matrix. For a relatively small 3D system with 1,000 particles, the stiffness matrix will have 36×10^6 entries including the 0 valued terms as each particle has 6 degrees of freedom. Even if efficient algorithms to solve sparse systems of linear equations are used, the sequence of calculations will be very computationally expensive, both in terms of the number of operations required to solve the system and in terms of memory requirements. While some further consideration to this type of approach is given in Section 2.5 below, most geomechanics researchers use an alternative *explicit* approach that was originally outlined by Cundall and Strack (1979a,b).

In Cundall and Strack's distinct element approach, and in molecular dynamics, solution of the global system of equations is avoided by considering the dynamic equilibrium of the individual particles rather than solving the entire system simultaneously. This approach also avoids creation and storage of the large global stiffness matrix and, as highlighted by Potyondy and Cundall (2004), relatively modest amounts of computer memory are then required to consider large populations of particles. The implementation is somewhat similar to the implementation used for finite difference continuum analysis. Referring to Zhu et al. (2007) probably the most general format for expressing the equation governing the translational dynamic equilibrium of a particle p with mass m_p is

$$m_p \ddot{\mathbf{u}}_p = \sum_{c=1}^{N_{c,p}} \mathbf{F}_{pc}^{\text{con}} + \sum_{j=1}^{N_{nc,p}} \mathbf{F}_{pj}^{\text{non-con}} + \mathbf{F}_p^f + \mathbf{F}_p^g + \mathbf{F}_p^{\text{app}} \qquad (2.3)$$

where $\ddot{\mathbf{u}}_\mathbf{p}$ is the acceleration vector for particle p, $\mathbf{F}_{pc}^{\text{con}}$ are the contact forces due to contact c when there are $N_{c,p}$ contacts between particle p and either other particles or boundaries, and $\mathbf{F}_{ck}^{\mathrm{non-con}}$ are non-contact forces between particle p and $N_{nc,p}$ other particles (or boundaries). From a geomechanics perspective, the most likely origin of non-contact forces would be capillary forces in unsaturated soil. $\mathbf{F}_p^f$ is the fluid interaction force acting on particle p, $\mathbf{F}_p^g$ is the gravitational (body) force and $\mathbf{F}_p^{\mathrm{app}}$ is a specified applied force (for example this may arise where a "stress-controlled membrane" is used as discussed in Section 5.4). Comparing Equations 2.1 and 2.3, there is no explicit consideration of damping in Equation 2.3, rather the contribution from damping is included in the calculation of the contact force (refer to the viscous dashpots described in Chapter 3 and also to Section 2.7 below).

The torque generated at each contact point is calculated as the cross-product of the contact force and a vector from the centre of the particle to the contact point. The dynamic rotational equilibrium is given by

$$\mathbf{I}_p \frac{d\boldsymbol{\omega}_p}{dt} = \sum_{j=1}^{N_{\mathrm{mom}}} \mathbf{M}_{pj} \qquad (2.4)$$

where $\boldsymbol{\omega}_p$ is the angular velocity vector and $\mathbf{M}_{pj}$ is the moment applied by the jth moment transmitting contact forces involving particle p and there are a total of N_{mom} moment transmitting contacts. As will be discussed in more detail in Chapter 3, at each contact point there will be a component of the contact force that is normal to the contact and a second component that acts along or tangential to the contact. The tangential forces will always impart a moment; however, the normal forces will only impart a moment if their line of action does not pass through the centroid of the particle (i.e. if the particles are non-circular or non-spherical). Moment transmitting contact models, e.g. rotational springs or the parallel bond model, have also been proposed.

During the deformation of a granular material the particle positions and the forces acting on the particles continuously evolve. In a DEM simulation time is discretized; this means that the system is examined at specific points in time and the real, continuously changing physical system is not accurately captured. As illustrated in Figure 1.8 in Chapter 1, at each time step there are two main sequences of calculations. The contact forces are calculated based on the most recently updated particle positions. This means that the applied forces and torques in Equations 2.3 and 2.4 are assumed to be known. Then Equations 2.3 and 2.4 can therefore easily be manipulated to give the particle translational and rotational accelerations, $\dot{\omega}_p$ and $\ddot{\mathbf{u}}_p$, i.e. equilibrium equations generate two sets of ordinary differential equations for each particle.

2.2   Updating Particle Positions

Knowing the resultant forces acting on the particles we can calculate the accelerations for particle p from the equation of dynamic equilibrium for the particle. If the translation motion of the particle is isolated, this equation is simply given by:

$$\mathbf{m}_p \mathbf{a}_p^t = \mathbf{F}_p^t \qquad (2.5)$$

where $\mathbf{m}_p$ is the inertia (mass) matrix, $\mathbf{a}_p^t = \ddot{\mathbf{u}}_p^t$ is the acceleration vector at time t, and $\mathbf{F}^t$ is the resultant force vector. Note that the acceleration vector, $\mathbf{a}^t$ considers only the translational degrees of freedom and has 2 components in two dimensions and 3 components in three dimensions. The force vector $\mathbf{F}^t$ also has 2 components in two dimensions and 3 in three dimensions. In the two-dimensional case the mass (inertia) matrix is given by

$$\mathbf{m_p} = \begin{pmatrix} m_p & 0 \\ 0 & m_p \end{pmatrix} \qquad (2.6)$$

where m_p is the particle mass, calculated as the particle density times the volume. In three dimensions the mass matrix, $\mathbf{m_p}$ is a 3×3 matrix, with the diagonal terms equal to m_p, and the off-diagonal terms equal to 0.

The next stage in the analysis involves using these acceleration values to obtain incremental displacements and hence update the particle positions. In numerical analysis, the techniques used to update parameters given their first and second derivatives with respect to time (i.e. to get displacements from accelerations), are called time integration methods. Many time integration algorithms exist (reference to Wood (1990b) may be useful for readers specifically interested in this topic). It is important to appreciate that for general 3D particles, analysis of the rotational motion is significantly more complex than the translational motion.

In most DEM codes a time integration approach similar to the central-difference method with a time increment Δt is used. This approach can most easily be understood by considering the relationship between the acceleration and velocity vectors, as follows:

$$\mathbf{a}_p^t = \frac{1}{\Delta t}(\mathbf{v}_p^{t+\Delta t/2} - \mathbf{v}_p^{t-\Delta t/2}) \tag{2.7}$$

where $\mathbf{v}_p^{t-\Delta t/2}$ and $\mathbf{v}_p^{t+\Delta t/2}$ are the velocity vectors at $t - \Delta t/2$ and $t + \Delta t/2$ respectively for particle p. Rapaport (2004) terms this time integration approach a "leap-frog" method as the velocities and displacements are calculated with a time lag of $\Delta t/2$. Other authors (e.g. Munjiza (2004)) refer to it as the position Verlet time integration scheme. As with $\mathbf{F}_p$ and $\mathbf{a}_p$, the $\mathbf{v}_p$ vector has 3 components in two dimensions and 6 components in three dimensions. The velocity at time $t + \Delta t/2$ is then calculated as:

$$\mathbf{v}_p^{t+\Delta t/2} = \mathbf{v}_p^{t-\Delta t/2} + \Delta t \mathbf{m}_p^{-1}(\mathbf{F}_p^t) \tag{2.8}$$

The velocity at time $t + \Delta t/2$ is taken to equal the average velocity over the time increment t to $t+\Delta t$. Then we can calculate the updated particle position $\mathbf{d}_p^{t+\Delta t}$ as:

$$\mathbf{x}_p^{t+\Delta t} = \mathbf{x}_p^t + \Delta t \times \mathbf{v}_p^{t+\Delta t/2} \tag{2.9}$$

where the particle position vector $\mathbf{x}$ gives the particle Cartesian coordinates and the total rotation about the principal axis (axes in 3D).

For two-dimensional discrete element simulations there is no coupling between the three degrees rotational of freedom. This means that the particle's rotational or angular velocities can be calculated by considering the following dynamic rotational equilibrium equation:

$$I_{p,z}\dot{\omega}_{p,z} = M_{p,z} \tag{2.10}$$

where $\omega_{p,z}$ is the angular velocity about an axis through the centre of the particle orthogonal to the analysis plane. For a circular or disk particle the moment of inertia $I_{p,z}$ equals $\frac{\rho\pi r_p^4}{2}$ where r_p is the particle radius and ρ is the particle density. The central-difference time integration approach can easily be applied to incrementally solve this equation as follows:

$$\omega_{p,z}^{t+\Delta t/2} = \omega_{p,z}^{t-\Delta t/2} + \Delta t \frac{M_{p,z}^t}{I_{p,z}} \qquad (2.11)$$

This angular velocity is used to calculate the tangential component of the contact force (refer to Section 3.7). It is also used to update the position of the edges of non-spherical particles, and to calculate the total particle rotations (rotations are important as an indicator of localizations within the material (Chapter 8). A key decision to be made by the analyst is to choose the value of the time increment, Δt, to be used in the simulation.

2.3 Time integration and Discrete Element Modelling: Accuracy and Stability

In their description of the distinct element method Cundall and Strack (1979a) proposed the use of the computationally efficient, explicit, central-difference type time integration scheme. A limitation of this scheme is that it is only conditionally stable, so small time steps must be used. However, this restriction on the size of the time increment due to numerical stability considerations is not as limiting as it might initially appear. To successfully capture the inherent non-linearity of the problem (changing contact conditions and non-linear contact response) the incremental changes in the particle positions and contact forces in a given time-step must be small. This translates into a constraint on the time increment to be small to capture the non-linearity of the system. Ideally the time increment chosen in a DEM simulation should be small enough that the motion of a particle over a given time step is sufficiently small to only influence its immediate neighbouring particles. Cundall and Strack (1978) stated that a fundamental idea of DEM is that the time step chosen be sufficiently small that in a single time step disturbances cannot propagate from a disk further than its nearest neighbours.

In the context of analysis of physical systems, a numerical algorithm is a procedure involving a sequence of calculations developed to model the response of the system. In DEM there is a set of calculations where information about the current configuration of particles is used to step forward and predict the system state at a future time. This prediction will be approximate, rather than exact. It is important to carefully consider the limitations and approximations involved in the numerical model. In DEM it is important to consider the accuracy, stability and robustness of the time integration algorithm used. Sutmann (2002) considers these issues from a molecular dynamics perspective. During each cycle in a DEM simulation the dynamic equilibrium equation is solved for each particle in the assembly. The system of differential equations is an idealization of the real physical system, limiting accurate prediction. Specific approximation errors are introduced when the equation is solved numerically. The round-off error introduced in calculations using computers is considered briefly in Section 1.9. A second, much larger, error is introduced as a consequence of the approximations used to calculate the particle incremental displacements from the calculated accelerations. This error is called the truncation error.

In any numerical model that simulates the response of a transient or dynamic system there will be truncation errors introduced at each time step. The truncation error can be understood by reference to the Taylor series expansion. The Taylor series expansion provides an estimate for the value of a parameter, say the position, at time $t + \Delta t$ as given by $\mathbf{x}_{t+\Delta t}$, in terms of the position at time t and the temporal derivatives of the position at time t, as

$$\mathbf{x}_p^{t+\Delta t} = \mathbf{x}_p^t + \Delta t \left(\frac{d\mathbf{x}_p}{dt}\right)^t + \frac{\Delta t^2}{2!}\left(\frac{d^2\mathbf{x}_p}{dt^2}\right)^t + \frac{\Delta t^n}{n!}\left(\frac{d^n\mathbf{x}_p}{dt^n}\right)^t + O\left(\Delta t^{n+1}\right)$$

(2.12)

The term $O(\Delta t^{n+1})$ is the truncation error. This is the error introduced in the approximate, calculated value of $\mathbf{x}_p^{t+\Delta t}$ by considering only the first n derivatives of $\mathbf{x}_p$ at time t in the prediction. This truncation error is a measure of the amount by which the exact solution to the differential equation describing the particle motion differs from the approximate solution. The error is proportional to Δt^{n+1}. As Δt will be a small number, i.e. $\Delta t \ll 1$, then the higher the value of n, and hence the greater the number of derivatives that are included in the approximation, the smaller the error will be. The error will also be reduced using a smaller Δt value, with the resultant improvement in accuracy being much greater for large values of n. In a transient simulation, where we are calculating the values of $\mathbf{x}_t$ over many time increments, this error is considered to be a "local" truncation error that is introduced at each time increment.

Most DEM codes used in geomechanics use either the central-difference time integration algorithm or a slightly modified version of the central-difference method. As noted by Wood (1990b) there is more than one expression available for the central-difference time integration approach. The Verlet equations used in DEM are given by

$$\begin{aligned} \mathbf{v}_p^{t+\Delta t/2} &= \mathbf{v}_p^{t-\Delta t/2} + \Delta t \mathbf{a}_p^t \\ \mathbf{x}_p^{t+\Delta t} &= \mathbf{x}_p^t + \Delta t \mathbf{v}_p^{t+\Delta t/2} \end{aligned} \qquad (2.13)$$

In an alternative form of the central-difference method, the incremental displacement is calculated directly from the particle accelerations at time t (Wood, 1990b), so that

$$\Delta \mathbf{x}_p^{t \to t+\Delta t} = \Delta \mathbf{x}_p^{t-\Delta t \to t} + \Delta t^2 \mathbf{a}_p^t \qquad (2.14)$$

where $\Delta \mathbf{x}_p^{t \to t+\Delta t}$ is the incremental displacement over the time increment from t to $t + \Delta t$, i.e. $\Delta \mathbf{x}^{t \to t+\Delta t} = \mathbf{x}_p^{t+\Delta t} - \mathbf{x}_p^t$ and $\Delta \mathbf{x}_p^{t-\Delta t \to t} = \mathbf{x}_p^t - \mathbf{x}_p^{t-\Delta t}$. This means that Equation 2.14 gives the acceleration as

$$\mathbf{a}_p^t = \frac{\left(\mathbf{x}_p^{t+\Delta t} - 2\Delta \mathbf{x}_p^t + \Delta \mathbf{x}_p^{t-\Delta t} \right)}{\Delta t^2} \qquad (2.15)$$

In either form, the central-difference algorithm is a second-order scheme, i.e. the accuracy of the calculated displacement depends on the square of the time increment, Δt^2. This time integration scheme has also been implemented for consideration of structural dynamics problems, and reference to Chopra (1995) may be useful to aid in developing an understanding of this method. One author who discusses the issue of accuracy arising from the truncation error explicitly is Cleary (2000) who stated that for this method between 20 and 50 time increments are needed to accurately resolve each collision in his simulations, resulting in very small time increments.

When choosing a method to integrate the particle accelerations and calculate the updated particle coordinates, it is important that the method chosen be both *consistent* and *convergent*. If the local truncation error at step i is τ, then the method is consistent if $\lim_{\Delta t \to 0} |\tau| = 0$ for all steps in the calculation sequence. A method is convergent if $\lim_{\Delta t \to 0} |\mathbf{x}^{\text{exact},t} - \mathbf{x}^t| = 0$ where $\mathbf{x}^{\text{exact},t}$ is the exact solution to the differential equation describing the particle motion at time t, and x^t is the calculated (approximate) value at the same time. The truncation error will be magnified as the analysis proceeds, so at time $t = n\Delta t$ the error will be magnified n times.

The algorithm must also be "stable." There are a number of ways of explaining what is meant by "stability" in the context of numerical modelling. In general, for a stable system if there are small changes in the initial data input to the model, the resultant changes in the output will also be small. If an error, E_0, is introduced at a given point in time, the error after n subsequent calculations, E_n, is the global error. As noted by Burden and Faires (1997) it is difficult to determine the global error, but there is a close correlation between the local error and the global error. Typically a linear growth in the global error will be unavoidable, meaning that if a local error E_0 is introduced at some point in the calculation, the cumulative effect of the error after n time increments is $E_n = CnE_0$, where C is a constant. If the relationship between the local and global truncation errors is $E_n = CnE_0$ then the algorithm is typically stable, however if $E_n = C^n E_0$ where $n > 1$, then there will be an exponential growth in error and the method is considered to be unstable. In mechanics applications analysts sometimes monitor the stability of a numerical model by calculating the total energy of the system. The components of the total energy include the strain energy stored in the contact springs and the particles' kinetic energy (refer to Section 2.6). Where the numerical integration is stable there will be no drift in the energy of the system. In an unstable system there will be a non-physical increase in energy in the system, i.e. energy is not conserved.

2.4 Stability of Central Difference Time Integration

The stability of the central-difference time integration approach is outlined in many basic numerical analysis texts (e.g. Burden and Faires (1997)). The basic idea of any time integration is that knowing the position and acceleration of a body we can predict its future displacement. Typically in numerical analysis/dynamics courses the concept is introduced by considering the free vibration of a particle of mass, m, suspended on a simple, elastic sphere with stiffness k. The dynamic equilibrium equation for this single degree of freedom system is then given by $a = -kx$, where $a = \ddot{x}$. For this simple system, if the central-difference approach is used, the maximum time increment that can be used is $\Delta t = \frac{T}{\pi}$, where T is the period for free oscillation of the system. This period is calculated as $T = 2\pi\sqrt{\frac{m}{k}}$. If predictions are made using a time increment that exceeds this critical value the results quickly become physically unreasonable and the analysis is said to be unstable. These restrictions on the choice of time increment that occur when using the central-difference approach to this simple, single degree of freedom system also apply in the multi degree of freedom simulations in DEM.

The critical time increment for stable analysis can be calculated using linear stability analysis by considering the *amplification matrix*, Zienkiewicz and Taylor (2000a). In general the amplification matrix, $\mathbf{A}$, is defined such that $\mathbf{x}_{t+\Delta t} = \mathbf{A}\mathbf{x}_t$. If any eigenvalue μ_i of $\mathbf{A}$ has a magnitude exceeding 1 (i.e. if $|\mu_i| > 1$) any initially small errors will increase without bound and the analysis will be unstable. Note that the spectral radius of $\mathbf{A}$, $\rho(\mathbf{A})$, is the maximum magnitude of an eigenvalue of $\mathbf{A}$, i.e. $\rho(\mathbf{A}) = max\,(|\mu_i|)$ Munjiza (2004) adopts a slightly different approach and defines an amplification matrix $\mathbf{A}^*$ for a single degree of freedom system with position x so that

$$
\begin{pmatrix} \dot{x}_{t+\Delta t}\Delta t \\ x_{t+\Delta t} \end{pmatrix} = \begin{pmatrix} 1 & -\frac{\Delta t^2 k}{m} \\ 1 & 1 - \frac{\Delta t^2 k}{m} \end{pmatrix} \begin{pmatrix} \dot{x}_t \Delta t \\ x_t \end{pmatrix} = \mathbf{A}^* \begin{pmatrix} \dot{x}_t \Delta t \\ x_t \end{pmatrix}
$$
(2.16)

Munijza shows that where $\frac{\Delta t^2 k}{m} \leq 4$ the spectral radius of $\mathbf{A}^*$, $\rho(\mathbf{A}^*)$ will be 1, however once $\frac{\Delta t^2 k}{m} > 4$ the spectral radius will increase beyond 1 and the simulation of the single degree of freedom system will be unstable. Stability analyses are completed by considering the undamped dynamic equilibrium equation, as is the case here. Wood (1990b) states that for simple algorithms, this assumption is valid.

Accepting this limitation of the central-difference method, it is necessary to examine the implications of the stability limitation for the multi degree of freedom systems encountered in DEM analyses. A DEM system is significantly more complex than the simple, single degree of freedom system. Each particle will have multiple contacts and multiple contact springs. At each contact there are two orthogonal springs acting normal and tangential to the contact. There will also most likely be a range of particle inertia values. O'Sullivan and Bray (2003b) proposed an approach to calculate a bound on the critical time increment for DEM simulations by drawing an analogy between a discrete element framework, and a finite element framework. In their analysis, the discrete element particles correspond to finite element nodes and that the interparticle contacts correspond to the finite elements, as illustrated in Figure 2.1. A global stiffness matrix can be assembled as in a finite element analysis, with the contact between particle i and particle j forming an "element" stiffness matrix, $\mathbf{K}_{ij}^e$ and the mass matrix including the inertia of the particles. Itasca (2008) give an alternative derivation for the stiffness at a contact point that also accounts for translational and rotational motion.

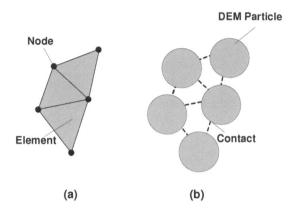

Figure 2.1: Analogy between a DEM model and a finite element mesh

In dynamic FEM (finite element method) or structural analysis an estimate for the critical time increment for linear systems can be considered by considering the global stiffness matrix. If it is assumed that this linear stability analysis also holds for nonlinear cases, then the maximum stable time increment (Δt_{crit}) is a function of the eigenvalues of the current stiffness matrix (e.g. Belytschko (1983)). For a linear, undamped system the relationship is given by:

$$\Delta t_{crit} = \frac{2}{\omega_{\max}} \qquad (2.17)$$

The maximum frequency, $\omega_{\max}$, is related to the maximum eigenvalue ($\lambda_{\max}$) of the $\mathbf{M}^{-1}\mathbf{K}$ matrix as

$$\omega_{\max} = \sqrt{\lambda_{\max}} \qquad (2.18)$$

Calculating the eigenvalues of a large matrix is an expensive operation. In explicit finite element analysis, the following relationship is frequently used (this is an extension of Rayleigh's theorem; a proof is given by Belytschko (1983)):

$$\lambda_{\max} \leq \lambda_{\max}^e \qquad (2.19)$$

where $\lambda_{\max}^e$ is the maximum eigenvalue of the $\mathbf{M}^{e-1}\mathbf{K}^e$ matrix for element "e", ($\mathbf{M}^e$ = element mass matrix, $\mathbf{K}^e$=element stiffness matrix). An estimate for the critical time increment can then be made by applying Equations 2.17 and 2.18, once $\lambda_{\max}^e$ is known.

O'Sullivan and Bray (2003b) compared two contacting DEM particles to a strut element in structural analysis. Referring to Figure 2.2 the effective stiffness of the contact between two disk particles, both with radius r, (considering both translational and rotational degrees of freedom), and a coordinate system parallel and orthogonal to the strut axis is given by:

$$\mathbf{K}^{e,local} = \begin{bmatrix} K_n & 0 & rK_s & -K_n & 0 & -rK_s \\ 0 & K_s & 0 & 0 & -K_s & 0 \\ rK_s & 0 & r^2K_s & -rK_s & 0 & -r^2K_s \\ -K_n & 0 & -rK_s & K_n & 0 & rK_s \\ 0 & -K_s & 0 & 0 & K_s & 0 \\ -rK_s0 & 0 & -r^2K_s & rK_s & 0 & r^2K_s \end{bmatrix}$$

(2.20)

In Figure 2.2, the particle nodal displacements in the local coordinate system are denoted by $(\bar{u}_1, \bar{v}_1)$ and $(\bar{u}_2, \bar{v}_2)$. To use this stiffness matrix in an eigenvalue analysis of the entire disk system, this local stiffness matrix must be rotated to the global Cartesian coordinate system using the following transformation matrix:

$$\mathbf{T} = \begin{bmatrix} \cos\theta & -\sin\theta & 0 & 0 & 0 & 0 \\ \sin\theta & \cos\theta & 0 & 0 & 0 & 0 \\ 0 & 0 & 1 & 0 & 0 & 0 \\ 0 & 0 & 0 & \cos\theta & -\sin\theta & 0 \\ 0 & 0 & 0 & \sin\theta & \cos\theta & 0 \\ 0 & 0 & 0 & 0 & 0 & 1 \end{bmatrix}$$

(2.21)

The stiffness of the strut element in the global coordinate system is then given by:

$$\mathbf{K}^{e,global} = \mathbf{T}^{-1}\mathbf{K}^{e,local}\mathbf{T}$$

(2.22)

This equation can be understood in a static sense by considering that

$$\mathbf{K}^{e,global}\Delta\mathbf{d} = \Delta\mathbf{F}$$

(2.23)

where $\Delta\mathbf{d}$ is a vector of displacement increments:

$$\Delta\mathbf{d} = \begin{pmatrix} \Delta u_1 \\ \Delta v_1 \\ \Delta \phi_1 \\ \Delta u_2 \\ \Delta v_2 \\ \Delta \phi_2 \end{pmatrix} \qquad (2.24)$$

where Δu_1, Δv_1, Δu_2, Δv_2 are the incremental (translational) centroidal displacements and $\Delta\phi_1$, $\Delta\phi_1$ are the incremental rotations of disks 1 and 2 about their centroids. These incremental rotations will introduce additional forces acting on particles 1 and 2, where

$$\Delta\mathbf{F} = \begin{pmatrix} \Delta F_{x,1} \\ \Delta F_{y,1} \\ \Delta M_1 \\ \Delta F_{x,2} \\ \Delta F_{y,2} \\ \Delta M_2 \end{pmatrix} \qquad (2.25)$$

and $\Delta F_{x,1}$ is the x-component of the increment in the force acting on particle 1, ΔM_1 is the increment in the moment acting on particle 1, etc.

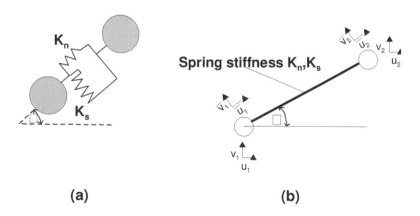

(a)   (b)

Figure 2.2: Analogy between a DEM model and an FEM mesh

Using this approach, a local stiffness matrix can be determined for each pair of contacting disks and rotated to a global reference system. To calculate the system eigenvalues a mass matrix is needed. In contrast to the strut in Figure 2.2, where the nodes have no inertia, and the mass of the strut is distributed to the nodes, DEM disks have inertia and they may be participating in many other contacts. In a simple first estimation it can be assumed that the mass of each disk is uniformly distributed to all of its contacts. Using this assumption our element mass matrix for contact between particles i and j is:

$$
\mathbf{M}_{ij}^{e,local} = \begin{bmatrix}
\frac{M_i}{n_c^i} & 0 & 0 & 0 & 0 & 0 \\
0 & \frac{M_i}{n_c^i} & 0 & 0 & 0 & 0 \\
0 & 0 & \frac{I_i}{n_c^i} & 0 & 0 & 0 \\
0 & 0 & 0 & \frac{M_j}{n_c^j} & 0 & 0 \\
0 & 0 & 0 & 0 & \frac{M_j}{n_c^j} & 0 \\
0 & 0 & 0 & 0 & 0 & \frac{I_j}{n_c^j}
\end{bmatrix}
\tag{2.26}
$$

where M_i is the mass of particle i, I_i is the moment of inertia of particle i about an axis through its centroid, n_c^i is the number of contacts surrounding particle i. Note that $\mathbf{M}_{ij}^{e,local} = \mathbf{M}_{ij}^{e,global} = \mathbf{M}^e$.

The calculation of the eigenvalues of the $\mathbf{M}^{e-1}\mathbf{K}^{e,global}$ matrix can be simplified as follows. As $\mathbf{M}^e$ is a diagonal matrix, then $\mathbf{M}^{e-1}$ is also diagonal and

$$
\mathbf{M}^{e-1}K^{e,global} = \mathbf{M}^{e-1}\mathbf{T}^{-1}\mathbf{K}^{e,local}\mathbf{T} = \mathbf{T}^{-1}\mathbf{M}^{e-1}\mathbf{K}^{e,local}\mathbf{T}
\tag{2.27}
$$

$\mathbf{T}$ is orthogonal so

$$
\mathbf{T}^{-1} = \mathbf{T}^T
\tag{2.28}
$$

and the matrices $\mathbf{M}^{e-1}\mathbf{K}^{e,local}$ and $\mathbf{T}^{-1}\mathbf{M}^{e-1}\mathbf{K}^{e,local}\mathbf{T}$ are similar and have the same eigenvalues (Golub and Van Loan, 1983). Then, for each contact element, the calculation of the eigenvalues of $\mathbf{M}^{e-1}\mathbf{K}^{e,local}$ is equivalent to calculating the eigenvalues of $\mathbf{M}^{e-1}\mathbf{K}^{e,global}$. Recognizing this equivalence and assuming that the normal and shear spring stiffnesses are equal (i.e. $K_n = K_s = K$)

the eigenvalues were calculated for a number of symmetrical configurations of uniform disks and spheres. The critical time steps were then determined using the above procedure. Full details of these calculations are given by O'Sullivan and Bray (2003b). The key conclusion of their study was that as the number of contacts per particle increased, n_c increased, and the magnitude of the terms in the element mass matrix decreased, thus increasing the maximum frequency and hence reducing the critical time increment Δt_{crit}.

To explore the implications of this stability limit for analyses consider a simple study, the case of a regular assembly of uniform spheres (with a face-centred-cubic lattice packing) subject to plane strain compression. Thornton (1979) provided an analytical solution for the peak strength of this assembly, knowing the inter-particle friction, so it is suitable for DEM validation (as discussed further in Chapter 11). The specimen considered contained 150 spheres and is illustrated in Figure 2.3, while the simulation input parameters are indicated in Table 2.1. The simulation was repeated a number of times, with different time steps, and the resulting responses are illustrated in Figure 2.4, with the theoretical strength clearly indicated for reference. The timestep values, Δt indicated are normalized by the ratio $\sqrt{\frac{m}{K}}$ where m is the particle mass and K is the spring stiffness, i.e.

$$\Delta t = \frac{\Delta t^{\text{sim}}}{\sqrt{\frac{m}{K}}} \tag{2.29}$$

and Δt^{sim} is the simulation time-step. Referring to Figure 2.4, the simulation with $\Delta t = 0.75$, is clearly unstable. However, for $\Delta t = 0.45$, while the results are incorrect, it would be difficult to determine that the results are erroneous in the absence of knowledge of the correct theoretical solution. When $\Delta t = 0.05$ and $\Delta t = 0.35$, the response is close to the theoretically correct value of peak strength. Note that an $\Delta t = 0.35$ is larger than the minimum critical time step for the face-centred-cubic packing case when both translation and rotation are allowed (O'Sullivan and Bray, 2003b). However, the lattice configuration considered in this analysis does not allow significant particle rotation, possibly explaining the lack of an apparent instability.

As discussed by Belytschko et al. (2000), numerical instabilities in explicit simulations can be detected by an energy balance check, as an instability results in the spurious generation of energy, which leads to a violation of the conservation of energy. From a molecular dynamics perspective, Rapaport (2004) also discusses the need to consider energy drift in order to ensure the reliability of DEM simulations. Rapaport noted that tests on angular momentum and energy conservation provide partial checks on the correctness of the calculation. O'Sullivan and Bray (2003b) demonstrated that the simulations exhibiting instability were associated with excessive errors in the energy balance of the system. (Calculation of the various components of energy in a DEM system is considered in Section 2.6).

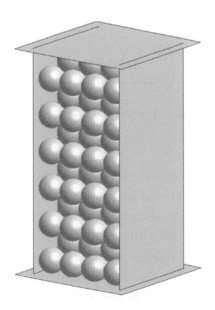

Figure 2.3: Illustration of specimen configuration for FCC plane strain simulation

The issue of numerical stability of DEM has not been widely

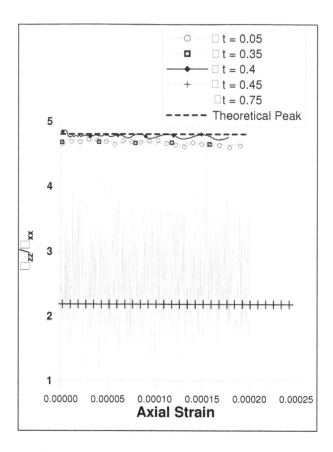

Figure 2.4: Specimen response in time-step parametric study

discussed in the DEM literature. Itasca (2008) discuss numerical stability in detail and present an alternative approach to determine the stability limit for an assembly of contacting spheres. Amongst those who have discussed this issue, Tsuji et al. (1993) used energy considerations to decide upon a value of $\Delta t_{\text{crit}} = \frac{\pi}{5}\sqrt{m/k}$ for their 2D simulations. An additional challenge is posed where a non-linear contact model is used (i.e. where stiffness varies). Itasca (2004) suggest that where the Hertz-Mindlin contact model is used, the safety factor on the time-step be reduced "especially under rapidly changing conditions." The Hertzian contact model relates the contact stiffness to the continuum shear stiffness of the material and is outlined in Chapter 3. Thornton and his colleagues

Parameter	Value
Normal Spring Stiffness	$1.0 \times 10^{11} M/T^2$
Shear Spring Stiffness	$1.0 \times 10^{11} M/T^2$
Density	$2{,}000 \ M/L^3$
Radius	20 L
Coefficient of Friction	0.3

Table 2.1: Analysis parameters for sensitivity analysis.($M =$ unit of mass, $L =$ unit of length, $T =$ unit of time)

(e.g. Thornton (2000) or Thornton and Antony (2000)) state that they select their simulation time-step by considering the minimum particle size and the Rayleigh wave speed. The Rayleigh wave velocity (v_r) for an elastic material with shear stiffness G and density ρ is (e.g. Sheng et al. (2004)):

$$v_r = \alpha \sqrt{G\rho} \qquad (2.30)$$

and α is given by the roots of

$$(2 - \alpha^2)^4 = 16(1 - \alpha^2) \left[1 - \frac{1 - 2\nu}{2(1 - \nu)} \alpha^2 \right] \qquad (2.31)$$

with an approximate value of α being given by $\alpha = 0.1631\nu + 0.876605$, where ν is the Poisson's ratio for the material. The critical time increment for DEM simulations with spheres and a Hertizan contact model is then given by (Sheng et al., 2004):

$$\Delta t_{\mathrm{crit}} = \frac{\pi R_{\mathrm{min}}}{\alpha} \sqrt{\rho G} \qquad (2.32)$$

where R_{min} is the minimum particle radius. Li et al. (2005) use a slightly different version of Equation 2.32, using the average particle radius rather than the minimum particle radius.

From the perspective of a DEM user the key points to note from the material presented in this section are:

1. The critical time-step is a function of the spring stiffness. The stiffer the spring the smaller the allowable time increment.

2. The critical time-step is a function of the particle mass, and hence density. The greater the density, the greater the allowable time increment.

3. The critical time-step is a function of the number of contacts. The greater the number of contacts the greater the overall, effective stiffness of the system and the smaller the allowable time step.

4. Where we use a non-linear contact model we need to consider the variation of stiffness during the simulation; the current tangential stiffness will govern the stability at any given time-step.

5. One way to monitor the development of instabilities is to consider the energy balance in the system; refer to O'Sullivan and Bray (2003b) for further details.

2.4.1 Density scaling

The critical time step is proportional to the particle mass, DEM users then often scale their particle densities to artificially increase the particle masses and hence the critical time increment to achieve results in reasonable run times. This approach is called "mass scaling" or "density scaling." Considering this option from a general computational mechanics perspective, Belytschko et al. (2000) suggest that mass scaling should be used in problems where high frequency effects are not important. When mass scaling is used it is assumed that the response of the system is not sensitive to inertia effects. Mass or density scaling is frequently used in discrete element analyses, including the simulations presented in this work and the studies described by Thornton (2000). Itasca (2004) propose the use of a differential density scaling coefficient so that the particle mass is increased to achieve a time increment of 1. They advise that this approach should be used with caution and that only the final steady state solution will be valid. Obviously the use of this approach in geomechanics will typically not be viable as the response observed and material behaviour tend to be path-dependent(i.e. the final answer depends on the details of how you get to that point). In a DEM simulation when a force or displacement is applied on a boundary, the response will propagate through the system. The speed of propagation of this "disturbance" will be a function of the properties of the system, most notably the contact spring stiffnesses, the particle mass, and the contact density.

Thornton and Antony (2000) scaled the particle density to ensure a quasi-static deformation state with "reasonable" simulation times. They argue that where this scaling of density is used the velocities and accelerations will be affected. However the simulations are quasi-static, and as no body force is applied, the contact forces and displacements are insensitive to the density value used.

In general, the use of density scaling should be approached with caution, and it is questionable whether its use can ever be recommended. To reduce the run time of simulations it seems preferable to maximize the rate of deformation in the simulations while ensuring that the simulations remain quasi-static by carefully monitoring the applied and internal stresses in the specimen.

2.5 Implicit Time Integration in Discrete Element Algorithms

Given the restrictions imposed by the central-difference time integration algorithm and the conditional stability of the central-difference and leap-frog algorithms, one might be tempted to use an implicit, unconditionally stable time integration approach in DEM. The idea of an implicit DEM algorithm is not new; in their original National Science Foundation (NSF) report describing the development of particulate DEM, Cundall and Strack (1978) cite the earlier work of Serrano and Rodriguez-Ortiz (1973) who developed an implicit DEM code. Other implicit approaches have been proposed or described by Ai (1985), Ke and Bray (1995), Zhuang et al. (1995), Tamura and Yamada (1996) and Holtzman et al. (2008). This section explores the implications of choosing this alternative approach.

In the absence of damping, any discrete element algorithm is concerned with solving the equilibrium equation at discrete time intervals for the system of particles. We can represent this at each time step as:

$$\mathbf{Ma} + \mathbf{Ku} = \mathbf{f} \tag{2.33}$$

where $\mathbf{M}$ is the (global) mass matrix, $\mathbf{a}$ is the acceleration vector (considering all the degrees of freedom), $\mathbf{K}$ is the (global) stiffness matrix, $\mathbf{f}$ is the force vector, and $\mathbf{u}$ is the incremental displacement vector. The elements of the stiffness matrix represent primarily the shear and normal springs that are present at the contact points, the "element" sub-matrices of the global stiffness matrix are the $\mathbf{K}^{e,\text{global}}$ matrices described above. $\mathbf{K}$ changes during the analysis as the system is non-linear.

In addition to the memory requirements considered above, Cundall and Strack (1978) drew attention to the fact that the matrix describing the contact stiffnesses must be reformulated every time a contact is made or is broken. The inherent geometric non-linearity restricts the time-increment to be very small so that changes in geometry can be captured accurately. The formation of new contacts, breakage of existing contacts, and initiation of sliding all need to be captured. A second consideration is that, as the particle rotations will contribute to the incremental tangential displacements and hence the shear force and linearization is involved, the accuracy of the calculation of the cumulative shear displacement will be improved where a small time increment is adopted. The use of a non-linear contact model will also require use of a small time step. Each of these considerations restricting the time-step to be small holds whether the time integration is implicit or explicit.

A numerical time stepping method is required in order to solve Equation 2.33 and obtain a history of the particle displacements over the specified analysis period. The system described in Equation 2.33 can be manipulated to form a set of simultaneous equations that can be solved for the unknown displacements, accelerations and velocities. In manipulating these equations, for simplicity, it is useful to consider a single degree of freedom system. The

Newmark's equations are used to develop relationships between the particle displacements, velocities and accelerations at times t and $t + \Delta t$.

$$\mathbf{x}^{t+\Delta t} = \mathbf{x}^t + \Delta t \mathbf{v}^t + \frac{\Delta t^2}{2}(1 - 2\beta)\mathbf{a}^t + \Delta t^2 \beta \mathbf{a}^{t+\Delta t}$$
$$\mathbf{v}^{t+\Delta t} = \mathbf{v}^t + \Delta t(1 - \gamma)\mathbf{a}^t + \gamma \Delta t \mathbf{a}^{t+\Delta t} \qquad (2.34)$$

where Δt is the time increment, $\mathbf{x}^t$ and $\mathbf{x}^{t+\Delta t}$ are the displacement vectors at time t and time $t+\Delta t$, respectively, $\mathbf{v}^t$ and $\mathbf{v}^{t+\Delta t}$ are the velocity vectors at time t and time $t+\Delta t$, and $\mathbf{a}^t$ and $\mathbf{a}^{t+\Delta t}$ are the acceleration vectors at time t and time $t + \Delta t$. The parameters β and γ define the variation of acceleration over the time-step and determine the stability and accuracy characteristics of the method. With a Newmark-type approach, for linear systems the requirement for unconditional stability is

$$2\beta \geq \gamma \geq 0.5 \qquad (2.35)$$

For example, referring to Ke and Bray (1995), the time integration scheme used in the implicit particulate discrete element analysis code DDAD is as follows:

$$\mathbf{x}^{t+\Delta t} = \Delta t \mathbf{x}^t + \frac{\Delta t^2}{2}\mathbf{a}^{t+\Delta t} + \mathbf{x}^{t+\Delta t}$$
$$\mathbf{a}^t = \frac{2}{\Delta t^2}(\mathbf{x}^{t+\Delta t} - \mathbf{x}^t) - \frac{2}{\Delta t}\mathbf{v}^t \qquad (2.36)$$

If Equation 2.36 is manipulated to take the form of Equation 2.34 it can be shown that the time integration scheme in DDA uses $\beta = 0.5$ and $\gamma = 1.0$; this is equivalent to taking the acceleration at the end of the time step to be constant over the time step (i.e. $a_t \equiv a_{t+\Delta t}$). This approach is implicit, meaning that to calculate the displacement up to time t the forces at time t are needed. Therefore, a large system of linear equations is required to be solved, at least once for each time increment. Using linear stability analysis it can also be shown that this approach is unconditionally stable. Consequently, theoretically a large time increment can be used. It is assumed that this linear stability also holds for non-linear cases.

In the DDA algorithm Thomas (1997) suggested limiting the time increment used to $\sqrt{8}\sqrt{(\frac{m}{k})}$ to ensure diagonal dominance of the DDAD stiffness matrix, and convergence of the solver used. Comparing the DDA and DEM algorithms, the critical time increments would need to differ by probably one order of magnitude in order for the computational costs associated with both methods to be equivalent. A key element of the implicit approach is that it requires the stiffness matrix at the end of the time increment to be determined. The stiffness matrix at the start of the time increment is used as an initial guess and the prediction is refined based on estimates of the particle displacements. This means that in an implicit approach the global set of equations must be solved numerous times in each time step in an iterative process to establish the stiffness matrix. A convergence criteria is difficult to establish. The source of the non-linearity in a DEM simulation renders the use of a Newton–Raphson approach to achieve convergence impractical. Shi (1988) proposed a complex system of open and close iterations for use in his DDA code. While the DDA method is viable for use in block DEM codes, where the number of bodies is relatively small, its applications to particulate systems with tens of thousands of particles is not viable.

The restrictions on implicit approaches noted here mean that the explicit Distinct Element Method algorithm proposed by Cundall and Strack (1979a) is the most commonly used approach for DEM. DEM is not unique in using an explicit time integration to model a highly non-linear problem. Belytschko et al. (2000) argue that explicit time integration is well suited to dynamic contact/impact problems as the time steps are small because of stability requirements, so the discontinuities due to contact-impact pose fewer challenges and neither linearization nor a non-linear solver is needed. From a general computational mechanics perspective, explicit methods are easier to implement than implicit approaches. They allow for element-by-element evaluation (or in the case of DEM, particle-by-particle evaluation) of the internal force vector, and hence do not require a global stiffness matrix with associated management and storage requirements. In explicit algorithms, for each time step the equations of motion for each particle are first integrated completely independently, as if not in contact. The uncoupled update indicates which parts of the body are in contact at the end of the time step, and then the contact conditions are imposed. Within each time step the iterations required in an implicit scheme are therefore avoided. Based upon these considerations, and realizing that to be realistic simulations of granular materials will need to include millions of small discrete particles, there is a strong case to adopt an explicit time integration scheme in particle-based discrete element codes, and consequently the DEM algorithm, as proposed by Cundall and Strack (1979a) is the dominant approach to discrete element modelling used currently in geomechanics.

More comprehensive discussions on the issues associated with time integration approaches specific to discrete element modelling can be found in Wang et al. (1996), Bardet (1998), O'Sullivan (2002) and Munjiza (2004). A comprehensive general reference on issues related to time integration is Wood (1990b). Doolin (2002) includes a detailed examination of the DDA time integration algorithm.

The conclusion that we can draw at this point is that while it is possible to create an implicit DEM formulation to overcome the numerical stability issues discussed in Section 2.4, implicit approaches are not easy to implement, they require computationally expensive iteration within each time increment, proving convergence is non-trivial, and the resulting global stiffness matrix, while sparse, will be very large. Whether an implicit or explicit approach is used, the time step in a discrete element simulation must be small to capture the changing contact conditions and to accurately calculate the increments in shear force caused by particle rotation.

2.6 Energy

In the analysis of any physical system it is useful to consider the energy of the system and the conservation of energy. Guidance on the calculation of energy terms in DEM simulations is given by Itasca (2004), Kuhn (2006), Bardet (1994) and O'Sullivan and Bray (2003b). During the simulation energy will be input via the boundary forces or body forces (i.e. gravity) and energy will be dissipated in frictional sliding and the rupture of contact springs. There will also be a continuous transfer or conversion of strain energy in the contact springs to kinetic energy and vice versa.

At any point the total translational kinetic energy, W_{kin}, in the system of particles is given by:

$$W_{\text{kin}} = \frac{1}{2} \sum_{p=1}^{N_p} \mathbf{v}_p m_p \mathbf{v}_p \tag{2.37}$$

where N_p is the number of particles in the domain, $\mathbf{v}_p$ is the velocity vector for particle p, and m_p is the mass of particle p.

In an individual spring, the stored strain energy is given by the area under the force-displacement curve. Therefore, where a linear force-displacement relationship is used the strain energy stored at each contact is given by

$$W_{\text{strain}}^c = \frac{F_n^{c2}}{2K_n} + \frac{F_s^{c2}}{2K_s} \tag{2.38}$$

where F_n^c is the normal component of the contact force, F_s^c is the tangential or shear component of the contact force for contact c and the normal and shear stiffnesses are given by K_n and K_s respectively. Where a non-linear contact constitutive model is used, an appropriate integration should be applied. The total strain energy in the system is the sum of the strain energy stored at each active contact.

Once the frictional strength of a contact is exceeded, energy will be dissipated in frictional sliding. The frictional energy dissipation is calculated incrementally so that

$$W_{\text{friction}}^{c,t} = W_{\text{friction}}^{c,t-\Delta t} + F_s^c \Delta s^{c,t-\Delta t \rightarrow t} \tag{2.39}$$

where $W_{\text{friction}}^{c,t-\Delta t}$ and $W_{\text{friction}}^{c,t}$ are the frictional energies dissipated up to time $t - \Delta t$ and time t respectively and $\Delta s^{c,t-\Delta t \rightarrow t}$ is the increment in tangential displacement at the sliding contact between times $t - \Delta t$ and t.

The sum, $W_{\text{strain}} + W_{\text{friction}}$ gives the total internal energy in the system (W_{int}). While there can be a transfer in kinetic energy to strain energy, energy will be dissipated or reduced in friction and so while the term W_{strain} is positive, W_{friction} is negative. Note also that when two particles lose contact there is a release in the elastic strain energy stored in that contact.

The system energy can only be incremented by input of external energy, (W_{ext}). The main contributors to the external energy are body forces $W_{\text{bodyforce}}$, the external applied forces $W_{\text{appliedforce}}$ and the boundary forces generated by interaction with rigid wall boundaries, $W_{\text{rigidwall}}$. Each of these components of the external energy can be expressed in incremental form. The total energy input by the body forces up to time t ($W_{\text{bodyforce}}^t$) is given by

$$W_{\text{bodyforce}}^t = W_{\text{bodyforce}}^{t-\Delta t} + \sum_{p=1}^{N_p} m_p \mathbf{b}_p \Delta \mathbf{x}_p^{t-\Delta t \to t} \qquad (2.40)$$

where $\mathbf{b}_p$ is the body force (applied acceleration) acting on particle p and $\mathbf{x}_p^{t-\Delta t \to t}$ is the increment in displacement of particle p from time $t - \Delta t$ to time t.

In a similar manner the applied external forces will generate an increment in energy so that the total energy input to time t is given by

$$W_{\text{appliedforce}}^t = W_{\text{appliedforce}}^{t-\Delta t} + \sum_{p=1}^{N_p} \mathbf{f}_p^{\text{app}} \Delta \mathbf{x}_p^{t-\Delta t \to t} \qquad (2.41)$$

where in this case $\mathbf{f}_p^{\text{app}}$ is a specific force applied to particle p, a possible origin of such an external force is the stress-controlled membrane discussed in Chapter 5. This contribution to the energy is calculated as the product of the force and the displacement of a particle, i.e. the work done.

The energy input due to motion of rigid wall boundaries (see Chapter 5), $W_{\text{rigidwall}}^t$, is also calculated by considering the increment in work done as the product of the force acting on wall w, $\mathbf{F}_w$, and the incremental displacement of wall w between $t - \Delta t$ and t, $\Delta \mathbf{x}_w^{t-\Delta t \to t}$:

$$W_{\text{rigidwall}}^t = W_{\text{rigidwall}}^{t-\Delta t} + \sum_{w=1}^{N_{rw}} \mathbf{F}_w \Delta \mathbf{x}_w^{t-\Delta t \to t} \qquad (2.42)$$

This discussion on the sources of input energy, stored energy and dissipated energy in a DEM simulation is not comprehensive. It is possible for a simulation to use a normal contact force model that dissipates energy during deformation, it is also possible that a form of numerical damping may be introduced in the simulation to simulate sources of energy dissipation that are not explicitly considered in the DEM model. Damping is considered in Section 2.7, below. It is important to understand the energy balance in a DEM simulation to avoid the problems with numerical stability discussed above. Referring to Belytschko et al. (2000) and O'Sullivan and Bray (2003b), the requirement for energy balance is that

$$W_{\text{kin}} + W_{\text{int}} - W_{\text{ext}} \leq \epsilon_{\text{max}} \left(W_{\text{kin}}, W_{\text{int}}, W_{\text{ext}} \right) \qquad (2.43)$$

where ϵ_{max} is a tolerance. Alonso-Marroqun and Wang (2009) give a nice description of how they use the consideration of the energy balance of their code as a means to check the accuracy of their simulations. From a geomechanics perspective, tracing the energy components in the system in detail will give insight into the material response. For example in their consideration of buckling mechanisms within the strong force chains that dominate granular material response (see Chapter 8), Tordesillas (2007) consider the strain energy stored at inter-particle contacts.

2.7 Damping

Where contact models that are elastic prior to yielding are used, the energy dissipation that occurs in physical systems of granular materials is not captured. Here yield is taken to mean the rupture of the contact spring in the contact normal direction or the initiation of inter-particle frictional sliding in the tangential direction. As highlighted by Cavarretta et al. (2010) and discussed in detail by Cavarretta (2009), in real inter-particle contacts there will be damage to surface asperities and plastic yielding from the initial formation of the contact. Subsequent to asperity yielding the contact response will be largely elastic, however, as noted in Chapter 3, as the stresses continue to increase plastic strains will develop in the solid particle material (this is captured in a contact model proposed by Thornton and Ning (1998)). This damage and yielding will dissipate energy, consequently the completely elastic contact models that are frequently used to describe the contact normal response in DEM codes are unrealistic. Munjiza (2004) describes this as a lack of "material damping" in rigid particulate DEM codes. The consequences for a DEM simulation are that if there is no yield by contact separation or frictional sliding the particles will vibrate constantly like a highly complex system of connected elastic springs. To avoid this non-physical phenomenon, DEM analysts often introduce numerical or artificial damping in their simulations.

To gain an initial understanding of the vibratory nature of particles in DEM simulations, consider the particle illustrated in Figure 2.5. In a simulation of this ball-wall system, a 1 kg ball rests on a horizontal boundary, gravity is switched on at time t=0, the beginning of the DEM analysis. The resultant response is illustrated in Figure 2.6. In Figure 2.6 three contact constitutive models are considered, a linear elastic contact model (with K=100 N/m), a non-linear elastic (Hertzian) contact model (with $G = 100N/m^2$ and $\nu = 0.3$), and a linear-elasto-plastic Walton-Braun type model (with K_1=100 N/m and K_2=1000 N/m). Each of these contact models is discussed in Chapter 3. Initially all three models behave in a very similar manner. In the absence of damping it is seen that both the linear elastic and non-linear elastic models will vibrate with the same period and oscillation amplitude as long as no external force modifies the contact conditions. For the non-linear model after yielding the frequency of the oscillation increases and the amplitude of the displacement vibrations decreases. Note however that the system still vibrates and the amplitude of the force vibrations is unchanged. Considering the energy components of the system, illustrated in Figure 2.7, it is observed that there is a decrease in the potential, kinetic and elastic strain energies after yielding occurs, however no further decrease in any of these energy components occurs. DEM codes include a damping parameter to damp out these non-physical vibrations. Here the two most common approaches to damping, mass damping and "non-viscous" damping, are considered. Viscous damping is discussed in Chapter 3.

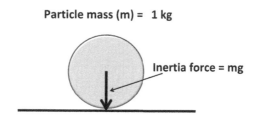

Figure 2.5: Single degree of freedom system: ball resting on horizontal boundary

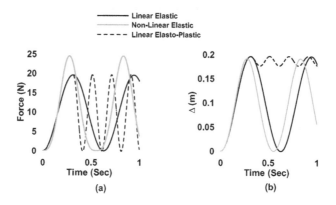

Figure 2.6: Single degree of freedom system: ball resting on horizontal boundary (a) Contact force versus time (b) Contact overlap versus time

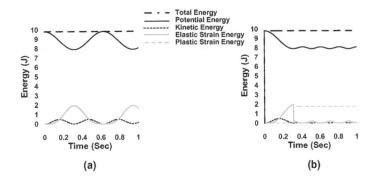

Figure 2.7: Single degree of freedom system: ball resting on horizontal boundary: energy considerations (a) Linear model (b) Walton-Braun model

Mass damping

Cundall and Strack (1979a) proposed a system of global damping, which "can be envisioned as the effect of dashpots connecting each particle to the ground." The amount of this damping that each particle "feels" is proportional to its mass. The way in which mass proportional damping is implemented in DEM analysis is summarized as follows (Bardet, 1998):

$$\mathbf{M}\mathbf{a}^t + \mathbf{C}\mathbf{v}^t = \mathbf{F}^t \tag{2.44}$$

where $\mathbf{M}$ is the mass matrix, $\mathbf{a}^t$ is the acceleration vector at time t, $\mathbf{C}$ is the damping matrix, $\mathbf{v}^t$ is the velocity vector at time t, and $\mathbf{F}^t$ is the force vector. Applying the Verlet time integration approach, with a time increment Δt:

$$
\begin{aligned}
\mathbf{a}^t &= \tfrac{1}{\Delta t}\left(\mathbf{v}^{t+\Delta t/2} - \mathbf{v}^{t-\Delta t/2}\right) \\
\mathbf{v}^{t+\Delta t/2} &= \tfrac{1}{\Delta t}\left(\mathbf{x}^t - \mathbf{x}^{t-\Delta t}\right) \\
\mathbf{v}^t &= \tfrac{1}{2}\left(\mathbf{v}^{t+\Delta t/2} + \mathbf{v}^{t-\Delta t/2}\right)
\end{aligned}
\tag{2.45}
$$

where $\mathbf{x}^t$ is the displacement vector at time t.

Substituting Equation 2.45 into Equation 2.44 gives

$$\frac{\mathbf{M}}{\Delta t}(\mathbf{v}^{t+\Delta t/2} - \mathbf{v}^{t-\Delta t/2}) + \frac{1}{2}\mathbf{C}(\mathbf{v}^{t+\Delta t/2} + \mathbf{v}^{t-\Delta t/2}) = \mathbf{F}^t \qquad (2.46)$$

assuming that the damping matrix is proportional to the mass matrix,

$$\mathbf{C} = \alpha\mathbf{M} \qquad (2.47)$$

then Equation 2.46 becomes

$$\mathbf{v}^{t+\Delta t/2}(1 + \alpha\Delta t/2) = \mathbf{v}^{t-\Delta t/2}(1 - \alpha\Delta t/2) + \Delta t\mathbf{M}^{-1}(\mathbf{F}^t) \quad (2.48)$$

which is equivalent to

$$\mathbf{v}^{t+\Delta t/2} = \mathbf{v}^{t-\Delta t/2}\left(\frac{1 - \alpha\Delta t/2}{1 + \alpha\Delta t/2}\right) + \left(\frac{\Delta t}{1 + \alpha\Delta t/2}\right)\mathbf{M}^{-1}(\mathbf{F}^t) \quad (2.49)$$

As outlined by Bardet (1998), Equation 2.49 is equivalent to the equation for dynamic relaxation.

Cundall (1987) discusses some of the limitations of mass proportional damping as follows:

1. This form of damping introduces body forces, which may be erroneous in flowing regions, and may influence the mode of failure.

2. The optimum proportionality constant (α) depends on the eigenvalues of the stiffness matrix.

3. The damping is applied equally to all nodes. In reality, different amounts of damping may be appropriate for different regions.

Local non-viscous damping

Cundall (1987) proposed an alternative damping system in which the damping force at each node is proportional to the magnitude of the out-of-balance-force with a sign that ensures that the vibrational modes are damped, rather than steady motion. The

66

"out-of-balance force" is the non-zero resultant force that acts on a particle to cause acceleration. Referring to Itasca (2004), the damping force is given by

$$\mathbf{F}_d = -\alpha^* |\mathbf{F}^p| \text{sign}(\mathbf{v}^p) \qquad (2.50)$$

where $\mathbf{F}_d^p$ is the damping force for particle p, α^* is the damping constant (default value of 0.7), $\mathbf{F}^p$ is the resultant or out-of-balance force acting on particle p, and $\mathbf{v}^p$ is the velocity vector for particle p. $\mathbf{F}_d$ acts in the opposite direction to $\mathbf{v}^p$, and sign($\mathbf{v}^p$) indicates the sign of the vector $\mathbf{v}^p$.

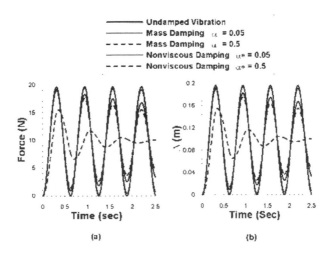

Figure 2.8: Single degree of freedom system: ball resting on horizontal boundary, linear spring. Comparison of response with mass damping and non-viscous damping

As argued by Itasca (2004), this form of damping has the advantage that only accelerating motion is damped, therefore no erroneous damping forces arise from steady-state motion. The damping constant is also non-dimensional and the damping is frequency independent. As proposed by Itasca (2004), an advantage of this approach is that it is similar to hysteretic damping, as the energy loss per cycle is independent of the rate at which the cycle is executed.

Figure 2.8 considers the response of the single degree of freedom ball-boundary system, with a linear contact model and $K = 100$ N/m as before. The undamped response is compared with the response observed with both mass damping and viscous damping. For both approaches it is seen that the contact force and deformation converge to a single value.

While damping is one means to overcome the non-physical nature of the contact constitutive models used in DEM simulations, it is difficult to select a physically meaningful value for damping or to relate the damping algorithms used to physical phenomena. In reality in a DEM simulation where particles are moving around each other, the dominant form of energy dissipation is in frictional sliding and contact breakages. Varying the damping may measurably effect the response, and it may be advisable to run DEM simulations with the damping parameter set to be very low or even zero. As highlighted in Chapter 7 it can be useful, however, to employ damping during specimen generation. Overdamping in quasi-static simulations can be avoided by checking that the system as a whole is in equilibrium, as discussed in Chapter 11.

2.8 Rotational Motion of Non-Spherical 3D Rigid Bodies

As noted in Section 2.2, consideration of particle rotations in 2D DEM simulations is trivial. However, in three-dimensional analyses the situation is more complex. For the general case, the translational and rotational degrees of freedom are coupled, and in addition there is a coupling between the three rotational degrees of freedom. If the reference point for the rotations is selected to be the body centre of mass, the coupling between the translational and rotational degrees of freedom is eliminated, and the

Newton-Euler equations can be used. There are a number of publications that consider the issues associated with calculation of the rotational motion of arbitrary 3D particles, and for further reading reference to Kremmer and Favier (2000), who provide further clarification of the mechanics involved, may be useful.

The inertia tensor for a 3D particle is given by

$$\mathbf{I} = \begin{bmatrix} I_{xx} & -I_{xy} & -I_{xz} \\ -I_{yx} & I_{yy} & -I_{yz} \\ -I_{zx} & -I_{zy} & I_{zz} \end{bmatrix} \tag{2.51}$$

The terms in this tensor are given by

$$\mathbf{I} = \begin{bmatrix} \int y^2 dm + \int z^2 dm & -\int xy dm & -\int xz dm \\ -\int yx dm & \int x^2 dm + \int z^2 dm & -\int yz dm \\ -\int zx dm & -\int zy dm & \int x^2 dm + \int y^2 dm \end{bmatrix} \tag{2.52}$$

where dm is an the mass differential. The integration is carried out with respect to a local Cartesian axis system, whose axes are parallel to the global Cartesian axes, but whose origin is at the particle's centre of mass. The diagonal elements I_{xx}, I_{yy} and I_{zz} are the moments of inertia of the particle, while the off-diagonal elements I_{xy}, I_{xz}, I_{yx}, I_{yz}, I_{zx}, and I_{zy} are called the products of inertia. The eigenvalues of this tensor, $\mathbf{I}$, give the principal moments of inertia $I_{x'}$, $I_{y'}$ and $I_{z'}$ and their orientations relative to the Cartesian axes are given by the eigenvectors.

For non-spherical particles, composed of aggregates of spheres, Itasca (2008) uses this inertia tensor directly when considering the rotational dynamic equilibrium using a local axis system centred on the particle centroid. The governing system of equations are given by

$$\mathbf{M} - \mathbf{W} = \mathbf{I}\dot{\omega} \tag{2.53}$$

where $\mathbf{M}$ gives the moments about the local Cartesian axes, i.e.

$$\mathbf{M} = \begin{pmatrix} M_x \\ M_y \\ M_z \end{pmatrix} \tag{2.54}$$

and

$$\mathbf{W} =$$
$$\begin{pmatrix} \omega_y\omega_z(I_{zz} - I_{yy}) + \omega_z\omega_z I_{yz} - \omega_y\omega_y I_{zy} - \omega_x\omega_y I_{zx} + \omega_x\omega_z I_{yx} \\ \omega_z\omega_x(I_{xx} - I_{zz}) + \omega_x\omega_x I_{zx} - \omega_z\omega_z I_{xz} - \omega_y\omega_z I_{xy} + \omega_y\omega_x I_{zy} \\ \omega_x\omega_y(I_{yy} - I_{xx}) + \omega_y\omega_y I_{xy} - \omega_x\omega_x I_{yx} - \omega_z\omega_x I_{yz} + \omega_z\omega_y I_{xz} \end{pmatrix}$$

$$(2.55)$$

The acceleration vector $\dot{\omega}$ includes the first derivatives with respect to time of the rotational velocities ω about the three axes:

$$\dot{\omega} = \begin{pmatrix} \dot{\omega}_x \\ \dot{\omega}_y \\ \dot{\omega}_z \end{pmatrix} \qquad (2.56)$$

Equation 2.53 therefore defines a system of three simultaneous equations with six unknowns. Itasca use an iterative approach to solve this system of equations, initially setting the angular velocities ω to be equal to the calculated velocities for the previous timestep, then calculating the accelerations $\dot{\omega}$. Itasca (2008) states that convergence is typically achieved within four iterations.

While Itasca's approach has been used in many geomechanics studies that have employed PFC, other DEM codes where non-spherical particles have been implemented use a slightly different approach. This alternative approach requires definition of three coordinate axis systems. Firstly there is the global, Cartesian coordinate system. There is also a local Cartesian coordinate system for each particle whose origin coincides with the particle centroid with axes parallel to the global Cartesian axes (as used by Itasca). Finally there is a local rotated coordinate system, whose origin is also at the particle centroid and whose axes coincide with the particle principal axes of inertia.

Consider an arbitrary point on the surface of a non-spherical particle with coordinates x_i^p in the global coordinate system. Then the coordinates in the local Cartesian coordinate system, x_i^{pb}, are given by $x_i^{pb} = x_i^p - x_i^c$, where the particle centroidal coordinates are x_i^c. This local Cartesian coordinate system is defined by axes (x_b, y_b, z_b). The axes of the local particle coordinate system are rotated relative to the local Cartesian coordinate system.

If the direction cosines defining the orientation of the principal axes of inertia of the particle relative to the local Cartesian coordinate system are given by $\left(n_{x'_b x_b}, n_{x'_b y_b}, n_{x'_b z_b} \right)$, $\left(n_{y'_b x_b}, n_{y'_b y_b}, n_{y'_b z_b} \right)$, $\left(n_{z'_b x_b}, n_{z'_b y_b}, n_{z'_b z_b} \right)$, then these direction cosines can be combined to give the following orthogonal matrix

$$T_{ij} = \begin{pmatrix} n_{x'_b x_b} & n_{x'_b y_b} & n_{x'_b z_b} \\ n_{y'_b x_b} & n_{y'_b y_b} & n_{y'_b z_b} \\ n_{z'_b x_b} & n_{z'_b y_b} & n_{z'_b z_b} \end{pmatrix} \qquad (2.57)$$

The coordinates of a point x_i^{pb} given in the local coordinate system can be rotated to obtain the coordinates in the particle-centred coordinate system $x_j^{'pb}$ by $x_j^{'pb} = T_{ij} x_i^{pb}$. If a series of coordinates in the local coordinate system define the edge of a particle, the particle can be rotated by multiplying all these edge coordinates by this matrix, without changing its shape or volume.

For an arbitrarily shaped body the relationship between the resultant moment acting on the particle about its centroid and its rotational motion is given by the Euler equations as

$$\begin{pmatrix} M_{x'} \\ M_{y'} \\ M_{z'} \end{pmatrix} = \begin{pmatrix} I_{x'}\dot{\omega}_{x'} + (I_{z'} - I_{y'})\omega_{z'}\omega_{y'} \\ I_{y'}\dot{\omega}_{y'} + (I_{x'} - I_{z'})\omega_{x'}\omega_{z'} \\ I_{z'}\dot{\omega}_{z'} + (I_{y'} - I_{x'})\omega_{y'}\omega_{x'} \end{pmatrix} \qquad (2.58)$$

where the subscripts x', y' and z' refer to a local coordinate system, centred at the particle centroid, and with the three orthogonal axes co-linear with the principal axes of inertia of the particle. The resultant moments about the three principal axes of inertia of the particle are denoted as $M_{x'}$, $M_{y'}$, and $M_{z'}$. The three rotational velocities are given by ω_x, ω_y, ω_z, and $\dot{\omega}_x$, $\dot{\omega}_y$, $\dot{\omega}_z$ are the time derivatives of these rotational velocities (i.e. the accelerations). For the simplest case of spherical particles $I_{x'} = I_{y'} = I_{z'}$ and Equation 2.58 becomes simply:

$$\begin{pmatrix} M_{x'} \\ M_{y'} \\ M_{z'} \end{pmatrix} = \begin{pmatrix} I_{x'}\dot{\omega}_x \\ I_{y'}\dot{\omega}_y \\ I_{z'}\dot{\omega}_z \end{pmatrix} \qquad (2.59)$$

Based upon earlier discussions, it is clear that Equation 2.59 can easily be integrated using the central-difference approach. This is a factor that contributes to the prevalence of spherical particles in 3D DEM simulations. However, for the general case (Equation 2.58) the coupling between the three rotational degrees of freedom remains and an alternative approach is needed.

Equation 2.58 can be rearranged to get expressions for the accelerations so that

$$
\begin{pmatrix} \dot{\omega}_{x'} \\ \dot{\omega}_{y'} \\ \dot{\omega}_{z'} \end{pmatrix} = \begin{pmatrix} \dfrac{M_{x'} - (I_{z'} - I_{y'})\omega_{z'}\omega_{y'}}{I_{x'}} \\ \dfrac{M_{y'} - (I_{x'} - I_{z'})\omega_{x'}\omega_{z'}}{I_{y'}} \\ \dfrac{M_{z'} - (I_{y'} - I_{x'})\omega_{y'}\omega_{x'}}{I_{z'}} \end{pmatrix} \tag{2.60}
$$

Thus while the complexity of the equations considered is reduced in comparison with Equation 2.53 the challenge of six unknowns distributed between three equations remains. Clearly the central-difference type, Verlet/leap-frog approach cannot be used. Vu-Quoc et al. (2000) describe a predictor-corrector approach, where they substitute an initial estimate of the $\boldsymbol{\omega}$ values into Equation 2.61, calculate the accelerations $\dot{\omega}$ and revise their estimate of $\boldsymbol{\omega}$. Kremmer and Favier (2000) adopt a slightly different linearized approach where the accelerations at time $t + \Delta t/2$ are calculated based upon the moments at time t and the rotational velocities at time $t - \Delta t/2$, so that

$$
\begin{pmatrix} \dot{\omega}_{x'}^{\,t+\Delta t/2} \\ \dot{\omega}_{y'}^{\,t+\Delta t/2} \\ \dot{\omega}_{z'}^{\,t+\Delta t/2} \end{pmatrix} = \begin{pmatrix} \dfrac{M_{x'}^{t} - (I_{z'} - I_{y'})\omega_{z'}^{t-\Delta t/2}\omega_{y'}^{t-\Delta t/2}}{I_{x'}} \\ \dfrac{M_{y'}^{t} - (I_{x'} - I_{z'})\omega_{x'}^{t-\Delta t/2}\omega_{z'}^{t-\Delta t/2}}{I_{y'}} \\ \dfrac{M_{z'}^{t} - (I_{y'} - I_{x'})\omega_{y'}^{t-\Delta t/2}\omega_{x'}^{t-\Delta t/2}}{I_{z'}} \end{pmatrix} \tag{2.61}
$$

Then the rotational acceleration at time t is calculated as $\dot{\omega}^t = \frac{1}{2}\left(\dot{\omega}^{t+\Delta t/2} + \dot{\omega}^{t-\Delta t/2}\right)$. Kremmer and Favier (2000) observe that the accuracy of this direct solution approach is limited in comparison with iterative approaches. Lin and Ng (1997) reduced the complexity of their system further by considering axisymmetric particles, resulting in a slightly different integration approach. Munjiza et al. (2003) proposed a method based upon the fourth-order Runge-Kutta method.

As outlined by Johnson et al. (2008), amongst other shortcomings, if the principal axis orientations are updated directly using these approaches there is a risk of loss of orthonormality, i.e. they may not remain mutually orthogonal. There is a growing consensus in the literature that quaternions are the most appropriate approach to adopt to deal with the challenges posed by time integration of non-spherical particles in three dimensions. The use of quaternions is considered by Zienkiewicz and Taylor (2000b), Sutmann (2002) and Rapaport (2004). Johnson et al. (2008) developed a time integration approach that fully integrates quaternions, while both Pöschel and Schwager (2005) and Vu-Quoc et al. (2000) use quaternions to update the principal axis orientations following calculation of the rotational velocities using the linear approach proposed above.

Quaternions were originally proposed by the Irish mathematician Hamilton. A complex number is represented as a sum of a real and an imaginary part $a + b \cdot i$. In an analogous manner a quaternion can also be represented as a sum with four terms $H = a + b \cdot i + c \cdot j + d \cdot k$. The fundamental formula of quaternion algebra is given by

$$i^2 = j^2 = k^2 = ijk = -1 \qquad (2.62)$$

Weisstein (2010) gives a number of examples of operations using quaternions.

The orientation of the principal axis of inertia of a particle can be related to the local Cartesian axis (centred at the particle centroid), by successive rotation through three angles, Φ, Θ, and Ψ about each of the coordinate axes. Adopting the notation of Pöschel and Schwager (2005), these angles are called the Euler angles. The corresponding quaternions are then given by

$$
\begin{aligned}
q_0 &= \cos\left(\frac{\Theta}{2}\right) \cos\left(\frac{\Phi+\Psi}{2}\right) \\
q_1 &= \sin\left(\frac{\Theta}{2}\right) \sin\left(\frac{\Phi-\Psi}{2}\right) \\
q_2 &= -\sin\left(\frac{\Theta}{2}\right) \sin\left(\frac{\Phi-\Psi}{2}\right) \\
q_3 &= -\cos\left(\frac{\Theta}{2}\right) \cos\left(\frac{\Phi+\Psi}{2}\right)
\end{aligned}
\qquad (2.63)
$$

The time derivatives of these quaternions relate to the rotations about the principal axes as follows:

$$
\begin{pmatrix} \dot{q}_0 \\ \dot{q}_1 \\ \dot{q}_2 \\ \dot{q}_3 \end{pmatrix} = \begin{pmatrix} q_1 & q_2 & q_3 \\ -q_0 & -q_3 & q_2 \\ q_3 & -q_0 & -q_1 \\ -q_2 & q_1 & -q_0 \end{pmatrix} \begin{pmatrix} \omega_{x'} \\ \omega_{y'} \\ \omega_{z'} \end{pmatrix} \tag{2.64}
$$

The solution of the set of 4 ordinary differential equations is then relatively straightforward, given that the vector ω is known. The Euler angles can be obtained by inverting the quaternion expressions.

In the simulation as a whole, a system of interacting particles whose locations are determined relative to a global Cartesian coordinate system is considered. All of the approaches noted above using the Euler equation are calculating the rotational velocities relative to a local axis and this local axis will differ for each particle considered. The DEM code then needs to keep track of the orientation of the local axis for each particle. In a rigid sphere cluster particle the coordinates of the constituent spheres will be updated relative to this local coordinate system and then related to the global coordinate system (using an orthogonal rotation tensor). Care must be taken in calculating the increments in the shear displacement at the contact points arising due to relative particle rotation and in using the rotation direction values in the analysis of shear bands or localizations.

The objective of the discussion on three-dimensional rigid body rotation given here is to highlight the complexity of the rotational motion, in comparison with the translational motion. Further details on the implementation of non-spherical particles in 3D DEM codes are given in the references cited in this section. Notably Munjiza (2004) describes some available approaches and Johnson et al. (2008) quantitatively compare different methods.

74

2.9 Alternative Time Integration Schemes

While the Verlet time integration algorithm is the most commonly used approach in particulate DEM, it is important to recognize that alternative, explicit time integration approaches exist. For example, Cleary (2000, 2008) states that he uses a second-order predictor-corrector time integration approach for his 2D simulations with super quadric particles. Cleary notes that his choice of time increment is such that about 15 time-steps can accurately resolve each collision. Xu and Yu (1997) also describe the use of a predictor-corrector approach to avoid overestimation of particle motion during the current time-step. Munjiza (2004) reviews a number of time integration schemes and compares their accuracy, stability and efficiency.

Predictor-corrector time integration schemes are multi-step integration methods (e.g. Burden and Faires (1997)). The central-difference method as implemented in DEM is essentially a single step method. The particle positions at time t give the accelerations at time t and these accelerations are double integrated to calculate the positions at time $t + \Delta t$. As noted by Munjiza (2004), this approach is a second-order time integration scheme, i.e. the accuracy of the method is proportional to the square of the time increment.

In a multi-step method information on the state of the system not only at time t, but also at previous times, is used to predict the particle positions at time $t + \Delta t$. Various multi-step methods are considered by Burden and Faires (1997), and many of them take a general form of

$$\mathbf{x}_p^{t+\Delta t} = \mathbf{x}_p^t + c_1 \mathbf{v}_p^{t+\Delta t} + c_2 \mathbf{v}_p^t + c_3 \mathbf{v}_p^{t-\Delta t} + c_4 \mathbf{v}_p^{t-2\Delta t} + c_{n+2} \mathbf{v}_p^{t-n\Delta t} \quad (2.65)$$

where (from the perspective of a particulate DEM simulation) $\mathbf{x}_p^{t+\Delta t}$, $\mathbf{x}_p^t$, $\mathbf{x}_p^{t-n\Delta t}$ are the particle positions at times $t + \Delta t$, t, $t - n\Delta t$, and $\mathbf{v}_p^{t+\Delta t}$, $\mathbf{v}_p^t$, $\mathbf{v}_p^{t-n\Delta t}$ are the corresponding particle velocities for n time-steps before time t. The parameters c_1, c_2, c_{n+2} are appropriate constants whose values can be determined using the Taylor series expansion. When the parameter c_1 is 0, the method is explicit, i.e. we can predict the value of displacement at time $t + \Delta t$ only considering the values at previous time increments. However, if the parameter c_1 is non zero the method is implicit. Typically, for the same number of steps (i.e. the same n value) an implicit approach will give a more accurate answer than an explicit approach. As discussed above there are many impediments to using implicit time integration in discrete element simulations, however the implicit multi-step methods can be adapted, so that an implicit multi-step method can be used to improve or "correct" the value of $\mathbf{x}_p^{t+\Delta t}$ calculated or "predicted" using an explicit approach. To assess the stability of a multi-step method, the characteristic equation of the method should be considered as described by Burden and Faires (1997).

Various predictor-corrector approaches exist. Pöschel and Schwager (2005) selected Gear's algorithm, and this method is also discussed by Munjiza (2004) and used by Garcia-Rojo et al. (2005). In this approach in the prediction phase the particle positions, and their derivatives, are calculated at time $t + \Delta t$ using a Taylor series expansion. Rather than considering the particle velocities at n previous time-steps, the higher-order derivatives of the displacement at time t are considered. Where a 5th order approach is used time derivatives up to $\frac{d^4}{dt^4}$ are considered in the expansion:

$$
\begin{aligned}
\mathbf{x}^{t+\Delta t} &= \mathbf{x}^t + \frac{d\mathbf{x}^t}{dt}\Delta t + \frac{d^2\mathbf{x}^t}{dt^2}\frac{\Delta t^2}{2} + \frac{d^3\mathbf{x}^t}{dt^3}\frac{\Delta t^3}{3!} + \frac{d^4\mathbf{x}^t}{dt^4}\frac{\Delta t^4}{4!} \\
\frac{d\mathbf{x}^{t+\Delta t}}{dt}\Delta t &= \frac{d\mathbf{x}^t}{dt}\Delta t + \frac{d^2\mathbf{x}^t}{dt^2}\Delta t^2 + \frac{d^3\mathbf{x}^t}{dt^3}\frac{\Delta t^2}{2!} + \frac{d^4\mathbf{x}^t}{dt^4}\frac{\Delta t^3}{3!} \\
\frac{d^2\mathbf{x}^{t+\Delta t}}{dt^2} &= \frac{d^2\mathbf{x}^t}{dt^2} + \frac{d^3\mathbf{x}^t}{dt^3}\Delta t + \frac{d^4\mathbf{x}^t}{dt^4}\frac{\Delta t^2}{2!} \\
\frac{d^3\mathbf{x}^{t+\Delta t}}{dt^3} &= \frac{d^3\mathbf{x}^t}{dt^3} + \frac{d^4\mathbf{x}^t}{dt^4}\Delta t \\
\frac{d^4\mathbf{x}^{t+\Delta t}}{dt^4} &= \frac{d^4\mathbf{x}^t}{dt^4}
\end{aligned}
\qquad (2.66)
$$

The contact forces and resultant forces on each particle at time $t + \Delta t$ are then calculated using the predicted particle coordinates, and from these forces the accelerations $\frac{d^2 x^{t+\Delta t}}{dt^2}$ can be calculated. The calculated accelerations are then used to correct or adjust the previously predicted derivative values by

$$
\begin{aligned}
\mathbf{x}^{t+\Delta t} &= \mathbf{x}^{t+\Delta t} + C_1 \frac{\Delta t^2}{2} \frac{d^2\mathbf{x}^{t+\Delta t}}{dt^2} \\
\frac{d\mathbf{x}^{t+\Delta t}}{dt} &= \frac{d\mathbf{x}^{t+\Delta t}}{dt} + C_2 \Delta t^2 \frac{d^2\mathbf{x}^{t+\Delta t}}{dt^2} \\
\frac{d^2\mathbf{x}^{t+\Delta t}}{dt^2} &= \frac{d\mathbf{x}^{t+\Delta t}}{dt} + C_3 \Delta t^2 \frac{d^2\mathbf{x}^{t+\Delta t}}{dt^2} \\
\frac{d^3\mathbf{x}^{t+\Delta t}}{dt^3} &= \frac{d\mathbf{x}^{t+\Delta t}}{dt} + C_4 \Delta t^2 \frac{d^2\mathbf{x}^{t+\Delta t}}{dt^2} \\
\frac{d^4\mathbf{x}^{t+\Delta t}}{dt^4} &= \frac{d\mathbf{x}^{t+\Delta t}}{dt} + C_5 \Delta t^2 \frac{d^2\mathbf{x}^{t+\Delta t}}{dt^2}
\end{aligned}
\tag{2.67}
$$

where C_1, C_2, C_3, C_4, C_5, are constant coefficients. When implementing Gear's algorithm different order schemes can be used, with the order of the scheme giving the highest derivative considered (Munjiza, 2004). The scheme presented above is a 4th scheme. If a different order scheme is used the coefficients in the corrector phase (Equation 2.67) will change. Consideration must also be given to the initialization.

There is a cost associated with the predictor-corrector scheme as the calculations required to determine the particle positions at each time-step are more complex, the information on the particle positions at more than one previous time-step needs to be calculated. However, Pöschel and Schwager (2005) completed a computer time profiling study and found that within each calculation cycle the calculations involved in updating the particle positions involves much less computing time than the calculations for the contact forces. They also argue that the increase in computing time associated with the predictor-corrector scheme is justified as, when comparison with the Verlet time integration scheme, a larger time-step can be used to achieve the same accuracy. In the context of the discussion on accuracy in Chapter 1 it is worth recalling that the number of approximations involved in calculating each of the higher-order derivative terms at time t can be very large.

Chapter 3

Calculation of Contact Forces

3.1 Introduction

In particulate DEM large numbers of interacting and potentially interacting bodies are considered. To determine the contact forces, or inter-particle reactions, the series of calculations must firstly identify which particles are contacting, then in a separate series of calculations the actual contact forces are calculated. These two stages are described as the *contact detection* and *contact resolution* stages of the analysis respectively (e.g. Hogue (1998)). Both sets of calculation involve predominantly geometrical calculations. The challenge with contact detection is to develop efficient algorithms to keep track of which particles are in contact or likely to come into contact and to form some type of "neighbour list". Contact resolution involves the accurate calculation of the contact geometry and kinematics, typically characterized by the overlap depth/separation and sometimes also relative tangential motion; however, in some cases the overlap area or volume may be considered. A contact constitutive model is then used to calculate the contact forces from this relatively simple description of the contact geometry. The sequence of calculations associated with calculating the contact forces is undoubtedly the most time-consuming aspect of a DEM simulation, Sutmann (2002) estimates that these calculations account for 90% of the DEM simulation time. The proportion of the analysis time taken up in contact resolution depends on the packing density of the system; when the void ratio is lower, the number of contacts per particle increases and the proportion of the analysis time taken up with contact resolution also increases.

To achieve computational efficiency, relatively straightforward analytical expressions to calculate the contact force are needed. In reality the load–deformation response of two contacting soil particles is highly complex. As noted by Zhu et al. (2007), it would be very difficult to accurately describe inter-particle contact as the distribution of the contact stress (or traction) depends on the particle geometry and material properties as well as the particle motion. At a sub-particle scale the asperities on the surfaces of the particles will initially contact, then after these deform and yield, the particles will interact over a (typically small) finite area. To facilitate the analytical description of their geometry, DEM particles are almost always analytically described as smooth surfaces and most DEM models simplify the contact to be a single point. The strains experienced by the contacting particles and the non-uniform stress distributions that are induced are not explicitly considered in the DEM simulation. Instead, the overlap between the rigid particles is considered to represent the deformation.

The contact forces in a DEM model are taken to represent the integral of the real stresses or tractions acting along a physical contact. The resultant inter-particle force is resolved into its two orthogonal components, normal and tangential to the contact point. Then, the stress-deformation response at the contact is represented using two orthogonal rheological models, acting in the normal and tangential directions respectively. These rheological models typically comprise a combination of springs, sliders and dashpots and are often called contact constitutive models or contact models. Adopting this simple contact modelling approach is one of the key fundamental aspects of DEM. It allows interactions between very large numbers of particles to be considered with good computational efficiency. The level of realism of the contact models can be advanced by specifying non-linear force-displacement relationships for the contact springs, or combining systems of springs and dashpots in various ways.

This Chapter introduces some of the more common contact models that have been adopted in geomechanics applications. The range of interaction models that have been considered to date extends beyond those considered here and this promises to be an interesting area of DEM research in future years. Readers interested in contact force modelling outside of geomechanics may wish to refer to Zhu et al. (2007). One challenge in this area is to accurately calculate the particle scale forces in micro-scale laboratory experiments. While contributions to meet this need have been made by researchers including Cavarretta et al. (2010), Yu (2004) correctly identified that better experimental micro-scale characterization is a central requirement necessary to improve DEM models.

3.2 Idealizing Contact for Particulate DEM Simulations

In particulate DEM contact forces are usually calculated by introducing "virtual" springs at the contact points, as illustrated in Figure 3.1. The particles considered in DEM simulations are completely rigid. However, as discussed in more detail later in this Chapter, in reality the particles acting in compression will deform at the contact points. In a DEM model, this deformation is simulated by allowing a small amount of overlap at the contact points. The springs are not restricted to being linear elastic. The equation that defines the force-deformation relationship for the contact spring is called a *contact constitutive model*. The calculation of the normal and tangential force components is largely decoupled, i.e. they are calculated separately. Two orthogonal rheological models are activated at the contact point to calculate the force components. These rheological models use combined systems of springs, sliders and dashpots with various complexities. The slider in the contact normal direction is used to prevent or limit tensile forces developing between the particles, while the slider in the tangential direction allows the particles to move relative to each other

81

when the contact frictional strength is exceeded (calculated using Coulomb friction). The symbol F_n is used to denote the normal component of the contact force, while the tangential component is denoted by F_t. Moments will also be transmitted to the particles, causing rotations. Irrespective of the particle geometry, moments will be imparted as a consequence of the tangential component of the contact forces. Moments will also be generated in the case of a finite contact area when the contact stress distribution is not symmetrical about the centre of the contact point. The axis of application of the moments is orthogonal to the contact normal. Further discussion on moment transmission at particle contacts is given in Section 3.9 below.

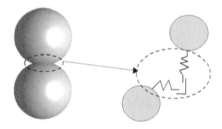

Figure 3.1: Schematic diagram of approach used to model contact in DEM

Figure 3.3 illustrates some of the geometrical features considered in a DEM simulation based upon the contact between two disks (although this diagram could also be considered to be a section through two contacting spheres). In Figure 3.3 the vector defining the contact normal has the same orientation as the vector joining the centroids of the two contacting disks as they are circular. The contact plane is at right angles to the contact normal and the contact coordinates are taken to be at the middle of the contact. If the contact normal orientation is described by the vector $\mathbf{n} = (n_x, n_y)$, the orientation of the tangent to the contact is given by $\mathbf{t} = (-n_y, n_x)$. Knowing the contact normal and the coordinates of the contact point, the equation of the contact plane (contact line in 2D) can easily be determined.

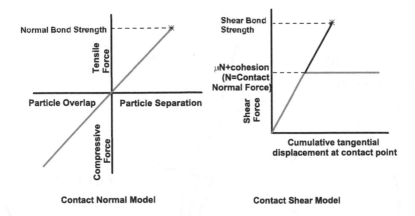

Figure 3.2: Diagram of normal and shear contact force models in DEM

Where the contact forces are non-zero there will be a contribution to the resultant forces acting on both contacting particles. These two contributions will be equal in magnitude and opposite in direction. In compression the normal inter-particle normal contact force calculated from the particle overlap acts to repulse the two contacting particles from each other. If there is a small separation between the particles, tensile forces will act to draw the particles towards each other, unless the limiting tensile force is exceeded. The limiting tensile force is most often specified to be zero in geomechanics, i.e. no inter-particle tension is allowed. The tangential forces will induce both relative rotation and translation. For non-circular or non-spherical particles the contact normal forces can impart a rotation and contribute to resist particle rotation.

The dominant approach used in particulate DEM can be classified as a penalty spring approach. Generally, in computational mechanics the approach adopted when the penalty method is used to model contact is to introduce a very stiff spring at the point of contact. Penetration then occurs at the contact point; however, a small amount of penetration yields a relatively large force acting equally (but in opposite directions) on the contacting bodies. As outlined by Munjiza (2004), there are two options in the penalty

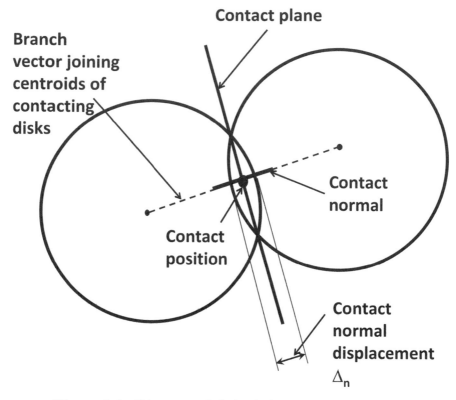

Contact plane

Branch vector joining centroids of contacting disks

Contact normal

Contact position

Contact normal displacement
Δ_n

Figure 3.3: Diagram of disk–disk contact geometry

formulation. One option is for the force to equal the magnitude of the overlap (i.e. the maximum overlap distance) times the penalty spring stiffness. Another option is to relate the area or the volume of the contact overlap to the contact force. This might be more appropriate in the case of a distributed contact force. Most particulate DEM implementations use the magnitude of the overlap to calculate the contact normal force.

It is useful to be aware of the other approaches used in general in numerical modelling contact (e.g. in the finite element method). Zienkiewicz and Taylor (2000b) consider three methods to model contact: the Lagrangian multiplier method, the penalty method and the augmented Lagrangian method. Munjiza (2004) extends this list to include the least squares methods. As outlined by Munijza, where Lagrange multipliers are used, an additional term is added to the global equilibrium equation so that the forces are solved for directly (refer to Munjiza (2004) or Zienkiewicz and Taylor (2000b), who use variational calculus to derive the modified dynamic equilibrium equations). Munijza identifies the two main drawbacks to the use of the Lagrange multiplier method over the use of a penalty spring approach to be the increase in the number of unknowns and the fact that in explicit analysis the impenetrability constraints are only approximately satisfied. As already noted in Chapter 1, the contact dynamics method proposed by Jean (2004) uses a slightly different approach such that there is no penetration at the contact points.

3.3 An Overview of Contact Mechanics

Before presenting the expressions for the contact constitutive models used in particulate DEM, it is useful to consider some of the theory associated with contact between two elastic spheres. The discussion presented here serves to highlight some of the basic concepts of contact mechanics relevant to particulate DEM. Readers interested in developing a more in-depth understanding of the load–deformation response observed at contact between solid bodies, and the stress distributions around the contact points, should refer in the first instance to the general text on contact mechanics by Johnson (1985).

Johnson clearly distinguishes *conforming* and *non-conforming* contacts. In a conforming contact the surfaces fit together closely prior to deformation. The two surfaces interacting at a non-conforming contact have dissimilar profiles and will initially contact only at a single point. As illustrated in Figure 3.4(a), the contact between two disks or two spheres will be non-conforming. Where the contacts are non-conforming the contact area is small relative to the particle dimensions and the stresses are highly concentrated in the contact zone. The stresses are not greatly influenced by the geometry of the body distant from the contact area. Most DEM models use spheres or disks as their basic particle types; therefore, the simulated contacts are non-conforming and the contact models have been developed based on a point contact assumption. One of the most commonly used contact models in geomechanics, the Hertzian contact model, was developed based upon the application of elastic theory to non-conforming spherical elastic bodies.

A theoretically perfectly conforming contact is illustrated in Figure 3.4(b), and, as can be appreciated from Figure 3.4(c), where real particles contact the complexity of their morphologies can generate a highly complex contact condition with more than one inter-particle contact. As noted by Fonseca et al. (2010), in a natural soil there are a variety of contact types, including many conforming contacts (refer also to Chapter 10). The real physical contacts will be three-dimensional, adding additional complexity to the contact geometry. Any particle surface of interest in geomechanics will have many surface asperities. As illustrated in Figure 3.4(d), the initial contact between surfaces is likely to be a non-conforming contact between two surface asperities, transitioning to a conforming contact as the asperities yield.

At the contact surface the surface pressures arising due to the contact forces are referred to as tractions. The term "traction" is often used in mechanics to describe the surface force per unit area acting along a boundary. The symbols f_n and f_t are used here to refer to the normal and tangential components of the surface tractions respectively. The contact forces are determined by integration of the tractions over contact area (A_c) as follows:

$$F_n = \int\limits_{A_c} f_n dA$$
$$F_t = \int\limits_{A_c} f_t dA \tag{3.1}$$

In Figure 3.5(a) the traction at the contact between two smooth spherical particles is illustrated. The traction is symmetric about the centre of the contact. The smooth, convex nature of the particles means that no rotational resistance is provided at the contact point and hence moment transmission cannot occur. Figure 3.5(b) illustrates the distribution of the normal tractions along the contact surface for a conforming contact. In this case the geometry of

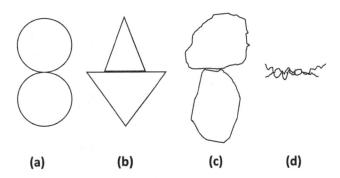

(a)   (b)   (c)   (d)

Figure 3.4: Contact types; (a) Non-conforming contact (b) Conforming contact (c) Soil particle contact (d) Contact between asperities on soil particle surface.

the particles provides a resistance to rotation and is transmitting both a resultant moment as well as a normal compressive force. The asymmetry in the traction distribution is a consequence of the moment loading.

3.4 Contact Response Based Upon Linear Elasticity

3.4.1 Elastic normal contact response

Anyone who has completed a basic undergraduate course in geotechnical engineering including shallow foundation design will have an appreciation of the complexity of the distribution of stresses beneath a contact. In shallow foundation analysis, the integration of the Boussinesq expression for a point load at the surface of an elastic half space is often used to derive an expression for the spatial variation in stress beneath the foundation. While the response

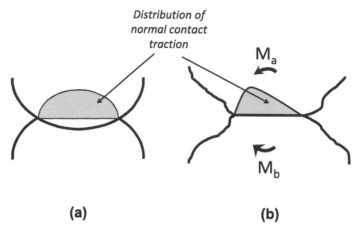

(a)   **(b)**

Figure 3.5: Illustration of normal traction distributions (a) Contact between smooth disks or spheres (b) General case with a non-conforming contact

of real soil is highly non-linear, this assumption of a linear elastic response is pedagologically useful when developing an understanding of the real stress distributions. Similarly, the use of continuum elasticity can help to develop an understanding of the response at the contact between two soil particles as compressive forces are transmitted between the particles.

Using elastic theory, expressions for the stress distributions and deformations within two contacting particles can be derived. These expressions can then give a description of the load–deformation response at the contact between two continuous, non-conforming bodies based on a contact mechanics theory proposed by Hertz. Hertzian contact mechanics assumes an initial point contact between the solid bodies and then provides expressions describing the growth of the contact area, the variation in surface tractions, the surface deformations and the stresses within the particles. This approach assumes that the solid contacting bodies are linearly elastic and then solves a boundary value problem, assuming that each contacting body is an elastic half space and that the contact area is elliptical in shape.

It is important to appreciate the assumptions underlying Hert-

88

zian contact mechanics so that the limitations of its applicability in geomechanics can be established. In the first instance the surface properties of the contacting particles are greatly idealized; the presence of surface asperities is neglected (i.e. the contacting surfaces are assumed to be perfectly smooth), and the spheres are assumed to be frictionless. The contact area is also assumed to be small relative to the dimensions of the contacting bodies and the strains induced are assumed to be sufficiently small that the material response remains linear elastic. It is also assumed that there is no interaction outside the loaded area (e.g. tensile forces). As long as the stiffnesses of the two contacting bodies are equal, friction is not brought into play as there is no relative tangential straining along the contact surface.

The relationships derived using Hertzian theory for the contact between two spheres s_1 and s_2 are expressed in terms of an effective particle radius, R^*, and an effective Young's modulus, E^*. These two parameters are given by

$$\frac{1}{R^*} = \frac{1}{R_{s_1}} + \frac{1}{R_{s_2}} \tag{3.2}$$

and

$$\frac{1}{E^*} = \frac{1 - \nu_{s_1}^2}{E_{s_1}} + \frac{1 - \nu_{s_2}^2}{E_{s_2}} \tag{3.3}$$

where R_{s_1} and R_{s_2} are the sphere radii, the Young's moduli are E_{s_1} and E_{s_2} and the Poisson's ratio are ν_{s_1} and ν_{s_2} for spheres s_1 and s_2 respectively.

This theory gives an expression for the radius of the circle defining the contact as follows:

$$a = \left(\frac{3F_n R^*}{4E^*} \right)^{1/3} \tag{3.4}$$

The maximum contact traction, $f_n^{\max}$, is

$$f_n^{\max} = \left(\frac{6F_n E^{*2}}{\pi^3 R^2} \right)^{1/3} \tag{3.5}$$

The deformation at the contact point, i.e. the increase in the proximity of the centroids of the two contacting particles, is given by

$$\delta = \left(\frac{9F_n^2}{16R^* E^{*2}} \right)^{1/3} \tag{3.6}$$

Figure 3.6 illustrates some aspects of contact response calculated using Hertzian theory. The parameters used to generate the results presented in Figure 3.6 were selected referring to Thornton (2000), as follows: $R_a = R_b = 0.258$, $E_a = E_b = 70$ GPa and $\nu_a = \nu_b = 0.3$. Figure 3.6(a) illustrates the variation in contact force as a function of the contact deformation for these values. Even though the material response is assumed linear elastic, a non-linear force-deformation response is observed, with the effective contact stiffness increasing as the contact force increases. This non-linearity can be understood by considering the variation in the contact area as the force increases. The rate of increase in surface area with deformation decreases as the deformation progresses. Figure 3.6(b) illustrates the radius of the contact surface at three discrete load levels (10 N, 50 N and 100 N). The variation in stresses within the particle along the line joining the centre of the contact surface and the particle centroid is illustrated in Figure 3.6(c). The normal stress within the particle σ_z is oriented along this line, while the radial stress σ_r is orthogonal to this line. The stress conditions will be symmetrical about this line.

While both σ_z and σ_r decrease monotonically from their maximum values at the contact surface, the variation in the deviator stress $\sigma_z - \sigma_r$ is more complex. It is important to consider the deviator stress as this value is used in both the von Mises and the Tresca failure criteria that can be applied to the solid particle material. The von Mises failure criterion states that

$$\frac{1}{6} \left[(\sigma_1 - \sigma_2)^2 + (\sigma_2 - \sigma_3)^2 + (\sigma_1 - \sigma_3)^2 \right] = \frac{Y^2}{3} \tag{3.7}$$

The Tresca failure criterion states that

$$\max \left(|\sigma_1 - \sigma_2|, |\sigma_1 - \sigma_3|, |\sigma_2 - \sigma_3| \right) = Y \tag{3.8}$$

where Y is the yield stress for the material. These equations then determine the onset of yield and plastic deformation in the particles. Thornton (1997a) and Thornton and Ning (1998) proposed that initially during loading, Hertzian elasticity can adequately describe the contact normal pressure distribution, however as loading progresses there is a "plastic" phase where the truncated Hertzian pressure distribution applies. The resultant force-displacement relationship for the plastic phase is linear (Thornton and Liu, 2000). For real soil particles, there will be flaws and cracks in the particles, resulting in inhomogeneous stress conditions. The (gross) particle failure mode is typically brittle.

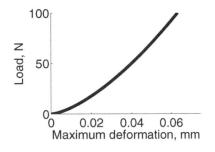

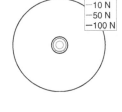

(a) Variation of Normal Force, F_n with contact deformation δ

(b) Variation in contact extent as F_n increases

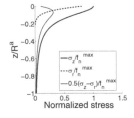

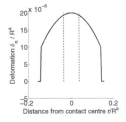

(c) Variation in stresses along the particle axis beneath the contact

(d) Variation in normal deformation along contact area

Figure 3.6: Illustration of normal contact force deformation relationships

3.4.2   Elastic tangential contact response

In comparison with the normal component of the contact force, the contact response along or tangential to the contact surface (i.e. orthogonal to the contact normal) is more difficult to understand. While (again) the key reference for this topic is Johnson (1985), the description of the response provided by Thornton (1999) (who considers the mechanics from a particulate DEM perspective) is (arguably) clearer. The discussion of tangential contact response presented here draws on these sources and is included here to facilitate a better appreciation of the assumptions inherent in the contact models used in DEM simulations.

The work of Mindlin (1949) and Mindlin and Deresiewicz (1953) forms the basis for some of the most important models of tangential response used in DEM simulations. A central assumption of this approach is that the tangential traction does not influence the normal traction distribution, which is assumed to follow the Hertzian response described above. This assumption holds true only for contact between two spheres with the same elastic properties. Mindlin (1949) showed that in the case where there is no variation in the normal force F_n, when a tangential force is applied, there will be a "slip" over part of the contact area, while over the remaining contact area there is "stick", i.e. there is no relative movement at the contact. From Hertzian theory, the contact area is circular. The slip region is then an annular region around a central circular "adhered" or "sticking" area. As outlined by Johnson (1985), slip is irreversible, consequently, there is an added complexity introduced as the contact state (i.e. the relationship between force and displacement) depends on the history of loading.

The laws of friction proposed by Amontons and Coulomb give the relationship between the normal and tangential tractions in the slip region to be as follows:

$$f_t(r) = \mu f_n(r) \tag{3.9}$$

where, as in the case of Hertzian normal contact, r represents the distance from the centre of the circular contacting area along the contact surface and f_n and f_t are the normal and tangential tractions respectively. This is illustrated in Figure 3.7(a), at the limiting condition where contact sliding is about to occur and $f_t(r) = \mu f_n(r)$ for $0 \leq r \leq a$, where a is the contact radius. Then the distribution of tangential tractions is simply calculated by multiplying the normal traction by the coefficient of friction. The entire extent of the contact area is then said to be slipping (Figure 3.7(b)). However, prior to this point of "gross yield", the modelling of the tangential response is non-trivial and the response depends on whether the contact is being loaded in the tangential direction for the first time or whether the contact is already experiencing a tangential force that experiences a change in direction.

Initial tangential loading

For the initial tangential loading, it is assumed that a contact subject to a normal force F_n experiences a tangential force that increases steadily (monotonically) from a value of 0 to F_t. If the radius of the adhered or stuck area is b and, as before, a is the radius of the total contact area (calculated using Hertzian theory), then the tangential traction at a distance r from the centre of the contact area is given by

$$f_t(r) = \tfrac{3\mu F_n}{2\pi a^3} \sqrt{a^2 - r^2} \qquad\qquad b \leq r \leq a$$

$$f_t(r) = \tfrac{3\mu F_n}{2\pi a^3} \left(\sqrt{a^2 - r^2} - \sqrt{b^2 - r^2} \right) \quad 0 \leq r \leq b$$

(3.10)

The resulting distribution of tangential tractions, assuming that $b = 0.5a$, is illustrated in Figure 3.7(c), and the extent of the slip area relative to the area in "stuck" mode can be appreciated by reference to Figure 3.7(d).

Mindlin (1949) gave the relative tangential displacement of the two contacting spheres as

$$\delta_t = \frac{3\mu F_n}{16 G^* a} \left(1 - \frac{b^2}{a^2} \right)$$

(3.11)

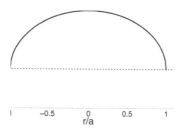

(a) Normalized tangential traction: First loading, fully sliding contact

(b) Extent of area experiencing slip: First loading, fully sliding contact

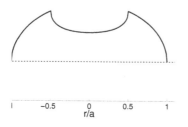

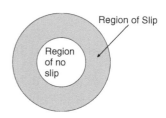

(c) Normalized tangential traction: First loading, partially sliding contact

(d) Extent of area experiencing slip: First loading, partially sliding Contact

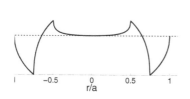

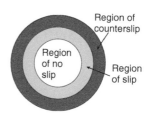

(e) Normalized tangential traction: First unloading, partially sliding contact

(f) Extent of area experiencing slip: First unloading, partially sliding contact

Figure 3.7: Illustration of tangential traction distributions and extent of slip areas for sliding of two spheres

where

$$\frac{1}{G^*} = \frac{2 - \nu_{s1}}{G_{s1}} + \frac{2 - \nu_{s2}}{G_{s2}} \tag{3.12}$$

where G_{s1} and G_{s2} are the shear moduli of the contacting spheres s_1 and s_2 respectively.

The tangential force can then be obtained from the following integral:

$$F_t = 2\pi \int_0^a f_t(r)r\,dr = \mu F_n \left(1 - \frac{b^3}{a^3}\right) \tag{3.13}$$

As the tangential force increases, the extent of the slip zone increases, i.e. b decreases and the slip zone progresses inwards. Eventually all of the contact area is slipping and, at this point $F_t = \mu F_n$ and $f_t(r) = \mu f_n(r)$ for $0 \le r \le a$. At any point during this period of monotonically increasing tangential loading, the relationship between the radius of the stick region (b) and the contact radius (a) is given by

$$\frac{b}{a} = \left(1 - \frac{F_t}{\mu F_n}\right)^{1/3} \tag{3.14}$$

Unloading - reversal of tangential force, F_t

The process of slip is dissipative and thus the slip region will not shrink once the direction of loading it reversed. Instead a region of micro-slip or counter-slip will begin at the edge of the contact. Then the response along the surface will be divided into three zones: an area with no slip, an area in slip and an area in counter slip. As explained by Thornton (1999) the energy required to produce the "annulus of counterslip" is twice that required to produce the original slip area, and generate slip in the opposite direction. The following equations now define the distribution of tractions along the contact surface

$$f_t(r) = -\frac{3\mu F_n}{2\pi a^3}\sqrt{a^2 - r^2} \qquad\qquad c \le r \le a$$

$$f_t(r) = -\frac{3\mu F_n}{2\pi a^3}\left(\sqrt{a^2 - r^2} - 2\sqrt{c^2 - r^2}\right) \qquad b \le r \le c$$

$$f_t(r) = -\frac{3\mu F_n}{2\pi a^3}\left(\sqrt{a^2 - r^2} - 2\sqrt{c^2 - r^2} + \sqrt{b^2 - r^2}\right) \quad 0 \le r \le b$$

$$(3.15)$$

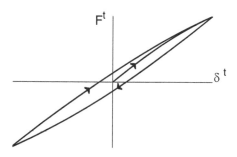

Figure 3.8: Illustration of hysteresis for a cycle of tangential loading with full reversal

Integration of the traction expressions in Equation 3.15, gives the following expression for the tangential force at the contact:

$$F_t = \mu F_n\left[1 - \left(\frac{b}{a}\right)^3\right] - 2\mu F_n\left(\frac{c}{a}\right)^3 \qquad (3.16)$$

Figure 3.8 illustrates the load–deformation response for the case of a contact where there is an initial monotonic tangential loading followed by a 180° load reversal to unload to a deformation of $-r$.

Excluding the contact constitutive models implemented by Thornton and his colleagues, (e.g. Thornton and Yin (1991)) most DEM models of tangential loading do not consider the details of the tangential contact stress distribution and do not distinguish between loading, unloading and reloading in the tangential direction. As discussed elsewhere in this book, DEM simulations and physical experiments indicate that granular material response is dominated by the contact normal forces, and so neglecting the details of the tangential stress distributions in DEM contact models may be acceptable. The description of tangential response has, however, been included here to highlight the complexity of the response in the tangential contact direction, even for the case of relatively simple elastic spheres.

3.4.3   Applicability of Hertzian contact mechanics to soil

Hertzian contact mechanics is attractive as it provides a rational basis for the development of contact models for application in DEM models. However, there are limitations to the application of Hertzian theory to real soils. The surface geometries assumed are highly idealized. Cavarretta et al. (2010) have demonstrated that even for relatively simple, manufactured materials, real particle contacts do not follow elastic theory; rather there is plastic yield of asperities prior to the development of an elastic Hertzian response. Experimental data for real soil response indicates that the Hertz-Mindlin contact model cannot correctly simulate the pressure-dependent nature of the small-strain modulus. As discussed by Goddard (1990), McDowell and Bolton (2001) and Yimsiri and Soga (2000), if the contact response in sands followed Hertzian theory, the small-strain shear stiffness of soil, G_{max} would be proportional to $p^{1/3}$, where p is the mean stress. However, as noted by McDowell and Bolton (2001), amongst others, experimental data indicate that (where the void ratio is kept constant) G_{max} varies approximately with $p^{1/2}$. This deviation from Hertzian behaviour may be a consequence of the non-spherical geometry of the particles, or the non-smooth nature of the particle surfaces.

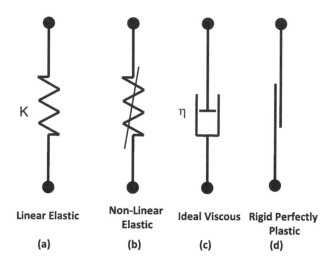

Figure 3.9: Graphical representation of basic rheological models

3.5 Rheological Modelling

Before discussing the details of contact modelling in particulate DEM, it is useful to review some "rheological" or "phenomenological" models that are applied in continuum mechanics. In this approach to modelling there are a limited number of base models, and these can be combined to capture various types of material response. DEM contact models often use this modelling framework to develop expressions for the load–deformation response at the contact points. Each of the base models has a graphical representation as illustrated in Figure 3.9 and these representations are often used in geomechanics-related DEM papers. In continuum mechanics these models describe a constitutive response relating stresses and strains. However, in the current context they may be used to relate a contact displacement (δ) to a contact force (F).

Figure 3.10 illustrates the load–deformation response captured by each of the basic rheological models. Figures 3.9 and 3.10(a) and (b) illustrate elastic response and a spring is used to represent this type of model. In a non-linear elastic spring (Figure 3.9(b)) the force–deformation response is given by a analytical expression (i.e. $F = f(\delta)$, where $f(\delta)$ is a non-linear function). A non-linear elastic model will not dissipate energy or capture plasticity in the response, i.e. the loading and unloading paths coincide. For the viscous response illustrated in Figure 3.10(c) the force is related to the rate of deformation or deformation velocity; as illustrated in Figure 3.9(c), this model is represented by a dashpot with damping η. For the rigid perfectly plastic response illustrated in Figure 3.10(d), there is no deformation (the rate of deformation, $\dot{\delta}$, is zero) until a yield point ($F = Y$) is reached, and after this point deformation continues at a constant load ($\dot{F} = 0$). This model is graphically represented as a slider that activates when the yield point is attained (Figure 3.9(d)).

The force–displacement response observed in each of these models can be expressed analytically as follows:

$$F = K\delta \qquad \text{Linear elastic, spring stiffnes K}$$

$$F = f(\delta) \qquad \text{Nonlinear elastic}$$

$$F = \eta\dot{\delta} \qquad \text{Viscous model, damping } \eta \qquad (3.17)$$

$$\begin{aligned} \dot{\delta} &= 0 \quad F < Y \\ \dot{F} &= 0 \quad F = Y \end{aligned} \qquad \text{Rigid perfectly plastic resonse}$$

The basic rheological models presented in Figure 3.9 can be combined to capture more complex response characteristics. While there is an almost infinite number of possible combinations, there are some common standard composite models, and these are illustrated in Figure 3.11. The linear Maxwell model illustrated in Figure 3.11(a) comprises a spring and dashpot arranged in series. With this arrangement both elements are restricted to experience the same force but can exhibit different deformations and the total deformation is the sum of the two components. The converse situation applies with the linear Kelvin model (Figure 3.11), i.e. the spring and dashpot experience different forces but are restrained to have equal deformations, and the total force is the sum of the two force contributions. The final model illustrated in Figure 3.11(c) is Burger's model, sometimes called Burger's fluid model or a four parameter fluid. This model is composed of a linear Maxwell model and a linear Kelvin model arranged in series.

The response of each of these models is given in Equation 3.18 below. For clarity in each case the response has been presented in the simplest format. However, for implementation in an explicit DEM code, the response function should give load as a function of displacement, i.e. it should be in the form $F = f(\delta)$.

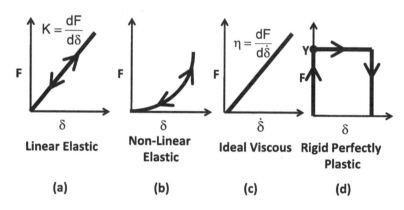

Figure 3.10: Load–deformation response of basic rheological models

$$\dot{\delta} = \frac{\dot{F}}{K} + \frac{F}{\eta} \qquad \text{Linear Maxwell model}$$

$$F = \eta\dot{\delta} + E\delta \qquad \text{Linear Kelvin model}$$

$$K_1\eta_1\eta_2\ddot{\delta} + K_1K_2\eta_1\dot{\delta} = \\ \eta_1\eta_2\ddot{F} + (K_1\eta_2 + K_2\eta_1 + K_1\eta_1)\,\dot{F} \qquad \text{Burger's model} \\ + K_1K_2F$$

$$(3.18)$$

where the terms $\dot{\delta}$ and $\dot{F}$ are the rates of change of deformation and force (i.e. first-order derivatives with respect to time), and $\ddot{\delta}$ and $\ddot{F}$ are the corresponding second-order derivatives.

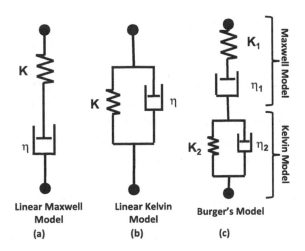

Figure 3.11: Graphical representation of composite rheological models

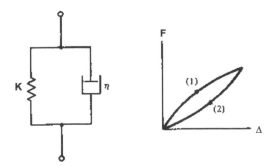

Figure 3.12: Load–deformation response of Kelvin rheological model

The load–deformation response of the Kelvin model is interesting as this model exhibits hysteresis. Referring to Figure 3.12 the force-deformation responses in loading and unloading are not co-linear and hence energy is dissipated. As discussed below, many

DEM modellers use this type of model to simulate energy loss due to plastic deformation at the contact points.

A key feature of these composite models is that because the viscous dashpot element is included in the model, they can capture a response that varies with time. This is of interest in geomechanics both to look at soil creep as well as to model the viscosity of the asphaltic binder in road pavements. The time–deformation response of the models illustrated in Figure 3.11 is illustrated schematically in Figure 3.13. Figure 3.13(a) illustrates the deformation of the linear Maxwell model under a constant load, F_0. It is clear that the model experiences a linear increase in deformation with time, i.e. it exhibits a linear creep. The creep response for the linear Kelvin model is given in Figure 3.13(b). In this case under a constant load, F_0, the deformation converges monotonically to a value of $\frac{F_0}{K}$. Burger's model, illustrated in Figure 3.13(c) is a combination of these two types of response. The analytical expressions for the time–deformation response captured by each of these three models to a force F_0 applied at a time $t = 0$, are given in Equation 3.19.

$$\delta(t) = F_0 \left(\tfrac{1}{K} + \tfrac{t}{\eta} \right) \qquad \qquad \text{Linear Maxwell model}$$

$$\delta(t) = \tfrac{F_0}{K} \left(1 - e^{-\left(\frac{K}{\eta} \right)t} \right) \qquad \qquad \text{Linear Kelvin model}$$

$$\delta(t) = F_0 \left[\tfrac{1}{K_1} + \tfrac{t}{\eta_1} + \tfrac{1}{E_2} \left(1 - e^{\frac{-tE_2}{\eta_2}} \right) \right] \qquad \text{Burger's model}$$

$$(3.19)$$

An appreciation of the range of responses that can be captured by use of rheological or phenomenological models comprising spring-dashpot combinations approach to modelling, as well as the approaches used to develop the load–deformation and time–deformation responses can be gained by reference to Shames and Cozzarelli (1997). The discussion here is limited to relatively simple spring-dashpot combinations; as discussed by Shames and Cozzarelli, the complexity of the response captured can be increased by adding additional components to the models presented here.

Itasca (2004) outlines the implementation of Burger's model

to capture contact response in a DEM code. A central-difference time integration approach was used to solve the second-order differential equation to describe the load–deformation response (refer to Equation 3.19).

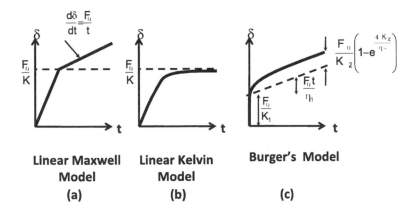

Figure 3.13: Time–deformation response of composite rheological models

3.6 Normal Contact Models

3.6.1 Linear elastic contact springs

The simplest type of contact model that can be used in particulate DEM to simulate the load–deformation response in the contact normal direction is a linear elastic spring. Where this model is used the contact normal force F_n is calculated as

$$F_n = K_n \delta_n \tag{3.20}$$

where K_n is the contact stiffness in the normal direction and δ_n is the overlap at the contact point, measured normal to the contact.

Typically in mechanics stiffness is a ratio of stress to strain and the units are those of stress (kPa). In this case however, the units of stiffness are force/length (e.g. N/mm). The calculated force is directed normal to the contact plane. The orientation of this force is along the line joining the centre of the two contacting particles, if disks, spheres or agglomerates of disks or spheres are used.

Given their widespread use in geomechanics, it is important to realize that in the Itasca PFC codes the user specifies a spring stiffness for each particle. So for two particles A and B in contact there are two spring stiffnesses $k_n^{p,A}$ and $k_n^{p,B}$ respectively in the normal direction and $k_s^{p,A}$ and $k_s^{p,B}$ in the tangential or shear direction. This results in effective normal stiffnesses K_n^{contact} and K_s^{contact} at the contact point of

$$
\begin{aligned}
K_n^{\text{contact}} &= \frac{k_n^{p,A} k_n^{p,B}}{k_n^{p,A} + k_n^{p,B}} \\
K_s^{\text{contact}} &= \frac{k_s^{p,A} k_s^{p,B}}{k_s^{p,A} + k_s^{p,B}}
\end{aligned}
\tag{3.21}
$$

In the case of uncemented, unbonded materials it is assumed that no tension is transmitted across the particle contact. Then when a gap develops between the particles the contact is considered ruptured or terminated.

The spring constants used in the linear elastic model cannot easily be directly related to the material properties of the solid particles, so where this model is used the springs should conceptually be considered to act as "penalty springs." In this way, as discussed in Section 3.2, they are stiff springs whose role is to minimize the amount of overlap that can occur at the contact point. As discussed further in Chapter 12, it is, however, possible to calibrate a DEM model and adjust the stiffnesses of these contact springs to match the overall response of an assembly of particles observed in the laboratory. Latzel et al. (2000) argue that use of a linear contact model is appropriate for 2D analyses; the extent of the simplification introduced by representation of a 3D material with a 2D model is such that the effort expended in using a more complex model is not worthwhile. The validity of this sentiment depends on the nature of the problem or the type of material behaviour being considered. Examples of geomechanics publications where useful insight into material response has been achieved using a linear contact model include Chen and Ishibashi (1990), Calvetti et al. (2004) and Rothenburg and Kruyt (2004).

3.6.2 Simplified Hertzian contact model

To overcome the non-physical nature of the linear spring stiffnesses, models have been developed to relate the spring parameters to the sphere material properties. Using the Hertzian theory of elastic contact that has been introduced in Section 3.3 above, an expression for the secant contact stiffness for the interaction between two spheres can be obtained. The Hertzian contact model is a set of non-linear contact formulations. This is sometimes called the Hertz-Mindlin contact model, as the approximate model used to describe the tangential force draws upon the work of Mindlin and Deresiewicz (1953), and this is discussed in Section 3.7 below. The normal contact stiffness is given by

$$K_n = \left(\frac{2\langle G \rangle \sqrt{2\widetilde{R}}}{3(1 - \langle \nu \rangle)} \right) \sqrt{\delta_n} \qquad (3.22)$$

The normal contact force is calculated as

$$F_n = K_n \delta_n \qquad (3.23)$$

where δ_n is the sphere overlap. For a sphere-sphere contact the coefficients $\widetilde{R}$, $\langle G \rangle$ and $\langle \nu \rangle$ are given by

$$\begin{aligned} \widetilde{R} &= \frac{2R_A R_B}{R_A + R_B} \\ \langle G \rangle &= \tfrac{1}{2}(G_A + G_B) \\ \langle \nu \rangle &= \tfrac{1}{2}(\nu_A + \nu_B) \end{aligned} \qquad (3.24)$$

and for a sphere-boundary contact, the coefficients are given by $\widetilde{R} = R_{\text{sphere}}$, $\langle G \rangle = G_{\text{sphere}}$, $\langle \nu \rangle = \nu_{\text{sphere}}$, where G is the elastic shear modulus, ν is Poisson's ratio, R is the sphere radius, and the subscripts A and B denote the two spheres in contact. This type of contact model has been used in many published DEM simulations, including Chen and Hung (1991), Lin and Ng (1997),

Sitharam et al. (2008), and Yimsiri and Soga (2010). As noted above, the relationship between mean stress and small-strain stiffness observed in physical experiments on sands differs from that which would be expected using Hertzian theory.

3.6.3   Normal contact models including yield

Walton-Braun linear model

In an elastic contact model, whether it is linear or non-linear, there is a unique relationship between the contact force and deformation and energy is conserved, i.e. the strain energy stored during loading equals the strain energy released in unloading. Walton and Braun (1986) proposed a linear contact model that dissipates energy, arguing that the particle interactions are non-conservative and that kinetic energy is dissipated in every collision. Their model is hysteretic and the normal force during first loading is given by

$$F_n = K_{1,n}\delta_n \qquad (3.25)$$

while the normal force during unloading or reloading is given by

$$F_n = K_{2,n}(\delta_n - \delta_{n,p}) \qquad (3.26)$$

where δ_n is the overlap normal to the contact point, $\delta_{n,p}$ is the plastic deformation. This plastic deformation depends on the maximum historical normal force, $F_{n,max}$, i.e. $\delta_{n,p} = \frac{F_{n,max}}{K_{2,n}}$. The stiffness during unloading is greater than during loading, i.e. $K_{2,n} > K_{1,n}$ and $K_{2,n}$ is either user-specified or given as function of $F_{n,max}$ (maximum normal force), i.e. $K_{2,n} = K_{1,n} + SF_{n,max}$. This model is simple to understand and relatively straightforward to implement, however, as with the liner elastic spring approach selection of suitable $K_{1,n}$ and $K_{2,n}$ values must be considered. The use of this model requires extra information to be stored at each contact point ($F_{n,max}$ and/or $\delta_{n,p}$), in comparison to the purely elastic models discussed above. Zhu et al. (2007) describe this to be a "semi latched" spring model. A version of this model is implemented in the commercial PFC codes and is referred to as the hysteretic damping model (Itasca, 2004).

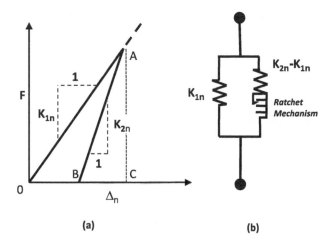

(a)   (b)

Figure 3.14: Illustration of Walton-Braun type contact model

Often when simulating the collision between two bodies, the collision is modelled by considering the relative velocities of the colliding bodies before and after impact, using a coefficient of restitution. This is the approach used in the event-driven simulations discussed in Chapter 1. The coefficient of restitution, e, quantifies the energy lost during collision. In a perfectly elastic collision there will be no energy loss and $e = 1$. Consider two particles a and b, that have velocities v_n^a and v_n^b in the direction normal to the contact prior to the collision, and post collision velocities v'^a_n and v'^b_n. The coefficient of restitution, e, relates these two velocities as follows:

$$e = \frac{v'^b_n - v'^a_n}{v_n^b - v_n^a} \tag{3.27}$$

As outlined in Chapter 2, the energy stored in an elastic spring equals the area beneath the force–displacement curve. During loading the kinetic energy of the particles is converted to strain energy and during unloading the stored strain energy is converted to kinetic energy. The reduction in kinetic energy can therefore be calculated by considering the force–displacement curve. Where the Walton-Braun model is used, referring to Figure 3.14, the coefficient of restitution is given by the square root of the ratio of the areas of triangles ABC and AOC (A_{ABC} and A_{AOC} respectively), i.e.

$$e = \sqrt{\frac{A_{ABC}}{A_{AOC}}} \tag{3.28}$$

which is equivalent to

$$e = \sqrt{\frac{K_{1,n}}{K_{2,n}}} \tag{3.29}$$

The energy balance for the Walton-Braun model is considered in Section 2.7 for a single-degree of freedom system.

The hysteretic normal contact model proposed by Thornton and Ning (1998) is similar to the Walton-Braun model, however the Thornton–Ning model uses a non-linear force-displacement relationship, based on Hertzian contact mechanics with a transition from elastic to plastic response being determined by the yield stress of the solid particle material. In the description of their model, Thornton and Ning give an expression for the coefficient of restitution that depends on the material Young's modulus and yield stress, the particle radii and densities, and the impact velocity of the particles.

Spring–dashpot model

The spring–dashpot model includes a dissipative viscous dashpot at the contact point to account for energy dissipation due to plastic deformation at the contact points. This model is equivalent to the Kelvin rheological model presented in Section 3.5 above. The force–displacement relationship is given by

$$F_n = K_n \delta_n + C_n \dot{\delta}_n \tag{3.30}$$

where C_n is the dissipative term. Examples of the use of this approach to contact modelling include Cleary (2000) and Iwashita and Oda (1998). Delaney et al. (2007) argue that the amount of energy dissipation is dependent on the velocity, and collisions at higher impact velocities dissipate less energy. They propose the use of an alternative dashpot formula where the contact force is given by

$$F_n = K_n \delta_n^{3/2} + C_n^* \dot{\delta}_n \delta_n^{3/2} \qquad (3.31)$$

where C_n^* is the modified dashpot term. Note that the viscous dashpot can be used with a both a linear spring formulation as in Equation 3.30 or with a Hertzian-type non-linear spring as in Equation 3.31.

As was the case for the Walton-Braun model, the visco-elastic parameters can be related to the coefficient of restitution for the contact (Pöschel and Schwager, 2005):

$$e = \exp\left(\frac{-\frac{\pi C_n}{2\langle m \rangle}}{\sqrt{\frac{K_n}{\langle m \rangle} - \left(\frac{C_n}{2\langle m \rangle}\right)^2}}\right) \qquad (3.32)$$

where $\langle m \rangle$ is the effective mass of the colliding particles and is given by

$$\langle m \rangle = \frac{m_a m_b}{m_a + m_b} \qquad (3.33)$$

Pöschel and Schwager (2005) include a discussion on the physical implications of using a coefficient of restitution in this way.

The inverse problem involves determining the contact parameters that will yield the required coefficient of restitution. As discussed by Cleary (2000), where a specific e value is required the dashpot constant C_n can be selected as follows:

$$C_n = 2\gamma \sqrt{\langle m \rangle K_n} \qquad (3.34)$$

where C_n is the viscous dashpot coefficient, K_n is the normal (linear) spring stiffness and the parameter γ is a function of the coefficient of restitution as follows:

$$\gamma = -\frac{\ln(e)}{\sqrt{\pi^2 + \ln(e)^2}} \qquad (3.35)$$

109

3.7 Calculating Tangential Forces in DEM

As discussed by Pöschel and Schwager (2005), there is a fundamental discrepancy in the typical approach used to model contact forces in DEM simulations. In the Hertzian model for normal force it is assumed that the spherical particle surface is completely smooth. Theoretically, frictional resistance cannot develop at the contact between two smooth perfectly spherical particles. However a sliding friction parameter is included in almost every DEM code and the frictional resistance is assumed to arise from the interlocking of asperities on the rough surfaces of the particles.

The terms "shear forces" and "tangential forces" are often used interchangeably to refer to the component of the force that acts along the contact surface, i.e. orthogonal to the contact normal. The tangential contact model must be able to describe the material response before gross sliding (i.e. when at least some of the contact surface is "stuck") and the response when the contact is sliding. The simplest approach to define yield, i.e. the initiation of gross sliding, is to assume a Coulomb friction model. Then a yield criterion is defined based upon μ, the coefficient of friction. This is always a positive number, i.e. $0 \leq \mu$ and usually $\mu \leq 1$. If F_t is the tangential force and F_n is the normal force, then $|F_t| \leq \mu F_n$ at all times. When $|F_t| < \mu F_n$ the contact "sticks", but when $|F_t| = \mu F_n$, sliding occurs and the tangential force acts opposite to the direction of slip and equals μF_n. In some cases a cohesion term may also be added to the failure criterion for the tangential force. Where this approach is used the contact remains "stuck" while $|F_t| < \mu F_n + C$, where C is a user-specified cohesion.

When a contact is first detected the tangential force and the cumulative tangential contact displacement are set to be 0. As long as the contact remains "stuck", the contact force is the product of the cumulative displacement in the tangential direction and the tangential spring stiffness. The cumulative displacement is the sum of the incremental relative displacements of the particles at the contact point that occur over each time increment from the time the contact is formed. Considering (for simplicity) a cohesionless contact, mathematically

$$F_t = -\min\left(|\mu F_n|, F_t\left(\delta_t, \dot{\delta}_t\right)\right) \frac{\dot{\delta}_t}{|\dot{\delta}_t|} \qquad (3.36)$$

where $F_t\left(\delta_t, \dot{\delta}_t\right)$ is the pre-sliding shear force calculated using the contact constitutive model. The parameter δ_t represents the cumulative relative deformation at the contact point, while the relative velocity at the contact point directed along the contact tangent is given by $\dot{\delta}_t$. Whether slipping or stuck, the tangential force acting on each particle will always act in the direction opposite to the apparent tangential sliding velocity $\dot{\delta}_t$. For a tangential displacement to exist at a the contact point, the particles must be moving at different rates, so the apparent tangential sliding velocity (or relative velocity of particle movement at the contact point) will be in opposite directions for each particle. This relative velocity at the contact point is a result of both relative translation of the particle centroids and rotation of the particles.

While a single coefficient of friction is used to model the sliding response in most DEM simulations, it is worth noting that physical test data for sliding along a range of interfaces indicates that the initial tangential force required to initiate sliding exceeds the tangential force that is measured once sliding begins. The ratio of the larger initial tangential force to the normal force gives a static coefficient of friction and the ratio of the force during sliding to the normal force gives the dynamic coefficient of friction. It is very difficult to measure accurately the coefficient of friction between two soil grains. The main difficulties arise from the small particle sizes involved and the non-conforming nature of the contact. A few customized apparatuses to measure inter-particle friction have been developed (e.g. Skinner (1969), Cavarretta et al. (2010)), however understanding of the tangential response at the contact between sand grains remains limited.

The most basic contact model assumes a linear relationship between the tangential contact forces and the cumulative tangential displacement prior to sliding. The cumulative deformation at the contact point is calculated by integration of the particle relative velocities at the contact point, so, for a linear spring with stiffness K_t, the pre-sliding shear force at time t is given by

$$F_t(\delta_t, \dot{\delta}_t) = K_t \int_{t_c^0}^{t} \dot{\delta}_t dt \qquad (3.37)$$

where t_c^0 is the time at which the two particles initially contact. In a DEM model the integral in Equation 3.37 is approximated by a summation, i.e. $\int_{t_c^0}^{t} \dot{\delta}_t dt \approx \sum_{t_c^0}^{t} \dot{\delta}_t \Delta t$. There will be an error, proportional to Δt, associated with the discretisation. This is an incremental force-displacement model, based upon the relative particle velocities in the tangential direction at the contact point. The need to use the cumulative displacement in the tangential direction to calculate the tangential component of the contact force is emphasized by Vu-Quoc et al. (2000) and O'Sullivan and Bray (2003a). Once sliding commences, a convenient way to calculate the sliding force in the appropriate direction is to use

$$F_t = |\mu F_n| \frac{F_t^*}{|F_t^*|} \qquad (3.38)$$

where F_t^* is calculated using Equation 3.37.

Referring to Itasca (2004) for the two-dimensional case, the tangential relative velocity of particle a relative to particle b at the sliding point, $\dot{\delta}^t$, is given by

$$\dot{\delta}^t = (v_i^b - v_i^a)t_i - \omega_z^b |x_i^C - x_i^a| - \omega_z^a |x_i^C - x_i^b| \qquad (3.39)$$

where t_i is the unit vector describing the orientation of the unit vector tangential to the contact, v_i^a and v_i^b are the translational velocities of particles a and b respectively in direction i, the positions of the particle centroids are given by $\mathbf{x^a}$ and $\mathbf{x^b}$ and the contact coordinates are given by $\mathbf{x^C}$. The rotational velocities ω_z^a and ω_z^b relate to rotations about axes through the centroids of the particles, orthogonal to the analysis plane (assumed to be $x - y$ plane).

The three-dimensional case is slightly more complicated (Itasca, 2008) and is given by firstly considering the relative velocity at the contact point $(\dot{\delta}_i)$:

$$\dot{\delta}_i = \left[v_i^b + e_{ijk}\omega_j^b(x_k^C - x_k^b) \right] - \left[v_i^a + e_{ijk}\omega_j^a(x_k^C - x_k^a) \right] \quad (3.40)$$

where e_{ijk}, the alternating tensor, is defined in Chapter 1. Here the rotational velocities are considered relative to a local Cartesian coordinate system, with origin at the particle centroids. The tangential component is then calculated by subtracting the normal component of the relative velocity vector:

$$
\begin{aligned}
\dot{\delta}_i^t &= \dot{\delta}_i - \dot{\delta}_i^n \\
\dot{\delta}_i^t &= \dot{\delta}_i - \dot{\delta}_j n_j n_i
\end{aligned}
\quad (3.41)
$$

The deformations used to calculate the normal contact forces could also be calculated by summing the incremental relative displacements, however it is best practice to calculate the normal forces based upon the contact geometry. This is clarified by Itasca (2004) who states that calculation of the normal contact force from geometrical considerations only makes the code less susceptible to problems from numerical round-off ("numerical drift").

3.7.1 Mindlin-Deresiewicz tangential models

The work of Mindlin and Deresiewicz (1953) indicates that the stiffness of the tangential contact spring should depend on the current normal load, the current tangential load, the load history and whether the tangential load is increasing, decreasing or increasing after a load reversal (i.e. loading, unloading or reloading). (Refer to Thornton (1999), Thornton and Yin (1991), Di Renzo and Di Maio (2004) and Vu-Quoc et al. (2000)). The path-dependent nature of the tangential force displacement relationship was already outlined in Section 3.4.2. Vu-Quoc et al. (2000) and Thornton and Yin (1991) both proposed contact constitutive models for implementation in particle DEM codes that capture the dependence of the tangential load response upon the load history.

Vu-Quoc model

Vu-Quoc et al. (2000) described their model to be a "highly simplified" version of the Mindlin-Deresiewicz model. In this model the tangential force at time $t + dt$, F_s^{t+dt}, is calculated as

$$F_s^{t+dt} = F_s^t + K_{s,t}\delta_s \tag{3.42}$$

where $K_{s,t}$, the tangential stiffness at time t, is given by

$$K_{s,t} = \begin{cases} K_{s,0}\left(1 - \frac{F_s^t - F_s^*}{\mu F_n^t - F_s^*}\right)^{1/3} & F_s \uparrow \\ K_{s,0}\left(1 - \frac{F_s^* - F_s^t}{\mu F_n^t + F_s^*}\right)^{1/3} & F_s \downarrow \end{cases} \tag{3.43}$$

where $K_{s,0}$ is the initial tangential stiffness and μ is the coefficient of friction. F_s^* is the value of the tangential force at the last turning point. The value of $K_{s,0}$ can be related to the $K_{1,n}$ parameter in the Walton-Braun contact normal model, as follows:

$$K_{s,0} = K_{1,n}\frac{2(1-\nu)}{2-\nu} \tag{3.44}$$

where ν is the Poisson's ratio of the solid particle material.

Thornton and Yin model

Thornton and Yin (1991) proposed a model for the interaction of spheres that contact obliquely. While the full implementation of the model also accounts for particle surface adhesion, here only the version of the model without adhesion is considered. Where this model is implemented to determine the tangential contact force, the contact normal force is calculated using Hertizan theory. Thornton and Yin developed their model based upon the experimental work of Mindlin and Deresiewicz (1953), and the tangential force (F_t) displacement (δ_t) relationship is illustrated in Figure 3.15 for a load–unload–reload cycle at a normal force F_n.

As illustrated in Figure 3.15 the tangential stiffness in this model is non-linear and dependent on the current normal force (F_n), the current tangential force, the load history and whether the

114

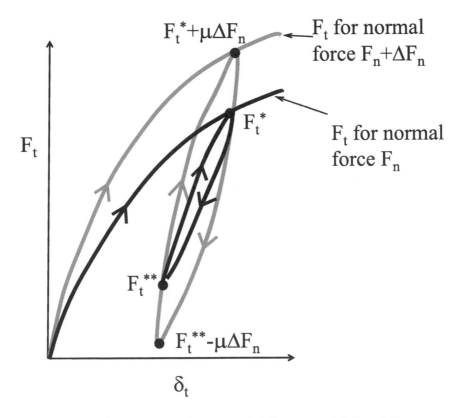

Figure 3.15: Illustration of tangential force model for oblique contact without adhesion proposed by Thornton and Yin (1991)

contact is in a state of tangential loading, unloading or reloading. This model therefore dissipates energy prior to the initiation of full tangential sliding.

The stiffness, K_t is then given by

$$K_t = 8G^*\theta\delta_n \pm \mu(1-\theta)\frac{\Delta F_n}{\Delta\delta_t} \qquad (3.45)$$

where the negative sign is invoked during unloading.

The parameter G^* is related to the Young's moduli (G_1, G_2) of the contacting spheres as follows:

$$\frac{1}{G^*} = \frac{2-\nu_1}{G_1} + \frac{2-\nu_2}{G_2} \qquad (3.46)$$

The normal displacement at the contact point is given by δ_n and the parameter θ depends on the loading state of the contact:

$$\theta^3 = 1 - \frac{F_t + \mu\Delta F_n}{\mu\Delta F_n} \quad \text{(loading)} \qquad (3.47)$$

$$\theta^3 = 1 - \frac{F_t^* - F_t + 2\mu\Delta F_n}{2\mu\Delta F_n} \quad \text{(unloading)} \qquad (3.48)$$

$$\theta^3 = 1 - \frac{F_t - F_t^{**} + 2\mu\Delta F_n}{2\mu\Delta F_n} \quad \text{(reloading)} \qquad (3.49)$$

As illustrated in Figure 3.15, the parameters F_t^* and F_t^{**} define the load reversal points. The second response curve for the normal load of $F_n + \Delta F_n$ is included in Figure 3.15 to illustrate that the reversal points must be continuously updated as the normal force changes, so that $F_t^* = F_t^* + \mu F_n$ and $F_t^{**} = F_t^{**} - \mu F_n$. To implement this model, consideration must be made for the case of small incremental displacements coupled with small increases in normal force, and the necessary details are given in Thornton and Yin (1991).

While the publications describing the implementations of the Mindlin-Deresiewicz type tangential models are not particularly recent, the linear and Hertz-Mindlin contact models seem to be the most commonly used tangential contact models in geomechanics DEM-related research. The importance of capturing the pre-sliding non-linearity of the tangential response is likely to depend

116

on the strain level of interest in the simulations. It would seem, for example, that capturing this aspect of contact response is very important where small-amplitude load reversals occur (e.g. during shear wave propagation). The contribution of Di Renzo and Di Maio (2004) is useful as they describe a comparison of different tangential force implementations by simulating oblique impact of a single sphere against an anvil using three different tangential force implementations.

3.8 Simulating Tensile Force Transmission

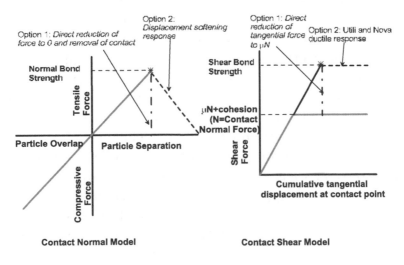

Figure 3.16: Basic bonding models in DEM

Conceptually, cement acts to provide a tensile resistance in the contact normal direction and a cohesion in the tangential direction in excess of the Coulomb frictional resistance. As noted above, bonding can be introduced into the model by specifying a tensile strength in the normal direction and a cohesion (or specified shear strength) in the tangential direction (Figure 3.16). In both cases the strength is usually specified in units of force.

Referring to Figure 3.16 the linear spring model can easily be extended to allow for transfer of tensile forces between the particles by restricting the slider action (and contact rupture) to occur when a finite tensile force is achieved. Where a tensile-capable contact model is used, this contact will be considered active if $|\delta_n| < \delta_n^{t,\max}$ where $\delta_n^{t,\max}$ is the separation distance at which the tensile strength of the contact $F_n^{t,\max}$ is mobilized. In a simple linear tensile model, if the bond strength is specified as $F_n^{\max}$, then $\delta_n^{t,\max} = \frac{F_n^{t,\max}}{K_n}$. In geomechanics this criterion will usually only apply to pre-existing contacts, i.e. if $|\delta_n| < \delta_n^{t,\max}$ for two particles that are not already in contact, no tensile force will be transmitted. If the normal tensile strength is exceeded, then the contact is removed. Should the two particles participating in the contact come into contact again later in the simulation, a new contact will be created and the tensile strength for the contact will be 0 and the basic notension, frictional contact model will govern the response.

Refinements to the tensile-capable contact normal model have been proposed. For example, a displacement softening response can be adopted as illustrated in Figure 3.16 (for the normal contact) where the normal tensile contact force reduces linearly with increasing displacement; an example implementation of this type of model is PFC's displacement-softening model (Itasca, 2004). Alternatively, in their simulations Utili and Nova (2008) chose to adopt the ductile-type model post yield as illustrated for the tangential contact in Figure 3.16. As the shear strength of the contacts in Utili and Nova's study is itself a function of the normal force and equal to $c + \mu F_n$ where c is a cohesion term, in the ductile case, post yield, the contact force will vary, while in the brittle case it reduces immediately to μF_n. An additional example of the use of a displacement softening model is the work of Hentz et al. (2004).

The linear tensile model has been used in a number of DEM related studies in geomechanics, including the 2D simulations of sand production by Cook et al. (2004), the 3D simulations of particle crushing by McDowell and Harireche (2002) and Cheng et al. (2003), and the 3D simulations of cemented sand response by Kulatilake et al. (2001). While Camborde et al. (2000) also adopt this simple approach to model tensile forces, they used non-circular particles and a non-linear, hysteretic model to simulate the response of contacts in compression.

3.8.1   Parallel bond model

Scanning electron microscope (SEM) images of the microstructure of both naturally and artificially cemented sands indicate that the cement at the particle contacts has a finite volume and it covers a finite area of the particle surface (e.g. Gutierrez (2007)). It seems reasonable to assume that the strength of the cemented bonds will depend on the volume of cement present at the contact. Furthermore, the finite area of the cemented bonds means that a moment can be transmitted in the contact normal direction and a resistance to rotation will be provided. The simple tension model described above cannot capture these facets of the contact response. These shortcomings are overcome in the parallel bond model described by Potyondy and Cundall (2004) and implemented within Itasca's commercially available codes PFC2D and PFC3D.

Where the parallel bond model is used, at each cemented contact a pair of parallel linear springs are effectively introduced to work in parallel with the conventional notension contact springs described above (Figure 3.17). In contrast to the simple bonding model, in the parallel bond model moments will be transmitted to the particles by both the normal and tangential contacts. As illustrated in Figure 3.18, the parallel bond has a finite size. The bond area is specified using a parallel bond radius multiplier, α, so that the radius of the bond, $R_{\text{bond}} = \alpha r_{\min}$ where $r_{\min}$ is the radius of the smaller of the two contacting particles and $0 \leq \alpha \leq 1$. The bond area (A_{pb}) is given by $A_{pb} = \pi R_{\text{bond}}$ for a unit thickness in 2D while in 3D $A_{pb} = \pi R_{\text{bond}}^2$.

Conceptually the size of the bond is a representation of the amount of cement, with a larger α value representing a case where there is more cement and a greater extent or degree of bonding. The physical volume of the cement is not, however, represented by the parallel bonds and the material void ratio will be unaffected by the size of the parallel bond. While simulations using the parallel bond model have been successfully calibrated against physical test data (see Chapter 12), it would be very difficult to directly link a physical volume of cement with a specific α value. When the bond size is 0 the parallel bond is effectively inactive, and the resistance in terms of strength and moment transmission increases as α increases.

In contrast to the simple linear contact bond described above, for the parallel bond the units of stiffness are given in units of $\frac{\text{stress}}{\text{displacement}}$ and the maximum strengths are specified in units of stress. In addition to the parameters that are required for the conventional linear contact stiffness model, the input parameters required to describe the parallel bond model also include the size of the parallel bond, α, the bond normal stiffness (K_N^{pb}), the bond tangential or shear stiffness (K_t^{pb}), the bond normal strength ($\sigma_N^{\max}$) and the bond shear strength ($\tau^{\max}$).

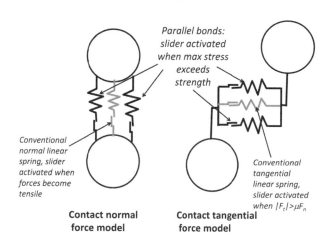

Figure 3.17: Parallel bonds acting in parallel with conventional linear contact model

The forces carried by the parallel bond in the normal and tangential directions (F_{pb}^N and F_{pb}^t) are given by

$$
\begin{aligned}
F_{pb}^N &= K_N^{pb} A_{pb} \delta_n \\
F_{pb}^t &= K_t^{pb} A_{pb} \Sigma \Delta \delta_t
\end{aligned}
\tag{3.50}
$$

where δ_n is the contact normal displacement and $\Sigma\delta_t$ is the cumulative tangential displacement. The bond breaks when either the maximum tensile or shear stress computed in the bond exceeds the defined strength.

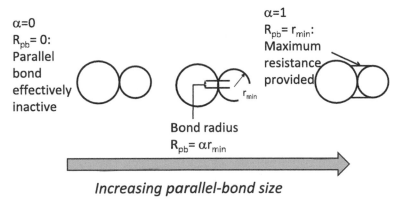

Figure 3.18: Parallel bond size variation with α

Two types of moment are transmitted by the parallel bond: a spin or twisting moment (M_{pb}^{spin}) and a bending moment (M_{pb}^{b}). The spin moment can only be calculated in a 3D implementation of this model as it relates to a moment caused by relative rotation about the contact normal. The increments in moment ($\Delta M_{pb}^{\text{spin}}$ and ΔM_{pb}^{b}) caused by an incremental rotation of the particles are given by

$$\Delta M_{pb}^{\text{spin}} = K_t^{pb} I_{pb} \Delta\theta_n$$
$$\Delta M_{pb}^{b} = K_N^{pb} I_{pb} \Delta\theta_s \tag{3.51}$$

where I_{pb} is the moment of inertia of the parallel bond, for 2D disks with unit thickness $I_{pb} = \frac{2}{3}R_{\text{bond}}^3$ while for 3D spheres $I_{pb} = \frac{1}{4}\pi R_{\text{bond}}^4$. The cumulative rotation about the contact normal is given by $\Sigma\Delta\theta_n$, while the cumulative rotation orthogonal to the contact normal is given by $\Sigma\Delta\theta_s$. The contact forces will add additional contributions to the moment given by the cross product of the resultant contact force and the vector directed from the contact to the particle centroid.

The breakage criteria, i.e. the stresses that will cause bond breakage of the normal and tangential parallel bond springs, are defined by the maximum tensile and shear stresses respectively. Expressions for the maximum normal stress ($\sigma_N^{\max}$) and the maximum shear stress ($\tau^{\max}$) were derived by consideration of beam bending theory to be:

$$\sigma_N^{\max} = \frac{-F_{pb}^N}{A_{pb}} + \frac{|M_{pb}^b|}{I_{pb}} R_{\text{bond}} \qquad (3.52)$$

$$\tau^{\max} = \frac{|F_{pb}^s|}{A_{pb}} R_{\text{bond}} + \frac{|M_{pb}^{\text{spin}}|}{J_{pb}} R_{\text{bond}} \qquad (3.53)$$

where J_{pb} is the polar moment of inertia of the parallel bond is only required in 3D simulations and is given by $J_{pb} = \frac{1}{2}\pi R_{\text{bond}}^4$.

Once $\sigma_N^{\max}$ exceeds the bond strength in the contact normal direction, the bond is considered to have failed in tension and the contact is removed. If subsequently these particles come into contact again, the contact between them will be governed by a conventional notension contact model, i.e. the particles will take on an un-bonded material response. In the shear direction once $\tau^{\max}$ is exceeded, the bond will be removed if there is a tensile force in the normal direction. Otherwise the contact response will revert to the notension contact model described above and relative slip or sliding of the particles at the contact point will be allowed.

Cheung (2010) carried out a simple two-particle analysis to understand the distribution of forces between the parallel bond and the contact model. Cheung showed that in tension the parallel bond takes all the tensile force, while in compression there is a share between the contact model and the parallel bond. At each time step a contribution to both the resultant force and the resultant *moment* acting on the two contacting particles from the parallel bond is calculated. The moment contribution is a function of the bond moment of inertia and the relative rotational velocities of the two contacting particles (refer to Itasca (2008)). Cheung demonstrated that to capture the brittle response typically observed in cemented sands, the relative stiffness and hence load share between the parallel bond model and the particle-particle model must be carefully considered. This model is available within the commercial DEM code, PFC 2D and has been used to model rock mass or cemented sand response in 2D (Wang et al. (2003), Fakhimi et al. (2006)) and 3D (Potyondy and Cundall (2004) and Cheung (2010)). The application of this contact model to simulate rock mass response is considered in Chapter 12.

Using an approach that is somewhat similar to the parallel bond model, Pöschel and Schwager (2005) proposed connecting their 2D, triangular particles using a contact model that is based on beam theory. In this approach the centroids of two contacting triangular particles are connected using an elastic beam that is fully fixed at each end. The deformation of this beam is then a combination of deformation due to elongation, bending and shearing, and elastic superposition is assumed so that the total deformation is the sum of these three contributions.

The rotational bonds proposed by Weatherley (2009) have some similarity with the parallel bond model; they transfer twisting as well as bending moments. In their model four spring stiffness parameters must be specified for the normal, tangential, bending and torsional responses. Then the breakage criterion is specified by considering the following summation:

$$\frac{|F_n|}{F_n^{\max}} + \frac{|F_t|}{F_t^{\max}} + \frac{|M_b|}{M_b^{\max}} + \frac{|M_t|}{M_t^{\max}} \qquad (3.54)$$

where F_n and F_t are the current normal and tangential components of the contact force. The current bending and torsion moments are given by M_b and M_t respectively and the superscript "max" is used to denote the breakage forces or moments, which are also input into the model.

3.9   Rolling Resistance

3.9.1   General discussion on resistance to rolling at contact points

The basic particulate DEM formulation was developed based upon smooth spherical or circular particles with non-conforming contacts that provide no resistance to rotation at the contact points. The non-convex, rough and often conforming surfaces that meet when real soil particles contact add a resistance to rotation at the contact points. For example, referring back to Figure 3.4(b) and (c), it is clear that particles who interact at these non-conforming are not free to rotate relative to each other. The terms "rolling resistance" or "rolling friction" are used to describe this phenomenon.

As illustrated in Figure 3.19, two types of rotation can occur. When two contacting particles roll, there is a relative angular motion about an axis that is parallel to their common tangent plane (i.e. the bending type moment in the parallel bond model). Particles also spin at their contact points i.e. rotate about an axis that is orthogonal to the contact plane and along the contact normal. While contact constitutive models have been proposed to account for rolling resistance at contact points, energy dissipation during spin or resistance to spin motion is rarely considered in DEM models. The parallel bond model discussed above considers both of these components of motion; however, in the parallel bond model the resistance to both bending moment and spin moment is assumed to be provided by a cement between the particles. The discussion on rolling resistance presented here considers the simulation of the resistance to rolling that arises from geometrical sources in unbonded materials.

The torque acting on particles arising from particle contacts has contributions from the tangential component of the force, the asymmetry of the normal stress (traction) distribution (if the contact has a finite area) and the contact normal force if the contact normal vector is not coincident with the branch vector (the branch vector is the vector joining the centroids of two contact-

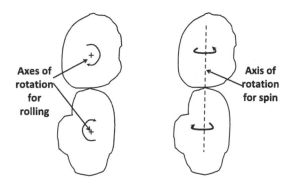

Figure 3.19: Rolling and spinning contacts

ing particles). The tangential moment can arise whenever there is non-zero inter-particle friction or a tangential (shear) cohesion. Determining the contribution of the normal component to the moment is non-trivial. This component is often called "rolling friction torque."

A limiting extreme case occurs when contacting particles can roll on each other without any sliding, their centroidal positions remain fixed and they act like cog wheels. The motion of one member is transmitted to the next (Figure 3.20). This mechanism requires the transmission of a moment at the contact point. Obviously the ideal spherical (or circular) convex geometries of the basic DEM particles (disks and spheres) will not automatically transmit this moment, and modifications to the contact constitutive models are required to replicate this geometrically derived mechanism. In contrast, the other ideal extremity is the case of free rolling, where contacting particles move relative to each other without generation of any resultant tangential force. A more detailed consideration of this topic is given by Greenwood et al. (1961) who discuss the issue of rolling friction (in relation to rubber) in detail from both an analytical and an experimental perspective.

Johnson (1985) explains that the sources of energy dissipation in rolling resistance include micro-slip and friction at the contact interface, the inelastic response of the material in the contacting particles and energy dissipation associated with the roughness of the rolling surfaces. Micro-slip will occur when the contacting materials have different elastic constants or where there are differences in the curvatures of the two contacting bodies. Energy dissipation can occur owing to both micro-slip in rolling and micro-slip in spin. From a mechanical perspective the resistance to rolling is associated with a couple that arises owing to the asymmetry of the contact pressure distribution. As two particles roll along each other there will be a higher pressure at the front of the contact than at the rear. If the pressure at the front of the contact becomes sufficiently high, there can be inelastic deformation within the contacting particles (and not just at the surface) and there will be an increase in the rolling resistance when this type of plastic zone develops (Johnson, 1985).

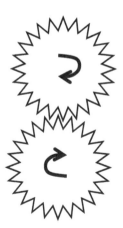

Figure 3.20: Illustration of rolling without sliding

The asperities along the surface play a key role in energy dissipation. In the initial loading they will intensify the real contact pressure so that some plastic deformation can occur along the contact, even if the nominal (bulk) stress is within the elastic limit of the material. Where a contact is subject to repeated loading, the asperities will deform plastically in the first loading, then the contact response will become more elastic in subsequent loadings. Roughness will also provide a resistance as energy is required to "surmount" the surface irregularities. Johnson (1985) compares this to the effort expanded when a wagon wheel rolls along a cobbled street.

Considering a cylinder of radius R rolling along a flat surface with an angular velocity ω, Johnson (1985) defines a coefficient of rolling resistance μ_R by equating the rate at which work is done to the rate of energy dissipation. If a resultant moment M_R acts on the cylinder then the rate at which work is done is $M_R\omega$. If the contact radius is a and the normal force acting on the surface is F_n, then the rate of energy dissipation is given by $\frac{2}{3\pi}\alpha F_n a\omega$, where α is the fraction of strain energy that is dissipated by hysteresis. The rolling resistance is then given by

$$\mu_r = \frac{M_r}{PR} = \alpha\frac{2a}{3\pi R} \tag{3.55}$$

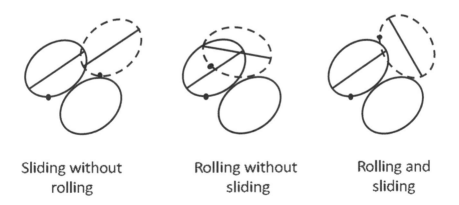

| Sliding without rolling | Rolling without sliding | Rolling and sliding |

Figure 3.21: Sliding and rolling in a granular material (after Oda et al. (1982))

In granular materials the rolling mechanism has a measurable significance for the bulk or overall material response. Oda et al. (1982) were amongst the earliest researchers to highlight the significance of rolling for bulk granular material response. On the basis of a series of experiments on photoelastic oval particles they identified three potential modes of contact deformation: pure rolling, pure sliding and simultaneous rolling and sliding (Figure 3.21). These experiments on two-dimensional materials provided clear evidence that in a granular material there is significant rolling at the contact points, and they concluded that micromechanical models should account for this mode of contact deformation.

3.9.2 Iwashita-Oda rotational resistance model

If disk or sphere particles are used the complexity of a DEM code and the computational cost of the simulations are both minimized. Two approaches have been proposed to account for rolling resistance in particulate DEM while continuing to model the system using geometrically simple spheres or disks. In the first approach, proposed by Iwashita and Oda (1998), an additional rotational spring-slider system is added in parallel with the normal contact spring. Iwashita and Oda (1998) originally developed this model in two dimensions, and a three-dimensional version of this model was implemented in the particle DEM code YADE (Belheine et al., 2009). This contact model is schematically illustrated in Figure 3.22. When the particles rotate relative to each other the combined spring-dashpot-slider system transfers a moment (M_r) to the contacting particles that is given by

$$M_r = -K_r\theta_r - C_r\frac{d\theta}{dt} \tag{3.56}$$

where K_r is the stiffness of the rotational spring, C_r is a rotational viscous dashpot and θ_r is the relative rotation of the two contacting particles. As detailed by Iwashita and Oda (1998), the incremental relative rotation is calculated by considering the incremental particle rotations as well as the change in orientation of the contact normal over the current time increment. The limiting value of M_r (i.e. the yield value, beyond which point there will be no increment in rotational resistance) is given by ηF_n, where F_n is the (compressive) contact normal force and the parameter η is the rolling friction. In their three-dimensional implementation Belheine et al. (2009) propose a linear relationship between η and the average radius for the two contacting particles. In both implementations the rotational resistance stiffness K_r is related to the linear tangential contact spring stiffness K_s.

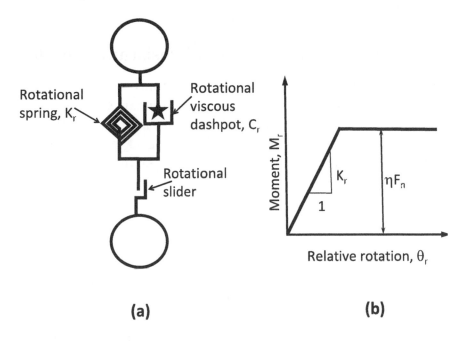

(a)

(b)

Figure 3.22: Rolling resistance model proposed by Iwashita and Oda (1998) (a) Schematic illustration of rotational component of contact model (b) Moment rotation response

Where this rolling friction model is used, the rolling resistance moment is added to the moment transmitted by the tangential contact force. Consequently the rotational dynamic equilibrium equation for particle p becomes

$$\mathbf{I}_p \frac{d\boldsymbol{\omega}_p}{dt} = \sum_{c=1}^{N_{ct}} \mathbf{F_t}\,^c r_p + \sum_{c=1}^{N_{cm}} \mathbf{M}_r^c \qquad (3.57)$$

where $\boldsymbol{\omega}_p$ is the particle's rotational velocity, the inertia tensor is given by $\mathbf{I}_p$ and the particle radius is r_p. A tangential force $\mathbf{F}_t^c$ acts at N_{ct} contacts and moment due to rotational resistance, $\mathbf{M}_r^c$ is induced at N_{cm} contacts. Iwashita and Oda (1998) described how they used their model in a two-dimensional parametric study to explore the influence of rotational resistance on both the macro- and micro-scale responses. The triaxial test simulations by Bel- heine et al. (2009) gave good correlations with physical test data for triaxial tests on Labenne sand.

3.9.3   Jiang et al. rotational resistance model

Jiang et al. (2005) proposed a rolling resistance model which ex- tends the earlier Iwashita-Oda rotational resistance model as the rotational resistance depends on the contact area (as do the nor- mal and shear contact springs). The Jiang et al. model also includes viscous damping. Jiang et al. (2009) extended this model to include roughness.

In both cases the model was derived from a conceptual model where systems of normal and shear springs acting in parallel sim- ulate the contact response (Figure 3.24). In the original imple- mentation by Jiang et al. (2005), both the normal force and the moment will depend on the rotation at the contact point, and an asymmetrical distribution of the contact normal traction is mod- elled. The contact normal stiffness k_n relates the normal stress and displacement at any point along the contact, then the normal force F_n is given by

$$F_n = \int\limits_{-B/2}^{B/2} \left[k_n(\delta_n + \theta z) + \nu_n(\dot{\delta}_n + \dot{\theta} z) \right] dz \qquad (3.58)$$

where the contact extends from $z = -B/2$ to $z = B/2$, the con- tact overlap is δ_n at the centre of the contact (where $z = 0$), the

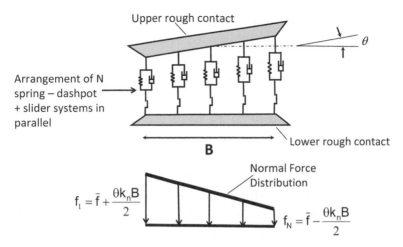

Figure 3.23: System of parallel springs and resultant distribution of normal forces in model proposed by Jiang et al. (2005)

rotation is θ and the viscous damping parameter is ν_n. (Here the contact overlap is taken to be positive in compression and counterclockwise rotation is positive.) The value of B depends on the grain shape it is sought to model, i.e. $B = \alpha\langle R \rangle$ where α is a dimensionless geometrical parameter and $\langle R \rangle = \frac{2R_1 R_2}{R_1 + R_2}$ with R_1 and R_2 being the radii of the two contacting particles. The moment at the contact point due to the contact normal traction can also be expressed in integral form as

$$M_n = -\int_{-B/2}^{B/2} \left[k_n(\delta_n + \theta z) + \nu_n(\dot{\delta}_n + \dot{\theta} z) \right] z\, dz \qquad (3.59)$$

In the two-dimensional formulation for rough contacts presented by Jiang et al. (2009) a rough contact is characterized by the geometrical parameters α as before, with a second parameter N being introduced that gives the number of contact points or asperities and these are assumed to be homogeneously distributed across the contact area. Each asperity will itself form a separate "sub-contact" point, with a spring stiffness k_n^N so that expressions for the total contact normal force and the total moment transmitted are derived by summing the contributions from the asperity forces. Two scenarios were proposed to calculate the moment induced at the contact point due to relative particle rotation. In the first case (Model 1 in Figure 3.24) a limiting rotation value is assumed, and once the rotation exceeds this value the rotation resistance is assumed to disappear as all the asperities are crushed. In the second case (Model 2 in Figure 3.24), termed the elasto-perfectly-plastic model, once a critical rotation value is exceeded the moment transmitted is assumed to be constant. Both the models proposed by Jiang et al.(2005 and 2009) have a conceptual advantage over the Iwashita and Oda approach as the width of the contact is explicitly considered in the model, and this better represents the real physical situation.

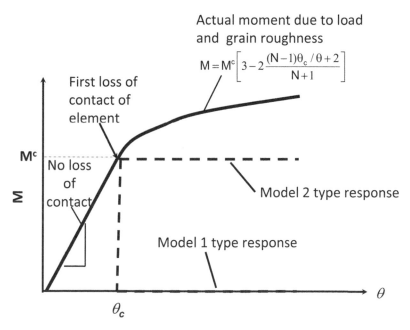

Figure 3.24: Moment rotation relationship for model proposed by Jiang et al. (2005)

3.10   Time-Dependent Response

Combinations of springs and dashpots can be used to model time dependent visco-elastic response, using the rheological modelling approach presented above. For example, the PFC codes include an option to use Burger's model to simulate soil creep (Figure 3.13(c)). Wang et al. (2008) simulated aging using a 2D DEM model with a Maxwell contact model.

While the time-dependent nature of soil response (including sand response) is now well established to be a complex phenomenon (e.g. Di Benedetto et al. (2005)) that certainly merits investigation at the micro-scale, there have been relatively few DEM simulations considering soil creep. An early example is the two-dimensional work of Kuhn and Mitchell (1992) who used an expression for the sliding velocity $\dot{\delta}_t$ at the contact points developed using rate process theory, and given by

$$\dot{\delta}_t = \lambda \frac{2kT}{h} \exp\left(-\frac{\Delta F}{RT}\right) \sinh\left(\frac{1}{2kT}\lambda n_1 \mu\right) \qquad (3.60)$$

where k is Boltzman's constant, h is Planck's constant, R is the universal gas constant, T is the absolute temperature, λ is the distance between successive equilibrium positions in the direction of the applied force, n_1 is the number of bonds per unit of normal contact force and ΔF is the activation energy. As discussed by Kwok and Bolton (2010) this is a thermally activated creep model that does not include the effects of inter-particle friction or damage to the contact asperities. Kwok and Bolton implemented the Kuhn-Mitchell model in a 3D DEM code with spherical particles. They then simulated a triaxial compression test and allowed creep to take place at various stages during the test and observed results that were qualitatively similar to the response observed in physical experiments on various soils.

The time-dependent response of rock mass was considered by Potyondy (2007) who proposed a modified version of the parallel bond model, called the parallel-bonded stress corrosion model. This model was developed to simulate the stress-dependent corrosion reaction that occurs in silicate rocks in the presence of water.

133

In this model an exponential function describes the rate of reduction in the parallel bond radius once a threshold stress is exceeded.

An additional motivation for the development of time-dependent contact models is the simulation of bituminous asphalt. Accurate modelling of this multiphase material which includes aggregate, filler, bitumen, and air phases, using DEM has not yet been achieved. However, Collop et al. (2007) demonstrated that the viscous response of the asphalt matrix between the soil particles could be simulated by including Burger's contact model, rather than explicitly simulating the asphalt phase. Their 3D simulations using spherical particles gave a good quantitative match to the development of strain with time observed in physical experiments. Abbas et al. (2007) also used Burger's contact model in their 2D simulations of asphalt pavement response.

3.11   Unsaturated Soil Response

Granular materials of interest to geotechnical engineers are not simply two-phase (particle-void) materials, the voids between the particles can be either dry, fully saturated, or contain a mixture of air and water. In petroleum engineering applications, the pore fluid may include both oil and water phases. The use of DEM to simulate a fully saturated system where one fluid completely fills the voids is considered in Chapter 6. Unsaturated or partially saturated soil response poses a particular challenge to geotechnical engineers and a number of research groups have been examining the fundamental mechanics of unsaturated soil response using DEM. Surface tension will develop at air-water or water-oil interfaces, effectively imparting a capillary force onto the individual particles. These particle-scale forces have a significant influence on the overall material response. In DEM, rather than explicitly modelling the interface between the different fluids, the influence of the capillary forces in unsaturated soil on the overall material response is modelled using specially developed contact constitutive models.

Restricting consideration to situations where air and water occupies the void space, the degree of saturation S_r is defined as the proportion of the void volume occupied by the water phase. The degree of saturation is related to the matric suction (i.e. the difference between the air and water pressures in the soil) via the soil water retention curve (SWRC) for a given soil. The sign convention adopted in soil mechanics takes the water pressure to be negative when the soil is partially saturated, and it is positive in a fully saturated soil. Likos (2009) provides an introduction to unsaturated soil concepts from a micromechanical perspective. The way the water is distributed in the soil is dependent on the degree of saturation. The situation where there is a low degree of saturation (roughly $< 20\%$) is called the *pendular* regime. In this case the water is found as a thin film on the particle surface and liquid bridges form between the particles. As the degree of saturation increases, and $20\% < S_r < 90\%$, the *funicular* regime develops where a network of liquid bridges forms in the partially filled pores and there are pockets of water-saturated pores. As S_r approaches 100% the air phase is present as isolated bubbles. In DEM studies of unsaturated soil response consideration has been limited to the pendular regime and contact constitutive models have been developed to determine the tensile forces imparted by the inter-particle liquid bridges. Gili and Alonso (2002), Jiang et al. (2004), Richefeu et al. (2008), El Shamy and Gröger (2008), and Scholts et al. (2009) have all proposed contact models to represent the inter-particle tensile and cohesive forces that arise owing to capillary tension in partially saturated soils. Zhu et al. (2007) list the capillary force as one of the non-contact forces that can exist between particles and they review implementations of contact models for unsaturated material response outside of the geomechanics research community.

All of the contact force expressions described in the literature consider a liquid bridge geometry similar to that presented in Figure 3.25. The liquid bridge itself is assumed to have a toroidal shape with principal radii r_1 and r_2 and separation distance a. The angle β is the half filling angle, θ is the contact angle and the volume of the liquid bridge is a function of r_1, r_2, β and θ. The half filing angle can be related to the degree of saturation and

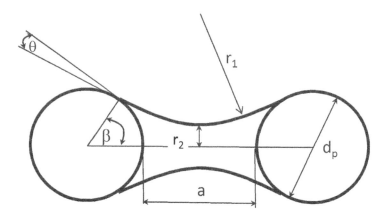

Figure 3.25: Illustration of pendular liquid bridge (using notation of El Shamy and Gröger (2008)

El Shamy and Gröger (2008) used an iterative procedure to determine β, which, at a particle scale, is a function of the liquid bridge volume, a and θ. By considering the surface tension acting on the liquid bridge neck and the matric suction or capillary pressure, an expression for the tensile force can be obtained. There is a slight variation in the expressions used to evaluate the force; however, referring to El Shamy and Gröger (2008), the tensile liquid bridge force is given by

$$F_l = \pi r_2 T_s \frac{r_1 + r_2}{r_1} \tag{3.61}$$

where T_s is the surface tension acting on the liquid bridge. This surface tension can be related to the matric suction as

$$P_c = T_s \left(\frac{1}{r_1} - \frac{1}{r_2} \right) \tag{3.62}$$

where the matrix suction, P_c relates to the air pressure (u_a) and water pressure (u_w) as $P_c = u_a - u_w$.

As noted by Gili and Alonso (2002) this force will act in the opposite direction to the compressive contact force when there is overlap between the particles, and will act as the only normal force when the particles separate. Consideration must be given to a separation distance where the liquid bridge will break and various options are considered. Both Richefeu et al. (2008) and El Shamy and Gröger (2008) describe 3D implementations of their unsaturated soil formulations, with 2D implementations given by Gili and Alonso (2002) and Jiang et al. (2004). In their implementation Jiang et al. (2004) account for merging of adjacent liquid bridges as the degree of saturation increases and adjacent voids become flooded with water.

3.12 Contact Detection

3.12.1 Identifying neighbours

The discussion on contact until now has focussed on contact resolution and determination of the contact forces. Referring back to Figure 1.7, the first step in calculating the contact forces is to identify the particles that are contacting or likely to contact in the current time increment, i.e. essentially to develop the list of contacts in the system. A variety of approaches can be adopted for contact detection. Irrespective of the approach chosen, for each particle a list of "neighbours", i.e. particles it contacts or particles it is very likely to contact in a given time increment, must be developed. Then each of these pairs of neighbouring particles will be considered in turn in the contact force calculation loop as illustrated in Figure 1.8.

To identify the set of neighbouring particles, a tolerance δ_n^{near} must be specified. Then the shortest distance between the two particles under consideration is calculated or sometimes estimated. The approach to calculate this distance will depend on the particle geometry. This distance will equal the contact normal overlap, δ_n when the particles are actually touching. Taking the overlap distance to be positive, when the particles separate and $\delta_n < 0$ and $|\delta_n| \leq \delta_n^{\text{near}}$, a and b can potentially contact this pair of contacts should be considered in the next time increment. If two particles were previously in contact, and $|\delta_n| > \delta_n^{\text{near}}$, then their contact has ended and no longer needs to be considered during contact resolution calculations. Pöschel and Schwager (2005) define a *Verlet distance*; if the distance between two particles is less than this Verlet distance, the particles are added to a *Verlet list* that lists the contacting, or potentially contacting particles in the system.

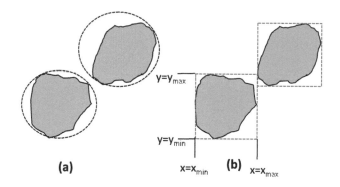

Figure 3.26: (a) Bounding sphere (b) Bounding parallelpiped used to assess potential contact between particles with irregular geometries

Referring to Figure 3.26, when the particles are highly irregular it is more efficient to assess whether either two bounding circles (spheres in 3D) or two bounding rectangles (parallelpipeds in 3D) intersect at the contact detection stage, rather than resolving the contact geometry in detail (see also Hogue (1998), Munjiza (2004), Vu-Quoc et al. (2000), Nezami et al. (2007)). The determination of the bounding box limits is more straightforward than determining the bounding sphere, as it is defined by the 4 lines (2D) or 6 planes (3D) given by $x = x_{\text{min}}$, $x = x_{\text{max}}$, $y = y_{\text{min}}$, $y = y_{\text{max}}$, $z = z_{\text{min}}$, and $z = z_{\text{max}}$, where x_{min} is the minimum x coordinate on the particle surface, x_{max} is the maximum x coordinate on the particle surface, etc. Alternatively the distance between the particle centroids can be considered and if it is within a limit (again called the *Verlet distance*) these particles are judged to be sufficiently close that contact is likely.

3.12.2   Contact detection searching strategies

The simplest thing to do from a coding/implementation perspective would be to check each particle against all other particles in the system at each time increment. This approach is naive and prohibitively computationally expensive. If this approach is used, the cost of the contact detection is proportional to N_p^2, where N_p is the number of particles in the system, and thus the simulation time will increase significantly as the number of particles increases. As highlighted by Munjiza (2004), in DEM codes it is important to develop a contact detection algorithm with minimal CPU and memory requirements. Munjiza (2004) lists the requirements for a contact detection algorithm, stating that it should be robust (i.e. reliable), easy to implement, CPU efficient, and RAM efficient. Grid based approaches are relatively easy to implement, and are commonly used, and so the basic idea of grid-based DEM contact detection is briefly outlined here. For more detailed discussions on implementation of contact detection algorithms refer to Pöschel and Schwager (2005), Munjiza (2004), or Munjiza and Andrews (1998). Bobet et al. (2009) also cite a number of additional references where DEM contact detection algorithms are described.

Binning algorithms

Where a "binning algorithm" is used to identify contacting particles, a regular grid is defined that covers the problem domain. The grid cell dimensions should be large enough that a cell in the grid can completely contain the largest particle. Each particle is mapped to a given cell in the grid and the distances between that particle and particles in the current cell and the adjacent 8 cells (2D) or 26 cells (3D) are determined when developing the neighbour list. Referring to Figure 3.27, particles in cell 1 will be tested for contact with the particles in cells 1 to 9. If these particles are sufficiently close and not already identified as contacting, a new contact will be created for consideration in the subsequent contact force calculations.

It is possible to omit this test for closeness and add all particles in the adjacent cells to the list of potential contacts (i.e. the neighbour list). Typically every particle in the domain is assigned an integer identification number (id number). If the id number of the particle under consideration is i and the id number of the particle that is being tested for a potential contact with i is j then to avoid duplication of contact information a convention is typically adopted, for example so that the neighbour list is only updated if j is less than i.

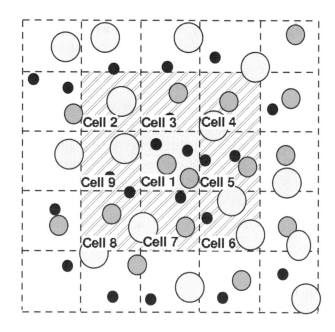

Figure 3.27: Grid used for 2D contact detection

Particles can easily be mapped to rows, columns and layers within a grid, by dividing the current particle coordinates by the grid size. Consider a grid with cell dimensions $\Delta x \times \Delta y \times \Delta z$, including $n_x^{\max}$ columns, $n_y^{\max}$ rows, and $n_x^{\max}$ layers. For a particle i with centroidal coordinates x_i, y_i, z_i we have:

- The column number: $n_x = \text{int}\left(\frac{x_i}{\Delta x}\right)$ where the function "int" converts real numbers to integers.

- The row number: $n_y = \text{int}\left(\frac{y_i}{\Delta y}\right)$.

- The layer number: $n_z = \text{int}\left(\frac{z_i}{\Delta z}\right)$.

- The cell number: $(n_z - 1)n_x^{\max}n_y^{\max} + (ny - 1)n_x^{\max} + n_x$.

Each cell then will have a list of particles mapped to it and each particle will be mapped to a particular cell.

A data structure to enable efficient searching through the contact information data following binning must then be selected. For consideration of this issue refer to Munjiza (2004) or Pöschel and Schwager (2005). One option is to create a list of contacts that is associated with each particle (avoiding duplication of contact calculations using the strategy presented above). Alternatively a list of contacts can be associated with each box, with the box id for a given contact being calculated by considering the contact coordinates (or an approximation of the contact coordinates).

The particles can be remapped to their appropriate cells at every time interval, after a specified number of time steps, or when a specified amount of deformation has taken place since the last remapping. For example, Vu-Quoc et al. (2000) and Pöschel and Schwager (2005) suggest the following criterion for updating their potential contact lists:

$$\sum_{t=t_i}^{t_j} \delta_{\max}^t \geq \frac{r_{nb}}{2} \tag{3.63}$$

where t_i is the time at which the lists were last updated, $\delta_{\max}^t$ is the maximum particle translation in time step t, and r_{nb} is the radius of the spherical "neighbourhood" region that surrounds the spheres.

The binning algorithm illustrated in Figure 3.27 becomes less effective where there is a wide range of particle sizes in the system.

The grid cell size must be at least as big as the largest particle and, if there is a broad particle size distribution, there may be many small particles within one cell, giving a large number of potential neighbours to consider. For systems with a broad range of particle sizes, it may be appropriate to use a hierarchy of box sizes for particle binning (e.g. Peters et al. (2009)). Reference to the discussion by Rapaport (2004) on calculating long-range interactions in molecular dynamics simulations may be useful for analysts contemplating implementation of this approach. To overcome this problem Pöschel and Schwager (2005) propose an alternative grid-based approach where, rather than saying that the lattice cells must exceed the size of the maximum particle, they require that each grid cell contain at most one particle centroid. Then the criterion for the cell size is based on the radius of the smallest particle, and referring to Figure 3.28, a suggested cell size is $\Delta x = \Delta y = R_{\min}$. The grid coordinates of each particle are determined using the approach proposed above and a search distance is determined based on the radius of the largest particle. In this way a significantly larger number of cells is considered; however, the maximum number of particles in a given cell is 1.

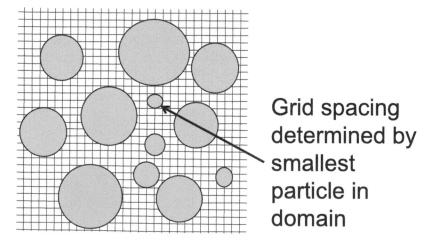

Grid spacing determined by smallest particle in domain

Figure 3.28: Lattice grid contact detection proposed by Pöschel and Schwager (2005)

Within a DEM code a number of links or pointers are required to create mappings between particles and their associated contacts, as well as between particles, contacts and their grid cells. A key challenge in DEM is the evolving nature of the contacts: during the simulation contacts will be created and deleted, and careful consideration must be given to the organization and management of the memory associated with the contacts. Authors who discuss the issues associated with this from the perspectives of developing DEM codes include Alonso-Marroqun and Wang (2009), Vu-Quoc et al. (2000), Munjiza (2004) and Pöschel and Schwager (2005).

Chapter 4

Particle Types

A key feature of particulate discrete element modelling is that the particles themselves are idealized, as already outlined in Chapter 1. All numerical models simplify the physical reality, and the users of a model should be aware of the extent of the simplifications and appreciate their implications. In a particulate DEM model the particles are assumed to be rigid and their geometries are restricted to shapes that can be analytically defined. Where rigid particles are used, only the translation of the particle centroids and the rigid body rotation of the particles need be considered in the governing equilibrium equations, i.e. there are three degrees of freedom for two-dimesional particles and six degrees of freedom for three-dimensional particles (as the rotations about the three principal axes of inertia are considered). If particle deformations were also to be considered, particle deformations (strains) would need to be included in the governing differential equations, effectively increasing the numbers of degrees of freedom in the system and the computational cost. This assumption of rigidity facilitates simulations involving large numbers of particles, and if it is assumed that the movement of particles relative to each other has a greater influence on the overall response than the deformation of individual particles, this is a reasonable approach to adopt. In the block DEM codes used in rock mechanics applications (e.g. UDEC and DDA), deformation is allowed and the implications of this for the DEM algorithm can be appreciated by reference to Shi

(1996) or Itasca (1998).

In reality, when two particles come into contact there is a deformation at the contact point. In a DEM simulation the amount of overlap that occurs at the particle contact points simulates this deformation. As noted in Chapter 1, it is because of this overlap that the term "soft-sphere" is used to describe this approach to modelling. The contact forces are very sensitive to the calculated overlap; hence the geometry of the contact must be very accurately determined, and this motivates the restriction of particle geometries to shapes that can be easily described analytically. The simplest type of geometry that can be considered is a disk (in two dimensions) or a sphere (in three dimensions). These are the most common types of particle geometry used in DEM codes. Hogue (1998) presents a coherent description of the issues associated with choice of particle geometry in DEM codes and Houlsby (2009) includes a more concise discussion. When choosing a particle type the analyst needs to assess the benefits of improvement against the geometrical and numerical challenges associated with adding complexity and computational cost. This Chapter summarizes of some of the issues associated with the various types of particle used in DEM simulations.

4.1   Disk and Sphere Particles

Disks and spheres are currently the most common type of particle considered in 2D and 3D DEM simulations respectively. These particles are popular as it is very easy to identify whether they are contacting, and if they are found to touch or almost touch (in the case of tensile force transmission), the geometry of the contact point, including the contact overlap or separation, can easily be accurately calculated. As noted in Chapter 1, in every time increment in a DEM simulation each contact is considered individually, and the geometry of that contact point is calculated. There will be many more contacts than particles in a DEM simulation, and contact resolution is usually the most computationally expensive part of the DEM algorithm. The more efficient this stage in the calculation is, the greater the number of particles that can be included. Achieving realistic numbers of particles is particularly important when simulating applied boundary value problems as discussed in Chapter 11.

Where disk or sphere particles are used, the contact overlap between two particles, a and b is simply calculated as

$$\delta_n = R_a + R_b - \sqrt{(x_a - x_b)^2 + (y_a - y_b)^2} \qquad \text{(2D)}$$

$$\delta_n = R_a + R_b - \sqrt{(x_a - x_b)^2 + (y_a - y_b)^2 + (z_a - z_b)^2} \quad \text{(3D)}$$
$$\text{(4.1)}$$

where radii R_a and R_b are the particle radii and the centroidal coordinates are given by (x_a, y_a, z_a) and (x_b, y_b, z_b) respectively. If the calculated overlap δ_n is positive then this contact is transmitting a compressive force, otherwise the contact is considered inactive (unless it can transmit tension). This calculation is also used in the contact detection phase of the simulation to assess whether the particles can potentially contact.

The contact location x_i^c is assumed to be at the midpoint of the contact overlap (refer to Figure 3.3), and for circular or spherical particles its coordinates can be determined by considering its location relative to either of the two contacting particles as follows (e.g. Itasca (2004)):

$$x_i^c = x_i^a + \left(R^a - \frac{\delta_n}{2}\right) n_i \qquad \text{(4.2)}$$

where $\mathbf{x}^c$ is the vector representing the contact coordinates, while the particle coordinates are represented by $\mathbf{x}^a$. The contact normal $\mathbf{n}$ is defined by considering the position of b relative to a:

$$n_i^c = \frac{x_i^b - x_i^a}{|x_i^b - x_i^a|} \qquad \text{(4.3)}$$

The calculations of the contact overlap and contact position from Equations 4.1 and 4.2 are straightforward; no iteration is required and the accuracy of the contact overlap calculation is determined only by the precision of the computer used (i.e. truncation error of floating point arithmetic). This is not the case where more general analytical forms are used. The applicability of the central-difference method to update the rotations of spherical particles also contributes to the prevalence of spheres in 3D DEM simulations. A further contributing factor to the widespread use of disks and spheres in DEM simulations in geomechanics is the popularity of Itasca's PFC2D and PFC3D DEM programs amongst geotechnical engineers. Examples of research studies completed using the PFC codes include, amongst others, the work by Cheng and her colleagues on particle crushing (e.g. Cheng et al. (2003)) (PFC3D), the 2D simulations incorporating particle crushing by Lobo-Guerrero et al. (2006) (PFC2D), the study of particle geometry effects by Powrie et al. (2005) (PFC3D). These Itasca codes are restricted to disks (PFC2D) and spheres (PFC3D), however they allow the creation of rigid particle clusters, and particles can also be bonded to each other to form crushable agglomerates, as discussed in Section 4.2 below.

To determine the shear contact force that acts along a tangent to the contact, the incremental relative tangential displacement at each contact point is required. Again, in the case where disk or sphere particles are used this is relatively straightforward. Firstly the relative velocity of the two particles at the point of contact must be determined. The velocity vector for contact c, $\mathbf{v}^c$ is then given by

$$v_i^c = \dot{x}_i^{a,c} - \dot{x}_i^{b,c} \tag{4.4}$$

where $\dot{x}_i^{a,c}$ is the velocity vector for the contact point c on the surface of particle a and $\dot{x}_i^{b,c}$ is the velocity of contact point, c on the surface of particle b. This velocity has contributions both from the particle translational velocities and the particle rotational velocities.

In 2D the calculation of $\dot{x}_i^{a,c}$ and $\dot{x}_i^{b,c}$ is relatively straightforward as rotation can only take place in one plane, i.e. about an axis that is orthogonal to the analysis plane. Then, if the angular velocities for particles a and b are given by ω_3^a and ω_3^b respectively, and anticlockwise rotations are taken to be positive, the velocities

148

of the contact points are given by

$$\begin{pmatrix} v_1^c \\ v_2^c \end{pmatrix} = \begin{pmatrix} (\dot{x}_1^a - \omega_3^a R_a) - (\dot{x}_1^b - \omega_3^b R_b) \\ (\dot{x}_2^a + \omega_3^a R_a) - (\dot{x}_2^b + \omega_3^b R_b) \end{pmatrix} \qquad (4.5)$$

where $\dot{x}_1^a$, $\dot{x}_2^a$, $\dot{x}_1^b$ and $\dot{x}_2^b$ are the velocities of the particle centroids.

When giving the expression for the velocity of a point on the surface of a spherical particle, it is more convenient to make use of the alternating tensor, e_{ijk}, (refer to Section 1.6). Then the velocity at the contact point c on the surface of particle a is given by

$$\dot{x}_i^{a,c} = \dot{x}_i^a + e_{ijk}\omega_j^a \left(x_k^c - x_k^a \right) \qquad (4.6)$$

where $\dot{x}_i^a$ is the velocity of the centroid of particle a in direction i, x_i^a is the centroidal position of particle a, ω_j^a is the velocity of rotation about axis j (the local axis through the centroid that is parallel to the three Cartesian axes with direction j) and x_k^c is the position of the contact c. An equivalent expression is obtained for particle b, and so the relative velocity at the contact point is

$$v_i^c = \left[\dot{x}_i^a + e_{ijk}\omega_j^a(x_k^c - x_k^a) \right] - \left[\dot{x}_i^b + e_{ijk}\omega_j^b(x_k^c - x_k^b) \right] \qquad (4.7)$$

The next step is to resolve the relative velocity into components along and orthogonal to the contact normal, as follows:

$$v_i^{c,t} = v_i^c - v_i^{c,n} = v_i^c - v_j^c n_j n_i \qquad (4.8)$$

where $v_i^{c,t}$ is the component of the relative tangential velocity in direction i.

The masses (m_p) and moments of inertia (I_p) of disks and spheres can easily be calculated once the radius is known as follows:

$$\begin{aligned} m_p &= \rho \pi r_p^2 \quad \text{(2D)} \\ I_p &= \tfrac{1}{2} m_p r_p^2 \quad \text{(2D)} \\ m_p &= \rho \tfrac{4}{3} \pi r_p^3 \quad \text{(3D)} \\ I_p &= \tfrac{2}{5} m_p r_p^2 \quad \text{(3D)} \end{aligned} \qquad (4.9)$$

149

The symmetrical nature of a sphere means that the moment of inertia is the same about any axis passing through its centroid.

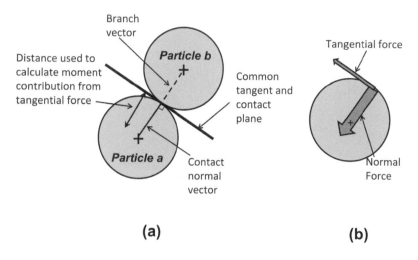

(a) **(b)**

Figure 4.1: Disk–disk contact

The drawback associated with using disk or sphere particles in geomechanics studies is that the rotations the particles experience greatly exceed the rotations particles experience in real soil under equivalent loading conditions. The issue of rolling resistance or rolling friction has already been considered in Section 3.9, where contact models to simulate rotational resistance were reviewed. The excessive rotations that occur in the case of disks and spheres arise because their geometry inhibits the transfer of moments to the particles by the normal component of the contact force. Compare the contact between two disks illustrated in Figure 4.1 and the contact between two ellipses illustrated in Figure 4.2. Where two disks contact, the branch vector (the vector connecting their centroids) and the contact normal are collinear (Figure 4.1(a)) and the normal component of the contact force passes through the centroid of the disk (Figure 4.1) without imparting any moment to the disk. On the other hand, when two ellipses contact, the branch vector and the contact normal are no longer collinear

(Figure 4.2(a)) and, as the line of action of the normal force no longer passes through the disk centroid, the contact normal force imparts a moment to the disk. This non-collinearity of the branch and contact vectors also means that when particle a tries to rotate in a counterclockwise direction, the normal contact force will increase, providing a greater resistance to rotation than the increment in tangential force provided when the circular particle a illustrated in Figure 4.1 rotates counterclockwise.

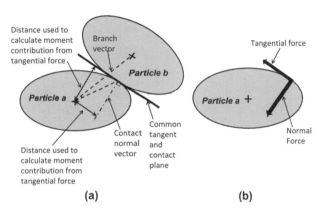

Figure 4.2: Ellipse–ellipse contact

While ellipses and ellipsoids have some advantages over disk and sphere particles, in contrast to rough, irregular soil particles, they have smooth and convex shapes. As illustrated in Figure 4.3(a) multiple points of contact can develop between real sand grains and the resultant action of the two normal forces illustrated in Figure 4.3 can be represented as a resultant force in combination with an equivalent moment. The pair of contacts between the irregular particles a and b in Figure 4.3 will provide a resistance to the rotation of particle a in both clockwise and counterclockwise directions; the contact between the elliptical particles in Figure 4.2 will provide no resistance to the rotation of particle a in the clockwise direction. As discussed by O'Sullivan and Bray

(2002), where non-convex particles are used, there is potential for a greater number of contacts to develop per particle, resulting in increases in strength and stiffness in comparison with convex particles. These particle-scale differences in rotation response have significant implications for the overall material response. A listing of the limitations of spherical/circular particles for modelling real materials includes the differences in shear strength, the differences in dilative response during shear, and the differences in distributions of void space (spherical/circular particles pack much more efficiently than real irregular grains) (Cleary, 2007).

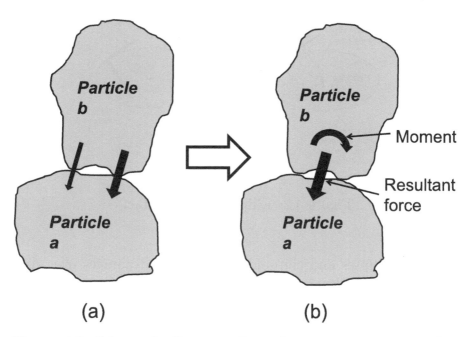

(a)   (b)

Figure 4.3: Schematic diagram of rough, non-convex contact between real soil grains

The rotational resistance models introduced in Chapter 3 provide DEM analysts with a means to incorporate rolling resistance in their simulations without incurring the additional computational costs associated with using non-circular or non-spherical particles. While it maybe difficult to relate the contact model parameters for these models to the characteristics of physical sand grains, at the very minimum these models are useful as they facilitate parametric studies to analyse the influence of rotation on the overall macro-scale response. Calvetti and his collaborators, (e.g. Calvetti et al. (2004)) consider an extreme condition and use spherical particles whose rotation is completely inhibited. Once appropriate values of inter-particle friction and contact stiffness are selected, simulations using this modelling approach agree closely with the behaviour of real sands, and Calvetti and his co-workers have succeeded in calibrating their non-rotating sphere models to capture the response of a range of laboratory sands (Calvetti, 2008). When rotation is completely inhibited in this way, the contact conditions for real soil particles, where rotation is geometrically inhibited, are not accurately modelled. Any detailed the interpretation of the particle scale interactions should consider this limitation. When they are completely artificially inhibited from rotating, the particles at rest will not be in a state of rotational equilibrium, consequently the stress tensor for the individual particles (calculated using the approach outlined in Section 9.4.2) will not necessarily be symmetric.

4.2 Rigid Disk and Sphere Clusters or Agglomerates

While contact detection and contact resolution for spheres and disks is straightforward, in the general case both sets of calculations become quite complex. Even considering ellipsoids, which have relatively simple geometries that can be analytically described, contact resolution involves the solution of a non-linear equation, which adds to the computational cost of the simulations. The accuracy with which the contact displacements are resolved must also be considered (see Favier et al. (2001) or Hogue (1998) for further discussion of this issue). These issues are avoided if disks and spheres are used as building blocks to create particles with more realistic geometries.

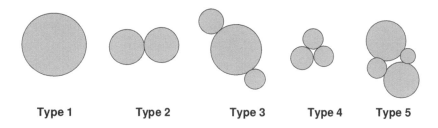

Figure 4.4: Touching disk clusters of the type considered by Thomas and Bray (1999)

A range of non-smooth, non-convex, non-spherical geometries can be modelled by "gluing" disk or sphere particles together to create rigid clusters. The disks or spheres may either touch or overlap. Thomas and Bray (1999) considered touching disks in their 2D simulations (Figure 4.4). The restriction on the particles not to overlap constrains the particles to have a limited range of geometries, in comparison with cluster configurations where over-lapping between the base particles is allowed. Favier et al. (1999) and O'Sullivan (2002) proposed the use of rather simple combinations overlapping spheres arranged in axisymmetric configurations (similar to the geometry illustrated in Figure 4.5). Vu-Quoc et al. (2000) also used clusters of overlapping spheres to approximate the shape of ellipsoidal particles.

More recently the level of sophistication in the use of over-lapping clusters has increased significantly. Various authors have proposed algorithms to create cluster particles from digital images of real sand particles. For example, the algorithm used by Das et al. (2008) is illustrated in Figure 4.6. In this approach a watershed-type algorithm is applied to a binary image of the particle to define the particle "skeleton." Then the disk that will cover the maximum area is inscribed within the particle outline and subsequent disks are each added to give maximum coverage to the remaining uncovered particle area. The number of disks required to accurately capture the shape depends on particle complexity.

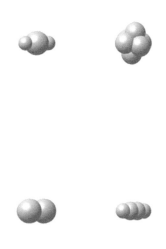

Figure 4.5: Overlapping sphere clusters of the type used by O'Sullivan (2002) and Favier et al. (1999)

Traditionally particle geometries were assessed in 2D; however, advances in optical microscopy and micro-computed tomography mean that 3D morphological characterization of particle geometry is now feasible. 3D algorithms to create sphere cluster particles have been developed by a number of authors including Das et al. (2008) and Garcia et al. (2009).

Whether the particles overlap or not, the cluster is treated as a rigid body. No contact forces are calculated between the disks or spheres making up the cluster even where the overlap would generate significant inter-particle compression forces, if these base particles were treated as separate degrees of freedom. The parallel axis theorem is used to calculate the contribution of each base

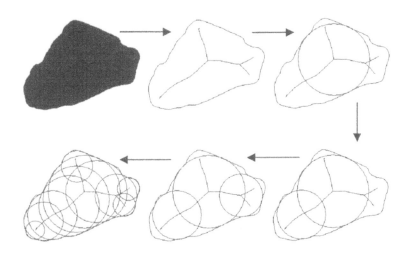

Figure 4.6: Approach used to create overlapping disk cluster particles proposed by Das et al. (2008) *(Figure supplied by B. Sukumaran)*

particle to the cluster moments of inertia about a local Cartesian axis centred at the cluster centroid I^{cluster}, as follows:

$$I_{ij}^{\text{cluster}} = \sum_{p=1}^{N_p} \left(I_{ij}^p \delta_{ij} + m^p a_i^p a_j^p \right) \qquad (4.10)$$

where the cluster contains N_p base disks of spheres, each with mass m_p, and inertia $\mathbf{I}^p$ relative to the disk/sphere local Cartesian axis, and δ_{ij} is the Kronecker delta. The vector $\mathbf{a^p}$ is given by $a_i^p = x_i^p - x_i^{\text{cluster}}$, where $\mathbf{x}^p$ and $\mathbf{x}^{\text{cluster}}$ are the vectors describing the base particle and cluster centroids respectively.

Where there is no overlap of disks or spheres, the total mass is the sum of the masses of the constituent disks or spheres. In the case where overlap occurs and the cluster geometry is relatively straightforward (e.g. as in Figure 4.5) the overlap volume can be analytically determined by integration and the mass and inertia values can be adjusted. For the more complex geometries described by Garcia et al. (2009) and Lu and McDowell (2008) the implications of the overlap volume must be considered. Gar-

cia et al. (2009) use an approach where the cluster is overlain by
a grid of cells. Where a cell is overlain by one or more spheres its
mass and position are used to calculate the mass and moments of
inertia. Lu and McDowell (2008) and Ashmawy et al. (2003) adopt
a less complex approach where the density of the base spheres is
scaled to have a value ρ^b_{scaled} so that

$$\rho^b_{\text{scaled}} = \frac{\rho_p V_p}{\sum_{i=1}^{N_p} V_i^b} \tag{4.11}$$

where ρ_p is the required density of the cluster particle, V_i^b is the
volume of base sphere i, N_p is the number of base spheres in the
cluster and V_p is the volume of the target particle for that cluster.

The motion of the particle is then calculated by summing
together the contact forces on its constituent disks or spheres.
The resultant moments will be calculated by considering the cross
product of each contact force and the vector directed from the
contact point to the cluster centroid. Where a body or external
force is applied to a base particle, this should also be accounted for
in both the translational and rotational equilibrium equations for
the cluster. Note, however, that in contrast to perfectly spherical
particles, the normal forces can impart a moment, as the contact
normals and vectors from the contact point to the cluster centroid will not be collinear. The shear forces also impart a torque
as in the single-disk/sphere particles and the lever arm for this
tangential force moment no longer equals the sphere radius.

Using this approach, the clusters themselves are rigid and each
cluster is itself a single particle, with three degrees of freedom
(in two dimensions) or six degrees of freedom (in three dimensions). The approaches for integrating the rotational motion of
non-spherical particles presented in Section 2.8 are required. As
noted by Das et al. (2008), it must be recognized that the resultant total mass of each cluster particle should be used to calculate
the critical time increment for stable analysis, rather than the
mass of the base particles. If the base particle mass were used, a
significantly smaller time increment for stable analysis would be
predicted. Once the centroidal position and the rotations of the
clusters have been updated, then the new positions of the base par-

ticles are calculated, prior to progressing to the contact resolution stage of the analysis and moving on to the next time increment. There is an increase in the simulation time when clusters are used in comparison with an equivalent simulation using the same number of spheres, as the number of contacts each cluster particle can have exceeds the potential number of contacts for disk/sphere particles.

4.3   Crushable Agglomerates

Particle crushing and damage has become an important research area in geomechanics (e.g. Coop et al. (2004)). Outside geomechanics, comminution is an important issue in many manufacturing processes and Cleary (2000) identifies analysis of comminution to be one of the main drivers motivating the uptake of DEM in industrial/process engineering. The fracture of even a single particle is a highly complex process. Using DEM, however, models that are conceptually simple can be developed to study the phenomenon of crushing.

One successful approach to simulate particle crushing has been to create breakable agglomerates of disks or spheres by forming bonds between particles in the cluster. The disks or spheres then are base units or elements within the agglomerate. To create these agglomerates the coordinates of the base disks or spheres are generated to be just touching, or overlapping slightly at the time of creation. Tensile and cohesive bonds are introduced at the contacts between these base particles. The resultant agglomerate will act as a coherent body until the forces between the base particles cause rupture of the bonds. If sufficient bonds rupture, the agglomerate will disintegrate into two or more smaller agglomerates. The comminution limit for the particle (i.e. the smallest size it will break down to) is determined by the size of the base particles. Figure 4.7(a) is a schematic diagram of a single particle crushing test simulation. In contrast to the rigid clusters described above, each agglomerate is itself a multiple-degree-of-freedom system and each base particle is a degree of freedom. Therefore, while the dynamic equilibrium of the rigid clusters considers the effective inertia of the cluster, the system of equations for a bonded agglomerate considers the individual base particles, and the contact forces acting on each of these base particles include contributions from contacts with spheres in adjacent agglomerates as well as contacts between the other spheres/disks in its own parent agglomerate. The rotational motion does not require consideration of non-spherical rigid body motion.

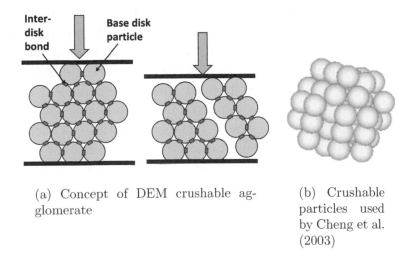

(a) Concept of DEM crushable agglomerate

(b) Crushable particles used by Cheng et al. (2003)

Figure 4.7: Crushable DEM agglomerates

Thornton and Liu (2004) and Kafui and Thornton (2000) used this approach to simulate the fracture of agglomerates upon impact or in collisions for powder processing applications. These ideas were then adapted for use in geomechanics by Robertson (2000) and McDowell and Harireche (2002) so that a solid soil particle is simulated as an bonded agglomerate of base sphere particles. Cheng et al. (2003 and 2004) then simulated isotropic compression and shearing tests on samples of these agglomerate particles to advance fundamental understanding of the role of particle crushing on soil response.

A representative crushable agglomerate used by Cheng et al. (2003) is illustrated in Figure 4.7(b). This agglomerate was created following an approach originally proposed by Robertson (2000). A dense, regular assembly of 57 spheres in hexagonal close packing (HCP), without initial overlap was initially generated. Then, in order to replicate the response observed in real experiments, and achieve a variability in strength and shape, about 20% of the spheres were removed in a probabilistic-based algorithm. Removing spheres from the agglomerates in these ways introduced flaws in the clusters. Cheng et al. (2003) calibrated the particle contact parameters to achieve a load–deformation response for compression of a single particle that matched a single-particle physical compression test and also found good agreement between the DEM simulations of isotropic compression tests and the real, physical tests. Lu and McDowell (2006 and 2007) proposed idea of using parallel bonded spheres to represent railway ballast particles, with small spheres being added to the particle surfaces to simulate abrasion of surface asperities. Pöschel and Schwager (2005) raised a slight concern that the smaller particles are effectively more rigid than the larger particles. While this is true, it may also be true in a physical system where the larger particles may have a greater number of deformable asperities and a greater number of internal flaws that could also contribute to their deformability.

A different approach to model particle crushing was adopted by Lobo-Guerrero and Vallejo (2005). They initially model the entire system using disk particles; when a certain pre-defined failure stress is experienced by the disk particle, each disk is then replaced by the assembly of eight disk particles illustrated in Figure 4.8. In this approach the overall mass of the system is not conserved. Pöschel and Schwager (2005) adopt a similar methodology where a fracture criterion is specified, the original particle is removed from the simulation and the new positions of fragments are randomly determined within an envelope defined by the original disk or sphere particle location. Their model differs from that proposed by Lobo-Guerrero and Vallejo (2005) as conservation of mass is achieved by setting a criterion that $\sum_{i=1}^{N^{P,f}} R_{fi}^2 = R^2$ (2D) and $\sum_{i=1}^{N^{P,f}} R_{fi}^3 = R^3$ (3D), where R is the original radius and the radii of the $N_{p,f}$ particle fragment disks or spheres are given by

R_{fi}. Initially the particle fragment disks or spheres will overlap, the resultant compressive contact force will cause the particles to be repelled from each other and additional "non-physical" energy will be introduced to the system. Refer to Pöschel and Schwager (2005) for a discussion on the breakage criterion and the approach used to control the particle size distribution of the fragmented particles.

One final approach to modelling crushable particles that is worth considering is that proposed by Ben-Nun and Einav (2010). These authors also proposed a criterion for particle crushing that is based on the forces acting on smooth ideal particles (disks). For each particle the sum of the contact forces acting on the particle is considered by summing the dyadic product of the contact normals and the contact forces to obtain a second-order tensor s_{ij}, where $s_{ij} = \sum_{k=1}^{N_c^p} n_i^k F_j^k$, where n_i^k and F_j^k are the normal and contact force vectors for contact k and there are N_c^p contacts involving particle p. The eigenvalues of the tensor s_{ij} are then used to calculate a nominal shearing force, S, and a nominal normal force, N such that

$$N = \frac{s_1 + s_2}{2}$$
$$S = \frac{s_1 - s_2}{2} \tag{4.12}$$

where s_1 and s_2 are the eigenvalues ($s_1 > s_2$). The particle is considered to have failed if $2S - N \geq F_{\text{crit}}$, a criterion is similar to the Brazilian test used in rock mechanics. If a particle is judged to have failed, it is replaced with an agglomerate particle that initially fits within the envelope defined by the original particle location and radius. Then the agglomerate particle is rotated and expanded, so that mass is conserved without inducing unreasonably large contact forces. This approach requires a "freezing" of the system while the adjustment is made. Readers interested in developing criteria for particle failure in DEM simulations may find the analytical study of Russell et al. (2009) useful as the stresses induced within particles by compressive contacts are considered and related to the particle failure.

4.4   Superquadrics and Potential Particles

Spheres, disks, ellipses and ellipsoids are all subsets of a general type of functions called superquatratics (in 2D) and superquadrics (in 3D). The general functional forms for these geometries are given by

$$f(x,y) = \left(\tfrac{x}{r_a}\right)^{m_a} + \left(\tfrac{y}{r_b}\right)^{m_b} \qquad \text{(2D)}$$

$$f(x,y,z) = \left(\tfrac{x}{r_a}\right)^{m_a} + \left(\tfrac{y}{r_b}\right)^{m_b} + \left(\tfrac{z}{r_c}\right)^{m_c} \quad \text{(3D)}$$

$$(4.13)$$

where the principal axis lengths are given by $2r_a$, $2r_b$, and $2r_c$, and in the case of a sphere the radius is given by $r_a = r_b = r_c$. The particle "squareness" or "blockiness" is controlled by the exponents m_a, m_b and m_c. Examples of superquadric geometries are presented in Figure 4.9.

Excluding disks and spheres from consideration (as their contact calculations are straightforward), the most commonly used superquadric/superquadratic geometries in geomechanics related DEM are as ellipses (2D) or ellipsoids (3D). These are convex shapes, and only one contact can exist between a pair of ellipses /ellipsoids; however, the non-collinearity of the normal and branch vectors enables moment transmission by the contact normal forces and consequently a much greater resistance to rotation in comparison with spheres or disks. Elliptical particles for two-dimensional analyses by were proposed by Rothenburg and Bathurst (1991) and Ting (1993). Ng has probably been the leading geomechanics researcher promoting the use of ellipsoids, with the approach used to simulate his axisymmetric particles being described by Ng and Dobry (1995) and Lin and Ng (1997). Examples of the use of this code to advance fundamental understanding of material response include the parametric studies considering the sensitivity of the particle scale and overall material response described in Ng (2001), while Ng (2004b) examined different failure criteria that have been proposed for soil (considering a full 3D stress

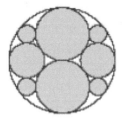

Figure 4.8: Crushable particles used by Lobo-Guerrero and Vallejo (2005)

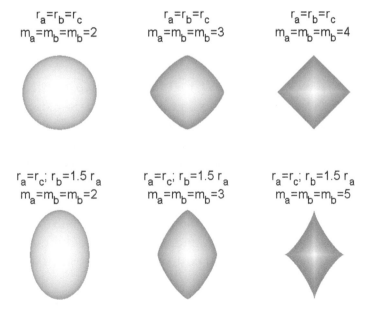

Figure 4.9: Example superquadric geometries

state). The use of more general superquadratic/superquadric particles in geomechanics has been limited. Examples of studies using these geometries outside geomechanics are given by Mustoe and Miyata (2001) and Cleary (2008) who both considered 2D process-engineering-type simulations.

Houlsby (2009) and Ng and Dobry (1995) consider contact detection between smooth, convex particles whose surfaces are analytically described. Then a point, o, in the problem domain (x_o, y_o, z_o) can be tested against the specific particle, p, whose surface is described by the function f_p. If $f_p(x_o, y_o, z_o) = 0$ then the point is on the surface of the particle, if $f_p(x_o, y_o, z_o) < 0$ then the point is inside the particle surface and if $f_p(x_o, y_o, z_o) > 0$ it is outside the particle surface. The contact between two particles P_1 and P_2 whose surfaces are described by two functions f_{P_1} and f_{P_2} can be calculated using Lagrange multipliers. The point on P_1 that is closest to P_2, $(x_{P_{1,2}}, y_{P_{1,2}}, z_{P_{1,2}})$, is determined by minimizing the sum $f_2 + \Lambda f_1$. This minimization can be achieved by differential calculus and Houlsby (2009) used an iterative Newton–Raphson approach to determine the minimum point. If $f_{P_2}(x_{P_{1,2}}, y_{P_{1,2}}, z_{P_{1,2}}) < 0$ then the two particles overlap, and there is contact. The point on P_1 that is closest to P_2 is then sought for calculation of the contact forces. Hogue (1998) outlined an alternative approach to contact detection and resolution where the surface of one of the particles is discretized and then each of these points is tested against the functions used to define the surfaces of adjacent particles. The contact detection between superquadratics/superquadrics becomes more expensive as the geometrical non-linearity increases and Hogue quantitatively compared the computational cost of contact resolution for a range of superquadric particles.

Just as disk and sphere particles can be combined to create clusters with more complex geometries, composite higher-order particles can be created. For example Kuhn (2003b and 2006) proposed the use of a composite, 3D convex particle termed an "ovoid" that comprises spherical "caps" and a toroidal middle section. Using similar concept, that avoids any iteration to resolve the contact geometry Pournin et al. (2005) proposed the use of a "spherocylinder" geometry comprising cylinders with a sphere inserted at each end.

A highly flexible approach was proposed by Houlsby (2009), who introduced the concept of "potential particles" (in 2D). Harkness (2009) implemented this approach in 3D. The idea of a potential particle is to use the contact detection for smooth, convex particles outlined above. The innovative aspect of this approach is the use of MacCauley brackets and Heaviside step functions to add functions together and create more complex geometries that are analytically described by "potential functions." This approach can be explained with reference to Figure 4.10(a) which illustrates the scheme adopted by Harkness (2009). Here a geometry comprising circles with flat sides is sought to simulate the long contacts that exist in locked sands and the composite geometry is constructed by taking the equation of a circle.

$$f_c = x^2 + y^2 - r_c^2 \qquad (4.14)$$

and the equation of a line (or "flat") given by

$$f_{\text{flat}} = a_i x + b_i y + c_i \qquad (4.15)$$

The potential function describing a sphere with k flats is then given by

$$f(x, y) = (1 - k) \left\{ \left\langle \sqrt{x^2 + y^2} - r_c \right\rangle^2 + \sum_{i=1}^{n_{\text{flat}}} \left\langle \frac{a_i x + b_i y + c_i}{\sqrt{a_i^2 + b_i^2}} \right\rangle^2 - s^2 \right\}$$
$$+ k(x^2 + y^2 - r^2)$$

$$(4.16)$$

where k defines the extent of circularity, there are a total of n_{flat} flats, and s^2 is a constant. The McCauley brackets $\langle \cdot \rangle$ are defined such that $\langle x \rangle = x$ if $x > 0$, while $\langle x \rangle = 0$ if $x \leq 0$. Figure 4.10(b) illustrates the extension of this concept to spheres with flats to create 3D particles.

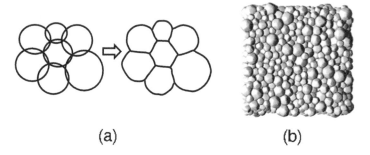

(a) (b)

Figure 4.10: Potential particles used by Harkness (2009): (a) Particle geometry generation (b); Image of 3D assembly of particles

4.5   Polygonal/Polyhedral Particles

As noted in Chapter 1, the "block" discrete element codes include polygonally shaped, simply deformable blocks. Similarly particle DEM codes exist that use rigid 2D polygonal particles, (e.g. Mirghasemi et al. (1997, 2002) and Matsushima and Konagai (2001)), or 3D polyhedral particles (Nezami et al., 2007). Where polygons are used the geometry is defined by the corner coordinates, the edges and the particle orientation. The dynamic equilibrium equation is used to update the coordinates of the particle centroid, then, similar to the cluster particles, the corner coordinates are updated (accounting for the incremental rotation). The volume of data required to describe a particle is approximately proportional to the number of corners (Hogue, 1998). By restricting consideration to a specific number of protoype shapes, the information can be reduced to the centroid location, the particle orientation and a shape identification index (Figure 4.11, Nezami et al. (2007)).

| Cube | Tetrahedron | Pyramid | 6 faces | 6 faces | 8 faces |

Figure 4.11: 3D polyhedra particles used by Nezami et al. (2007)

An appreciation of the additional effort involved with contact detection for polygonal/polyhedral particles can be gained by reference to Cundall (1988b). Cundall clearly identified 6 potential types of contact that can occur where 3D polyhedral particles are involved as follows: corner–corner, corner–edge, corner–face, edge–edge, edge–face and face–face. For the case of 2D polygonal particles, consideration can be restricted to corner–corner, corner–edge and edge–edge. In the case of 2D contacts involving edges, the contact normal is orthogonal to the edge. Cundall (1988b) proposes a "common plane" approach for contact identification. The essence of this approach is that the plane that bisects the space between two polygons is firstly determined, then contact between each of these particles and the common plane is considered. A comprehensive discussion on contact detection for their 2D triangular particles is given by Pöschel and Schwager (2005), where 6 contact scenarios are considered.

An interesting extension of polygonal particles has recently been proposed by Alonso-Marroqun and Wang (2009). In their approach they consider the Minkowski sum of a polygon and a disk; this operation generates a geometry where the disk is "swept" along the polygon edge, as illustrated in Figure 4.12. The overlap distance is then given by

$$\delta = r_a + r_b - \delta_{\text{polygon}} \tag{4.17}$$

where r_a and r_b are the radii of the disks used in the two contacting particles and δ_{polygon} is the shortest distance between the two polygons (e.g. a vertex–edge distance).

4.6 Achieving More Realistic Geometries

The geometries considered to date have been restricted to analytically described shapes. Real sand particles are highly irregular. Numerous studies have shown that grain shape influences the mechanical behaviour of sands, for example Cho et al. (2006) showed that shape influences the critical state friction angle, the critical state line intercept, and the slope of the critical state line, while Duttine and Tatsuoka (2009) considered how particle geometry influences the nature of the viscous response of sand. These studies, amongst many others, show that there is a need to accurately capture sand particle morphologies in DEM codes to achieve accurate, quantitative predictions of sand response. Considering the possibility simulating a broad range of morphologies, Hogue (1998) described the use of discrete functional representations to represent a range of particle types. In this approach a series of vertices or corners are defined by their polar or spherical coordinates relative to the particle centre. Then, the edges between these vertices can

168

be ascertained by interpolation. Hogue outlined how a bounding sphere can be placed around these particles to facilitate preliminary contact detection. The location of the point of intersection of these bounding spheres can be used to define the search region used to accurately resolve the contact.

From the perspective of finite element analysis, Zienkiewicz and Taylor (2000b) describe an approach to detect contact between irregular particles whose edges are defined by nodes and the surfaces between the nodes are defined by interpolation functions (i.e. shape functions). Contact identification is then based upon finding a value $\xi = \xi_c$ that minimizes the function

$$f(\xi) = \frac{1}{2}\left(\mathbf{x}_s^T - \mathbf{x}^T\right)(\mathbf{x}_s - \mathbf{x}) \tag{4.18}$$

where $\mathbf{x}_s$ is the position vector of a node and $\mathbf{x}$ defines a surface that is potentially contacting $\mathbf{x}_s$; the minimization can be achieved using the Newton–Raphson method. The equation of the surface is defined as

$$\mathbf{x} = N_\alpha(\xi)\mathbf{x}_\alpha \tag{4.19}$$

The function $N_\alpha(\xi)$ may be linear or nonlinear. The contact point location on the surface is then given by $\mathbf{x}_c = N_\alpha(\xi_c)\mathbf{x}_\alpha$.

As micro-computed tomography and imaging technologies develop and their use becomes more widespread in the characterization of granular materials, it is likely that composite triangular particles will become more widespread in DEM analyses, with the triangular element geometries being generated by applying the triangulation methods discussed in Section 1.8 to the image data. Pöschel and Schwager (2005) proposed the use of triangles connected by deformable spring beams. This approach has the potential to draw upon the extensive work on automatic mesh generation that is ongoing in continuum FEM analyses. It is also likely that analysts will build particles up using cubic base particles, as this type of base particle can be directly linked to the voxels (3D pixels) that are generated from micro-computed tomography scans. This approach is used to construct continuum models for biomedical engineering applications (e.g. Dobson et al. (2006)).

Zienkiewicz and Taylor (2000a) consider the challenge of modelling systems of bodies that are "pseudo-rigid" and proposed that a "faceted shape" for use in this type of analysis can be directly constructed from a finite element discretization. A "pseudo-rigid" system is defined as being a system containing many relatively small particles, each of which can experience large displacements and is restricted to uniform deformations. The block DDA and UDEC codes (Shi (1988), Itasca (1998)) are examples of pseudo-rigid body DEM models. In the case where particle deformation is introduced in the simulations, additional degrees of freedom (i.e. strain) will be included in the global equilibrium (balance) equations, thus increasing the computational cost.

4.7 Linking Ideal DEM Particles to Real Soil

Figure 4.13 illustrates images of real soils, and these shapes are clearly significantly more complex than the DEM particle geometries discussed above. Both the material characteristics (strength, stiffness) as well as the geometry or morphology of real particles will determine their mechanical response. The mechanical response of an individual particle in a granular material depends both on the material properties and the particle geometry (Cavarretta, 2009). For example, Cavarretta (2009) demonstrated the influence of surface roughness on the observed contact stiffness. In particulate DEM, the material response is captured through the contact model, either the inter-particle contact model or, in the case of crushable particles, the contact model used to determine the base particle interactions.

From a morphological perspective, when considering the simulation of real materials both the particle sizes and shapes should be considered. The level of complexity required to model the particle interactions and gain reasonable representations of reality depends on the particle inertia. If the particles are sufficiently large,

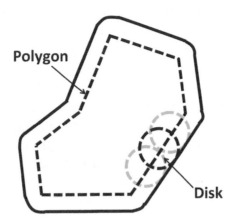

Figure 4.12: Spheropolygons created using the Minkowski sum as described by Alonso-Marroqun and Wang (2009)

(a) SEM image of Huisinish Sand

(b) Monterey No. 16 Sand

Figure 4.13: Images to illustrate the geometric complexity of real soils

171

the magnitude of the surface forces will be negligible in comparison with the particle inertia and these surface interactions will not noticeably influence the particle motion. It seems reasonable to restrict consideration only to particles whose diameter exceeds 100 μm, unless surface interaction forces are to be included in the model. In soil mechanics typically a diameter of 60 μm is taken to define the silt - sand boundary. More generally, Duran (2000) proposes that particles smaller than 100 μm be considered powders, while particles larger than 100 μm can be considered as granular solids. Painter et al. (1998) define a thermal condition for materials to be considered "granular" as

$$mgd >> k_B T \qquad (4.20)$$

where m is mass, g is gravity, d is the grain diameter, k_B is Boltzmann's constant, and T is the absolute temperature. This equation states that the thermal energies $k_B T$ are irrelevant compared to the gravitational energies (mgd). Duran (2000) also highlights that as the particle size decreases, the ratio of the particle surface area to the particle volume increases. Consequently, per unit mass, the area of the particle available to interact chemically with ambient liquids and gases is much greater. This consideration does not exclude the application of DEM to fine particle systems, rather the model becomes more complex. Zhu et al. (2007) outline the analytical expressions that can be used to calculate van der Waals forces and electostatic forces between particles in DEM simulations. When one considers the smallest sized soil particles, clay particles, the challenge posed to DEM analysts is not only the accurate characterization of the chemical interaction forces, but also the simulation of the relatively complex particle shapes.

The geometry of real particles encountered in geomechanics applications is highly complex. The extent of this complexity is evident and quantifiable at different scales. When we consider the geometrical characteristics, the term "form" is used to describe the overall particle geometry. It may, however, be useful, and indeed necessary, to include consideration of more detailed measurements of the particle geometry, i.e. the particle surface roughness. Traditionally geotechnical engineers have used qualitative terms to describe soil, using visual inspection to determine whether the soil is angular, sub-angular, rounded or sub-rounded. Krumbein and Sloss (1963) proposed a chart to provide guidance in assigning quantitative values to sphericity and roundness. Sphericity is defined as the ratio of the surface area of a sphere with the same volume as a particle to its actual surface area. The roundness is calculated by drawing inscribed circles within each corner of a two-dimensional image of a particle and then taking the ratio of the average radius of those circles to the radius of the largest circle that may be inscribed within the particle outline. Cho et al. (2006) demonstrated the use of this chart to relate the mechanical response of particles to their regularity (average of sphericity and roundness).

The greater availability of optical microscopes with built-in digital cameras, along with the maturity of image analysis techniques, has motivated the development of a number of approaches to automate shape quantification and enable more objective measurements to be made (e.g. Bowman et al. (2001)). There are also now apparatus with built in schemes to record binary images and quantify shape for a statistically representative number of particles, for example the Sympatec QicPic apparatus scans particles falling under gravity with a laser, recording an outline of random orientation; images obtained from the QicPic apparatus are illustrated in Figure 4.14. The experimental techniques to measure and quantify particle geometry are evolving on an ongoing basis, and the discussion presented here is far from comprehensive. However, what seems clear is that when tying DEM geometries to real soil particles, analysts have two choices. In the first instance they can work directly from the digitized images to generate their DEM particles, this approach is adopted by Lu and McDowell (2008) and Das et al. (2008) as discussed above. Alternatively analysts can describe their particles using analytical shapes (e.g. using potential particles) that capture the measured sphericity and roundness of their real soil particles. A particular challenge is posed in the case of the highly complex carbonate sands such as those considered by Coop et al. (2004) or illustrated in Figure 4.13(b) above.

(a) Mon-
terey
Sand

(b) Ottawa
Sand

Figure 4.14: Sample outputs using the QicPic particle analyser

Chapter 5

Boundary Conditions

5.1 Overview of DEM Boundary Conditions

In continuum numerical modeling the choice of boundary conditions plays a central role and boundaries are equally important in DEM. A key choice in setting up a DEM simulation is to decide on the spatial domain that will be considered. The boundaries to this domain must then be numerically described in the DEM model.

In continuum modeling there are displacement boundary conditions, along which the displacement is restricted or specified, and traction boundary conditions, along which stress is specified. In a DEM simulation displacement boundary conditions can be achieved by fixing or specifying the coordinates of selected particles. Similarly, force boundary conditions can be achieved by applying a specified force to selected particles. This applied, external, force is added to the contact forces acting on the particle and the resultant force is then used to calculate the particle accelerations and incremental displacements. These force boundary conditions cannot easily be directly used with systems that include thousands of particles as the analyst must apply these conditions to specific, selected particles. DEM is well suited to problems involving large deformations, and forces may need to be applied to different particles as the system deforms. Consequently, algorithms to select boundary particles are needed. Here four types of boundary condition are considered, in order of their popularity of use; these are rigid walls (Section 5.2), periodic boundary conditions (Section 5.3), membrane boundaries (Section 5.4) and axisymmetrical boundaries (Section 5.5).

5.2 Rigid Walls

The most widely employed boundary type is a rigid boundary. These rigid boundaries are simply analytically described surfaces and they can be planar or curved. Examples of DEM simulations using rigid wall boundary conditions in the simulation of element tests are illustrated in Figures 5.1 and 5.2. Rigid boundaries can also be used to simulate inclusions or machinery interacting with the granular material. For example Kinlock and O'Sullivan (2007) used rigid wall boundaries to represent the penetrating object in their simulations of pile installation and cone penetration testing (refer to Figure 1.2). These boundaries themselves have no inertia; the contact forces determined at particle-boundary contact are used to update the particle coordinates only; thus in some respects they are similar to the displacement boundary conditions used in FEM analyses. While the forces acting on the walls do not influence motion of the walls, the user can control the wall movement by explicitly specifying a wall velocity. Users can also specify wall velocities indirectly, by developing an algorithm to move the walls according to some criterion; for example the wall velocity can be related to the current stress conditions, as considered further below. In either case, when the walls are moved, deformations and forces are applied to the assembly of particles through the walls via the wall-particle contacts. In typical DEM simulations contacts are not generated between walls that intersect or touch.

In comparison with the particles, relatively little information is required to describe a rigid wall boundary. A planar rigid wall can be described by a point coordinate, fixing its position in space, and the normal vector describing its orientation. Similarly a circular, cylindrical or spherical wall can be described by specifying the centre of symmetry and radius, and the analyst may need to specify whether the particles will contact the outside or the inside of the boundary. As described by Weatherley (2009), the planar boundaries can be extended to consider more complex geometries by using a series of line segments in two dimensions or a triangular mesh to define the complex surfaces in 3D.

The normal contact forces are calculated by considering the distance from the particle centroid to the wall in a direction normal to the wall. If the normal vector defining the wall orientation is given by (a, b, c), the equation of the wall will be $ax+by+cz+d = 0$, where $d = -ax_w - by_w - cz_w$ and (x_w, y_w, z_w) are the coordinates of a known point on the wall. Then typically the distance between a particle centroid (x^p, y^p, z^p) and the wall will be calculated (in 3D) to be

$$D = \frac{ax^p + by^p + cz^p + d}{\sqrt{a^2 + b^2 + c^2}} \tag{5.1}$$

Note that this distance D is a signed distance, consequently the wall will have an active side and an inactive or "blind" side. In some DEM codes the contact normal direction is not explicitly input, rather the user specifies three or more coordinates (in 3D) to define the plane. Care must be taken as the contact normal direction will depend on the order in which the vertices are input, with the normal pointing in exactly the opposite direction if the vertices are input in a clockwise, rather than a counterclockwise order. If a particle is located on the inactive side of a wall, it will not "see" the wall, i.e. no contact will develop between the particle and the wall and it can simply move, unimpeded through the wall. If the wall stiffness is too low or the velocity of the particles adjacent to the wall is too high in comparison with the time increment chosen, in a given time step a particle centroid can move from the active to the inactive side of the wall and essentially "fall" through the wall. The shear forces are calculated by considering the relative displacement of the wall and the particle at the contact point, and orthogonal to the contact normal. These rigid boundaries are well suited to model the rigid top and bottom platens in the triaxial

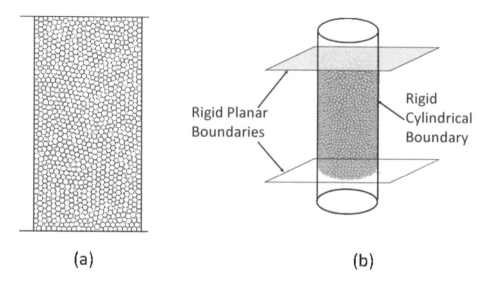

(a)   (b)

Figure 5.1: Rigid boundaries used to simulate soil mechanics element tests (a)Biaxial compression test (b) Triaxial compression test

Servo-controlled rigid boundaries are often used in published DEM analysis to simulate element tests such as the triaxial or direct shear test (e.g. Cheng et al. (2003)). The concept of servo-controlled rigid boundaries is illustrated schematically in Figure 5.3. A representative stress for the sample is measured. Then if this measured internal stress, $\sigma_{ii}^{\text{meas}}$, differs from the user-specified stress, σ_{ii}^{req}, the walls orthogonal to the direction i are slowly moved. If the $\sigma_{ii}^{\text{meas}}$ term exceeds σ_{ii}^{req} (compressive stress is positive), the walls are moved outwards, and if the $\sigma_{ii}^{\text{meas}}$ term is less than σ_{ii}^{req}, the walls are moved inwards. Two approaches can be adopted to measure or quantify the stress σ_{ii}. The contact forces along the relevant boundaries can be summed (integrated) and divided by the boundary area (length in 2D). Alternatively an internal volume can be specified and the average stress within this area can be calculated using the methods discussed in Chapter 9. Whichever approach is used, the wall velocity is proportional to the magnitude of the stress difference, $V_i^{\text{wall}} = \alpha |\sigma_{ii}^{\text{meas}} - \sigma_{ii}^{\text{req}}|$,

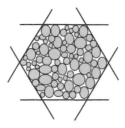

Figure 5.2: Boundaries facilitating controlled rotation of principal stress orientation (Li and Yu, 2009)

where the proportionality constant, α, is user-specified. The parameter α is referred to as the gain in control engineering. The wall velocity can be set to zero when $|\sigma_{ii}^{\mathrm{meas}} - \sigma_{ii}^{\mathrm{req}}| \leq \epsilon$, where ϵ is a user-specified tolerance. Appropriate selection of the gain parameter is essential to ensure that the required stresses are attained, and it is not always easy to find the optimal α value, especially where particle breakage is modelled (e.g. Carolan (2005)). It is good practise to carefully check the success of a servo control system.

The examples given in Figures 5.1 and 5.2 can all be used to simulate element tests in DEM in conjunction with a "servo-controlled" approach. Figure 5.1(a) illustrates the boundary conditions for a 2D biaxial test. In a strain-controlled biaxial compression test, the rigid lateral boundaries are adjusted to maintain a constant horizontal stress, while the top boundary moves downwards. This approach can be directly extended for 3D triaxial test simulations; in this case the sample is encased within six planar rigid boundaries and the position of the four vertical boundaries are adjusted to achieve the required horizontal stresses. Using these six planar boundaries it is possible to set up a fully three-dimensional anisotropic stress state with $\sigma_{11} \neq \sigma_{22} \neq \sigma_{33}$. Figure 5.1(b) illustrates the use of a rigid cylindrical boundary to maintain the axisymmetric stress state that typically exists in triaxial compression tests. In this case the vertical cylinder forms a rigid boundary and the radius of the cylinder is adjusted to maintain the required radial confining pressure. Figure 5.2 illustrates a hexag-

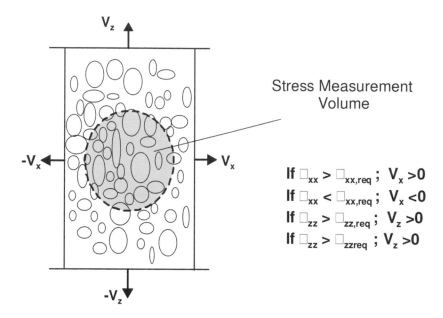

Figure 5.3: Schematic diagram of servo-controlled rigid boundaries used in DEM analyses, (in this case the representative stress, σ_{ij} is measured in a subvolume within the sample.)

onal type configuration of rigid walls used by Li and Yu (2009) to allow rotation of the principal stress directions; this configuration allows the loading direction to be varied in 15° increments.

Pöschel and Schwager (2005) propose an alternative implementation of rigid boundaries, where the boundaries are made up of particles. Their implementation has the advantage of generating walls with a geometrical roughness. These composite particle walls are not included in the main calculations, i.e. their positions are not determined in the main DEM calculation cycle, rather they are either held in place, or their motion is determined using a user-specified boundary velocity. Marketos and Bolton (2010) describe a detailed study on the use of planar boundaries in DEM simulations. They highlight the difference in packing geometry that exists close to the particle boundary and the influence this has on the contact force network and the implications for calculating strain.

5.3 Periodic Boundaries

Another type of boundary condition commonly used in DEM simulations is periodic boundaries (Figure 5.4). Many of the DEM simulations discussed in the geomechanics literature have used periodic boundary conditions (e.g. Thornton (2000) or Ng (2004b)). While the use of periodic boundaries is widespread, initially the concept can be slightly difficult to understand. Using periodic boundaries allows simulation of very large assemblies of particles by considering only a selected subdomain, called a periodic cell. This periodic cell is surrounded by identical copies of itself. This means that each particle, and the local geometry around the particle, is repeated at intervals of L_x in the x direction, L_y in the y direction and L_z in the z direction, similar to the repetition of a pattern in wall paper or fabric. Where periodic boundaries are used the granular material is then effectively infinite in extent. In the DEM simulation it is then assumed that the material response can be represented by considering the response of repeated, identical representative elements, filling this infinite space. Each periodic cell is then a representative volume element (RVE) for the material (the concept of an RVE is considered further in Chapter 9). The systems considered are thus spatially homogenous when observed at a scale that is greater than the cell dimensions.

A good overview of the use of periodic space in DEM simulations is provided by Cundall (1988a) and the description provided by Thornton (2000) is also very useful. The discussion by Rapaport (2004) on periodic boundaries from the perspective of molecular dynamics simulations is applicable to particulate DEM and the comments by Pöschel and Schwager (2005) on the use of periodic boundaries in event driven simulations are also informative. Section 11.5 considers the influence of cell size on simulation results. Referring back to the discussion on contact detection in Section 3.12, where a cell-based search approach is used to determine the list of neighbours, periodic boundaries can readily be accommodated (see also Rapaport (2004)).

5.3.1   Periodic cell geometry

The periodic cell, or "periodic solution space" (Cundall (1988a)), comprises a periodic cell that is almost always a parallelogram in 2D and a parallelepiped in three dimensions. Periodic cells are not strictly restricted to these geometries, and any space-filling convex-shaped region can be used (Rapaport, 2004). For example a hexagon could be used in 2D and this would allow consideration of a broader range of principal stress orientations (similar to the rigid wall approach of Li and Yu (2009)). There is a numerical connection between opposite faces in the periodic cell so that the material responds as if the cell repeats itself infinitely in the directions normal to each periodic cell face. The numerical connection is developed by allowing particles to contact across periodic boundaries and particles to move through the boundaries. Both these issues are considered here. An illustrative diagram of a periodic cell (for the 2D case) is given in Figure 5.4. As discussed by Rapaport (2004), a periodic system can be considered to be a topological remapping of the a 2D region enclosed within the cell boundaries onto a 3D torus (with a 3D cell being remapped to the 4D equivalent of a torus, a construct that is somewhat difficult to visualize).

In the periodic space, particles along the boundaries can contact both their immediately adjacent neighbouring particles and particles close to the opposite periodic boundary. These potential cross-boundary contacts must be considered both in the contact detection stage and in the calculation of the contact forces. When calculating the distance between particles contacting across a periodic boundary a means to identify that the contact crosses a periodic boundary is needed. The distance between the particles can be considered, and where it exceeds a specified distance (say a multiple of the grid size used in the contact detection stage; re-

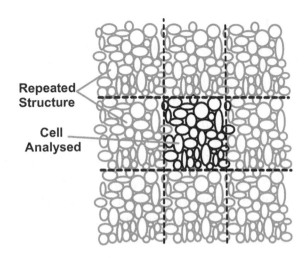

Figure 5.4: Periodic boundaries

fer to Section 3.12) the distance is reassessed while accounting for the possibility of a cross-boundary contact. For example, referring to Figure 5.5 considering two contacting particles A and B in a two-dimensional simulation, with centroids x_A, y_A, x_B, y_B and radii r_A, r_B, the distance between the particles in the x direction is given by

$$l_x = x_A - x_B \tag{5.2}$$

Then if $|l_x|$ exceeds the limit specified, and the periodic cell is bounded by the planes $x = 0$ and $x = x_{\max}$, the overlap at the contact point is

$$\Delta_n = \sqrt{\left(x_A - x_B - \frac{l_x}{|l_x|}x_{\max}\right)^2 + (y_A - y_B)^2} - (r_1 + r_2) \tag{5.3}$$

The equations in the y- (and in 3D z-) directions are similar. Care must also be taken in calculating the distance from the contact points to the particle centroid when determining the contribution of the shear forces to moment loading on the particle.

Particles will tend to move outside the boundaries of the periodic cell during the simulation as a consequence of the motion of the particles and the movement of the periodic boundaries. They are then "remapped" so that they re-enter the cell at the corresponding location on the opposite periodic face. The criterion for this remapping considers the particle centroidal coordinates. When the centroid of a particle is detected to move outside the boundary face 1-1, it is introduced at the boundary face 2-2 at the same elevation (refer to Figure 5.5). The arithmetic involved in the remapping is simple: for particle A if x_A exceeds x_{max}^{cell}, x_A is adjusted to become $x_A = x_A - x_{max}^{cell}$, etc.. If a particle is near a face, it can protrude from that face before it actually crosses it. In the contact force calculations we are effectively introducing an "image" of the particle into the opposite periodic face, and this image is offset from the "real" particle by a distance equal to x_{max}^{cell}.

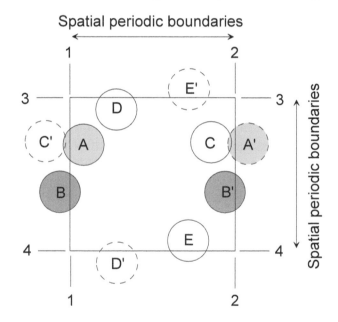

Figure 5.5: Consideration of boundaries in periodic cell *(Figure prepared by Cui (2006))*

The periodic cell itself can be deformed and its volume can change. This deformation is most easily achieved by specifying a strain rate (i.e. a strain-controlled test). The boundaries of the periodic cell will be then adjusted in accordance with the globally applied strain field. Typically the planes $x = 0$, $y = 0$ and $z = 0$ bound the periodic cell along three sides, while the remaining three sides are bounded by the planes $x = x_{\text{max}}^{\text{cell}}$, $y = y_{\text{max}}^{\text{cell}}$, and $z = z_{\text{max}}^{\text{cell}}$. The coordinates of the planes $x = x_{\text{max}}^{\text{cell}}$, $y = y_{\text{max}}^{\text{cell}}$, and $z = z_{\text{max}}^{\text{cell}}$ are adjusted during the simulation as follows:

$$
\begin{aligned}
x_{\text{max}}^{\text{cell}} &= x_{\text{max}}^{\text{cell}} + x_{\text{max}}^{\text{cell}} \dot{\epsilon}_{11}^{\text{grid}} \Delta t \\
y_{\text{max}}^{\text{cell}} &= y_{\text{max}}^{\text{cell}} + y_{\text{max}}^{\text{cell}} \dot{\epsilon}_{22}^{\text{grid}} \Delta t \\
z_{\text{max}}^{\text{cell}} &= z_{\text{max}}^{\text{cell}} + z_{\text{max}}^{\text{cell}} \dot{\epsilon}_{33}^{\text{grid}} \Delta t
\end{aligned}
\tag{5.4}
$$

where $\dot{\epsilon}_{11}^{\text{grid}}$, $\dot{\epsilon}_{22}^{\text{grid}}$, $\dot{\epsilon}_{33}^{\text{grid}}$, are the rates of axial strain specified in the periodic cell in the x, y and z directions respectively, and Δt is the time increment used in the simulations.

Almost all the published DEM simulations consider only triaxial or true triaxial conditions; however, as discussed by Barreto (2010) it is possible to deform the specimen in simple shear. In this case, if the sample is deformed with a strain rate of ϵ_{13} then the cumulative shear deformation of the cell would be calculated by monitoring the parameter $\Delta x_{\text{max}}^{\text{cell,shear}} = \Delta x_{\text{max}}^{\text{cell,shear}} + z_{\text{max}}^{\text{cell}} \dot{\epsilon}_{13}^{\text{grid}} \Delta t$. Adjustments then need to be made when calculating contacts and remapping particles across the periodic boundary defined by $z = z_{\text{max}}$ and across the boundary originally parallel to the plane $x = 0$. This implementation avoids modification of the rectangular grid system for contact detection. Zhuang et al. (1995) adopt an alternative approach where all the cells for contact detection deform as the specimen deforms in shear.

The stress within the periodic cell can be controlled by using a "servo-controlled" approach that is similar to the servo-controlled approach used in the rigid wall simulations considered above. The strain rate is adjusted to achieve a specified stress condition. Typically the average stress is calculated by considering the entire periodic cell volume, rather than sampling a subvolume. The strain rate to achieve the required stress state is then given by (in the x-direction, for example):

$$
\dot{\epsilon}_{11}^{\text{serv}} = \dot{\epsilon}_{11}^{\text{serv}} + \alpha \left(\sigma_{11}^{\text{req}} - \sigma_{11}^{\text{meas}} \right)
\tag{5.5}
$$

where σ_{11}^{req} is the required axial stress in the x-direction, $\sigma_{11}^{\text{meas}}$ is the measured axial stress in the x-direction. (Refer to Chapter 9 for details on how the stresses are calculated.) In this way tests along specified stress paths can be completed. The total strain rate in the system will be the sum of the servo-controlled strain rate and the user-specified strain rate, i.e. $\dot{\epsilon}_{ij}^{\text{grid}} = \dot{\epsilon}_{ij}^{\text{user}} + \dot{\epsilon}_{ij}^{\text{serv}}$.

5.3.2 Particle motion in a periodic cell

When a global strain field is applied in the periodic cell, a velocity is imparted to both the particles and the periodic cell boundaries. Then at each time increment the particle velocity includes a contribution from the solution of the dynamic equilibrium equation and a contribution from the grid strain rate. The incremental displacement of a point in the system with position x_i due to the grid motion $(\Delta u_i^{\text{grid}})$ is then given by

$$\Delta u_i^{\text{grid}} = \dot{\epsilon}_{ij} x_j \Delta t \tag{5.6}$$

These grid-induced displacements are calculated after considering the dynamic equilibrium of each particle for the current time step. Cundall (1988a) describes the particles as being "carried along" and moving "in sync" with the periodic cell deformation until collisions occur and the particles acquire velocities relative to the periodic space.

Cundall (1988a) notes the need to also consider the impact of the periodic cell deformation rate on the relative particle motion used in calculating the contact forces (refer to Section 3.7). When a periodic cell is deforming, the relative motion will be the sum of the velocities calculated from dynamic equilibrium considerations (Cundall (1988a) calls these the real velocities) and the velocity related to the periodic space, so that the relative velocity between two particles a and b $\dot{\delta}_i^{a,b}$ is given by

$$\dot{\delta}_i^{a,b} = \dot{u}_i^b - \dot{u}_i^a + \dot{\epsilon}_{ij} \left(x_j^b - x_j^a \right) \tag{5.7}$$

where the real velocities and position vectors of particles a and b are given by $\dot{u}_i^a$, $\dot{u}_i^b$ and x_i^a, x_i^b respectively.

5.3.3 Use of periodic cell

The periodic cell is taken to be a representative volume of the material and the strain of the periodic cell equals the average strain experienced by the volume. In a physical triaxial test a strain softening response is associated with the formation of localizations or shear bands, i.e. non-uniform strains. A strain softening type response has been observed in periodic cell simulations (e.g. Cundall (1988a) and Thornton (2000)). The periodic cell geometry prohibits formation of unique shear bands. As illustrated in Figure 5.6(a), when a localization forms across a sample it will intersect a pair of periodic boundaries. This will then mean that a second localization is introduced in the sample to preserve continuity of geometry across the periodic boundaries, as illustrated in Figure 5.6(b). Rather than seeing a single discrete shear band, Kuhn (1999) observed micro-bands in his discrete element simulations of biaxial compression tests. Kuhn's micro bands are chains of small regions (void cells) in which there is intensive slip deformation. Kuhn described how these microbands will wrap around a sample with periodic boundaries and join with themselves over an integer number of assembly widths and heights.

While periodic boundary systems can be used to achieve a triaxial stress state (e.g. Thornton (2000)), care should be taken when directly comparing periodic simulation results with results of physical triaxial tests, as the stress inhomogeneities present in real, physical triaxial tests are avoided in periodic cell simulations.

A particularly clear explanation of how to use periodic boundaries is given by Thornton and Antony (2000) who considered the simulation of stress-controlled tests in the periodic cell. The strain rate $\dot{\epsilon}$ used to achieve a desired isotropic stress p is given by

$$\dot{\epsilon} = \dot{\epsilon}_x = \dot{\epsilon}_y = \dot{\epsilon}_z = g(p_d - p_c) \tag{5.8}$$

where p_d is the required isotropic stress state and p_c is the calculated isotropic stress. Thornton and Antony (2000) determined the gain parameter by firstly specifying an initial strain rate $\dot{\epsilon}$ and then calculating g as:

$$g = \left(\frac{\dot{\epsilon}}{p_d - p_c} \right)_{\text{initial}} \tag{5.9}$$

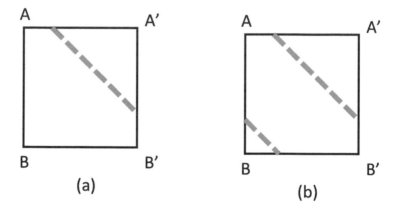

Figure 5.6: Illustration of localization in a periodic cell

Thornton and Antony (2000) clarify how they gradually increase the confining stress p_c value in stages or increments, for example if their target isotropic stress was 100 kPa, the p_c values considered would be 2, 5, 10, 20, 50, and 90 kPa. For each p_c value the strain rate decreases as the sample moves closer to the target stress level. They ensure that the system is stable at each stress level by monitoring the coordination number (i.e. the average number of contacts per particle) and the void ratio, and the calculation cycles continue until no changes in these parameters are observed. Thornton and Anthony also described the approach used to carry out a strain-controlled simulation. In this case the strain rate $\dot{\epsilon}_z$ is specified, while the strain rates $\dot{\epsilon}_x$ and $\dot{\epsilon}_y$ are calculated from Equation 5.3.3.

5.4   Membrane Boundaries

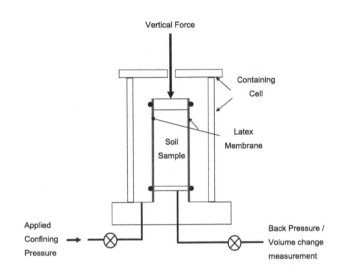

Figure 5.7: Schematic diagram of triaxial cell

The triaxial test is probably the most commonly used laboratory element test in soil mechanics to determine strength and stiffness parameters, and DEM simulations of the axisymmetric stress conditions encountered in the triaxial apparatus are common. Many published DEM simulations of element tests have used rigid walls as the test boundaries in combination with a servo-controlled system to control the stresses (e.g. Cheng et al. (2004)), while Thornton (2000) and Lin and Ng (1997), amongst others, simulated triaxial test stress conditions using a periodic cell. In comparison with periodic boundaries, specimens bounded by rigid walls are a closer approximation to real physical tests. Referring to Figure 5.7, the specimen in the triaxial cell is bounded above and below by steel platens that can be modelled using the rigid wall boundaries discussed above. The lateral boundary for the specimen is a highly flexible latex membrane that allows the pressure within the triaxial cell to be transferred to the sample; at the same time no constraint is imposed upon lateral deformation. The sample can bulge and localizations or shear bands can form. Where the test is simulated using rigid boundaries, while the overall stress state can be attained using the servo-controlled approach discussed above, the rigid walls will inhibit natural development of these localizations and there will be significant non-uniformities in the stresses applied along the boundary. The alternative is to develop a "stress-controlled membrane" to simulate the flexible latex membrane enclosing the sample.

There are two approaches to modelling flexible membranes for triaxial test simulations. In both cases the outermost particles that would touch the membrane in a real physical test must be identified. In the first approach, flexible contact springs with a high tensile capacity are inserted to link these outermost particles. Then a force is applied to these membrane particles that equals the product of the confining pressure and a representative length. Authors who have described the use of this approach include Mulhaus et al. (2001), Oda and Kazama (1998) and Wang and Leung (2008). In the second approach the force that must be applied to the outermost particles to achieve the required stress level are calculated and no special connection is created between these particles (Figure 5.8). The idea of creating a membrane in this way seems to have been first proposed by Cundall et al. (1982). This approach has been used in a number of element test simulations (e.g. Bardet (1994), Kuhn (1995) Powrie et al. (2005), O'Sullivan (2002) and Cui (2006)).

A limitation of the flexible spring approach, is that it is very difficult to relate the properties of discrete contact springs to a continuous membrane. For example Wang and Leung (2008) adopted a membrane particle contact spring stiffness that was 1/10 of the contact stiffness for the remaining particles and with a tensile strength of 1×10^{300} Pa. In a real, physical test when there are large localized displacements, new particles will touch the membrane and this cannot be simulated effectively where the flexible-bonded particle approach is used (as highlighted by Tsunekawa and Iwashita (2001)). However, for simulating membranes with a finite stiffness this approach is preferred.

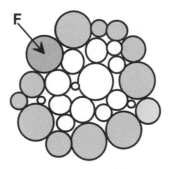

Figure 5.8: Concept of force membrane

Where a force-based numerical membrane is used, the two key stages in the membrane algorithm are the identification of the disks (2D) or spheres (3D) participating in the membrane and the calculation of the required forces. A representative length (in 2D) or area (in 3D) of the side of the sample is then associated with each particle along the specimen edge. A specified force is then applied to each of these side particles equal to the product of the required confining stress and the relevant length or area. As a consequence of the geometrical differences, the numerical algorithms required to identify the external particles and to calculate the representative lengths or areas differ for the 2D and 3D cases. An essential aspect of a constant stress membrane, both in the laboratory and in a DEM simulation, is that it allows the specimen to deform during loading. As the specimen deforms, the outermost particles must be re-identified and the equivalent forces recalculated in the DEM simulations; the criteria used to update the membrane are also considered here.

5.4.1 Two-dimensional implementation

Various approaches can be adopted to identify the outermost particles in the sample. Thomas and Bray (1999) and O'Sullivan et al. (2002) used rather inefficient approaches that considered only the

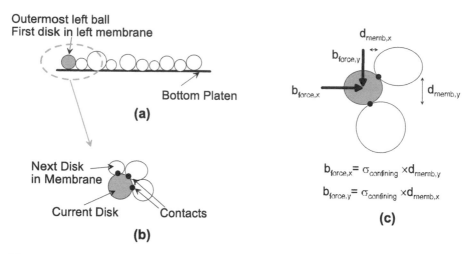

Figure 5.9: Two-dimensional stress-controlled numerical membrane: concepts

particle locations. A more effective and robust approach was developed by Cheung and O'Sullivan (2008) and implemented in a disk-based DEM code (PFC2D). Their algorithm considers the contact information as well as the particle coordinates. The procedure for identifying the outermost disks that would participate in the membrane is illustrated in Figure 5.9. The first step is to identify the disk contacting the bottom boundary whose centroid has the smallest x-coordinate (Figure 5.9(a)). This is the first disk in the left membrane. There is then a search through the linked list of contacts associated with this (current) disk, allowing all the disks touching this disk to be identified (Figure 5.9(b)). Then, the outermost contacting disk that is above the current disk is identified as being the next disk in the membrane. This procedure is repeated, by searching through the contacts associated with each new membrane disk, until contact with the top boundary is detected.

Having identified all the outer disks participating in the membrane, an external force is applied to each of these disks to achieve the specified confining pressure. For each disk, the vertical and horizontal distances, $d_{\mathrm{memb},x}$ and $d_{\mathrm{memb},y}$, between the contact with the membrane disk beneath and the contact with the membrane ball above are determined (Figure 5.9(c)). The horizontal and vertical forces applied to each disk are then given by $d_{\mathrm{memb},y} \times \sigma_{\mathrm{confining}}$ and $d_{\mathrm{memb},x} \times \sigma_{\mathrm{confining}}$, respectively, where $\sigma_{\mathrm{confining}}$ is the specified confining pressure. The vertical force orientation (positive or negative) is determined by considering the orientation of the vector normal to a line joining the two contacts (Figure 5.9(c)). The horizontal force will always be directed towards the centre of the specimen. Rotation of the disks participating in the membrane is inhibited.

Kuhn (2006) adopts a process that starts from a sample in a periodic cell. To apply the membrane (referred to by Kuhn as a "stress-controlled tight-fitting particle boundary"), the periodic boundaries are removed, or broken. Then the branch vectors that previously connected the boundary particles across the periodic boundary are used to apply the forces. Kuhn's implementation is more general than other membrane implementations, as he also includes in his code the capability to apply displacement control along these boundaries. The user then controls the rate at which the information on this boundary is updated as the sample subsequently deforms. Kuhn also allows for explicit user definition of the boundary forces.

5.4.2 Three-dimensional implementation

A comprehensive description of a membrane implementation in three dimensions is given in Cheung and O'Sullivan (2008) and some of the key ideas are presented in Cui et al. (2007). The general principles are outlined here. As before, the two main stages in the algorithm are identification of membrane spheres and the calculation of the applied force. Cheung and O'Sullivan (2008) implemented the algorithm in the commercially available sphere-based DEM code (PFC3D).

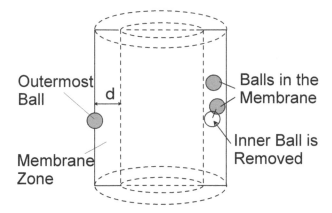

Figure 5.10: Schematic to illustrate membrane sphere identification scheme used by (Cui, 2006)

Identification of membrane spheres

The procedure for identifying the "membrane" spheres or balls is outlined in Figure 5.10. The external forces are applied to these particles. The first step in this sequence of calculations is the identification of a "membrane zone." This is a region containing all spheres that can possibly form part of the membrane. The membrane zone has a thickness d as shown schematically in Figure 5.10. The size of the membrane zone, d, is proportional to the mean sphere diameter a membrane zone thickness equal to the particle diameter is satisfactory.

In the second level of checking, a sphere is considered to be a membrane sphere only if it has no contact with another sphere that would prevent it touching the membrane in a real physical test (Figure 5.10). Spheres are removed from the membrane if there is another sphere outside them that would inhibit such contact.

In their implementation of a numerical membrane, Wang and Tonon (2010) adopt an alternative approach that may be easier to implement. As illustrated in Figure 5.11, firstly they create a cylindrical sample using rigid walls. Then a grid with cell size smaller than the minimum particle diameter is created along a cylindrical surface that is just inside the initial cylindrical wall. For each grid

194

cell, the outer most particle is identified. The advantage of the Cheung and O'Sullivan (2008)/Cui et al. (2007) approach is that the particles selected have an associated projected area that does not overlap with the projected areas from any other particles.

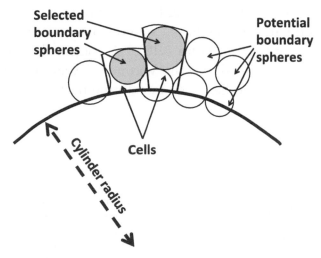

Figure 5.11: Schematic to illustrate membrane sphere identification approach proposed by Wang and Tonon (2010)

Calculation of applied forces

Having identified the membrane spheres, the confining pressure is maintained by applying calculated forces on these membrane spheres. The magnitudes of these forces are calculated by determining the areas of a set of Voronoi polygons created based on the centroids of the membrane spheres (refer to Section 1.8 for details on the Voronoi diagram). The Voronoi diagram is created on a planar surface by unfolding the membrane zone and projecting the coordinates of membrane spheres onto this 2D projection plane (as illustrated in Figure 5.12). As outlined by O'Sullivan (2002), for planar boundaries the Voronoi diagram is created on a plane through the middle of the membrane particles.

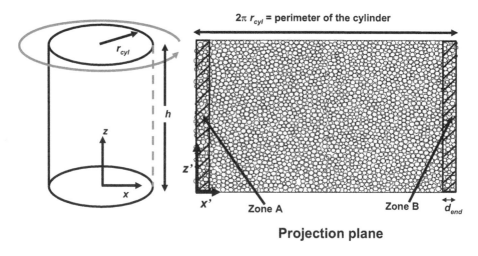

Figure 5.12: Projection plane used to construct Voronoi diagram

A Voronoi cell is then associated with the centroid of each membrane sphere, and this polygon represents the area of the membrane associated with that sphere. The product of that area and the specified boundary pressure gives the magnitude of applied

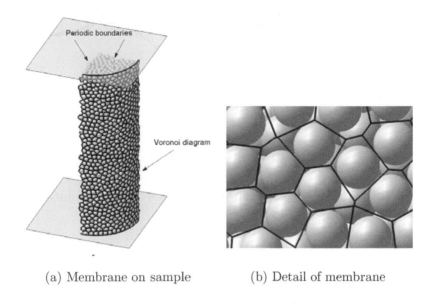

(a) Membrane on sample   (b) Detail of membrane

Figure 5.13: Illustration of 3D membrane Cui et al. (2007)

force required for that sphere. The force is directed along the vector connecting the sphere centroid to the centre of the specimen. In the three-dimensional implementations proposed by O'Sullivan (2002), Cui et al. (2007) and Cheung and O'Sullivan (2008), the vertical component is neglected and only horizontal forces are applied to the membrane spheres. Care needs to be taken along the vertical and horizontal boundaries to ensure that the area is completely covered, and that none of the Voronoi cells extends beyond the boundaries. The steps taken to achieve this are outlined in Cui and O'Sullivan (2006) and Cheung and O'Sullivan (2008); a sample membrane is illustrated in Figure 5.13.

A final point to note regarding the use of stress-controlled membranes in DEM simulations is that an interval for updating the membrane must be specified. As the specimen deforms, new disks will move closer to the outside of the specimen, especially when a localization develops. One possible "trigger" for updat-

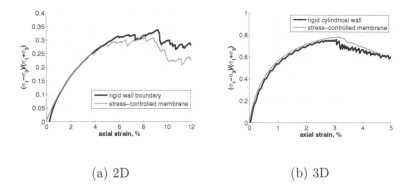

(a) 2D   (b) 3D

Figure 5.14: Macro-scale responses observed in simulations discussed by Cheung and O'Sullivan (2008)

ing the membrane is to measure the cumulative movement of each disk since the last update of the membrane and then to update once more if this value exceeds a specified tolerance. An additional check may be carried out at selected intervals during the simulation to "catch" any particles that start to escape from the membrane.

5.4.3 Comparison of rigid and membrane boundaries

Cheung and O'Sullivan (2008) compared 2D biaxial and 3D triaxial test simulations using both rigid walls and flexible membranes. Figure 5.14 illustrates the overall macro-scale responses obtained in these simulations. For the 2D simulation a specimen of 2377 disks with radii uniformly distributed between 0.48 mm and 0.72 mm was considered. The inter-particle coefficient of friction was 0.5 and the boundaries were assumed perfectly smooth, with a particle-boundary friction coefficient of 0.0. In both cases the specimen was brought to an isotropic stress state of $\sigma_1 = \sigma_3 = 1.0$ MPa, for servo-controlled simulation. The lateral (vertical) walls remained in place, while in the membrane simulation the lateral walls were removed and the membrane was inserted. During the simulation a constant lateral stress of $\sigma_3 = 1.0$ MPa was maintained. For the 3D triaxial test simulations the specimen comprised 12,622 spheres, with the sphere radii being uniformly distributed between 0.88mm and 1.32mm. In this case cementation was modelled using the parallel bond described in Chapter 3. As in the 2D case, two equivalent simulations were carried out. In the first simulation a rigid cylindrical wall enclosed the specimen for the duration of the test, while the stress-controlled membrane was used during the triaxial compression stage for the second test. In both simulations the specimen was brought to an initial isotropic stress state of 10 MPa. Figure 5.14 clearly indicates that in both the 2D and the 3D cases the overall specimen response was not very sensitive to the boundary condition used.

It is interesting to look at the patterns of specimen deformation, illustrated in Figure 5.15 for the three-dimensional case. In this figure the shading illustrates the cumulative particle rotation, with the darkest shade illustrating the particles that have experienced the most rotation. Rotations are used to indicate the position of localizations in the specimens and it is clear that the internal deformation patterns are very sensitive to the boundary conditions. Similar results were observed in 2D. Figure 5.16 compares the forces along the external boundaries at the beginning and end of the 2D simulation. It is clear that there is a far greater variation in the applied forces along the vertical boundaries where the rigid boundary is used. These results indicate that while the macro-scale response may not indicate a strong sensitivity to the lateral boundary conditions used, the internal material responses differ.

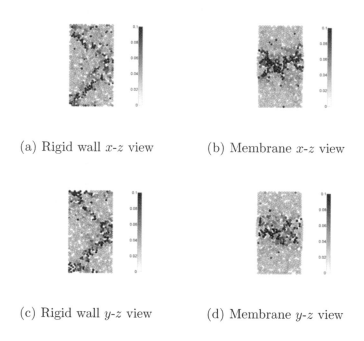

(a) Rigid wall x-z view   (b) Membrane x-z view

(c) Rigid wall y-z view   (d) Membrane y-z view

Figure 5.15: Particle rotations in radians at an axial strain of 4.5% for a (3D) triaxial test on bonded specimens with rigid and membrane boundary conditions

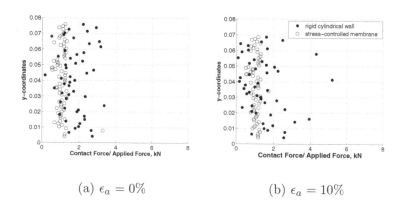

(a) $\epsilon_a = 0\%$ (b) $\epsilon_a = 10\%$

Figure 5.16: Comparison of forces along a rigid wall with applied external forces for 2D simulations

5.5 Modelling Axisymmetry in DEM

There are many axisymmetric systems of interest to geotechnical engineers and when analysing such systems it should be sufficient to model one "slice" of the specimen and thus gain significant computational efficiency. In a similar manner, the computational cost of 3D finite element analyses is greatly reduced if an axisymmetric framework is adopted. Weatherley (2009) proposed using rigid frictionless walls to simulate only a quarter of the domain and hence achieve symmetry. "Circumferential periodic boundaries" as proposed by Cui et al. (2007) allow for axisymmetrical simulations while maintaining a continuous internal system of particle-to-particle contacts throughout the specimen.

These circumferential periodic boundaries, illustrated in Figure 5.17(a), are conceptually similar to the rectangular periodic boundaries that are widely used in DEM simulations (e.g. Thornton (2000)). Particles with their centres moving outside one circumferential boundary (Oa) are re-introduced in at a corresponding location along the other circumferential boundary (Ob) (Figure 5.17(b)). Contact forces can develop between the particles close to each periodic boundary and particles along the other periodic boundary. These forces are calculated by using a rotation tensor when calculating the inter-particle distances (Figure 5.17(c)).

Cui and O'Sullivan (2006) describe the implementation and validation for a 90° segment. The most straightforward way to implement these boundaries is to centre the system on the z-axis. The x- and y-axes then form a periodic boundary pair. The position of a particle along one periodic boundary can be mapped to the other periodic boundary by an orthogonal rotation in the xy plane and the rotation tensor T is given by:

$$\begin{pmatrix} x' \\ y' \end{pmatrix} = T \begin{pmatrix} x \\ y \end{pmatrix} = \begin{pmatrix} \cos\theta & -\sin\theta \\ \sin\theta & \cos\theta \end{pmatrix} \begin{pmatrix} x \\ y \end{pmatrix} \qquad (5.10)$$

where x', y' are coordinates after rotation, x, y are the coordinates before rotation, and θ is the angle between the current periodic boundary and its partner periodic boundary (with anticlockwise rotation being positive). In the implementation of these boundaries, special care must be taken regarding particles that are located close to the origin. If a particle protrudes from both boundaries (Oa and Ob), then forces along both periodic boundaries must be considered. Furthermore, if the particle centroid is located exactly along the z axis, then the particle cannot move in the horizontal ($x - y$) plane. A more detailed description of the implementation of these circumferential periodic boundaries is provided by Cui (2006).

The implementation of axisymmetrical boundaries for "hollow" systems, e.g. the hollow cylinder apparatus, is relatively straight-forward. A problem arises when the system is continuous through the axis of symmetry. In that case, to avoid a local decrease in void ratio, particles need to be explicitly inserted along the central axis of symmetry.

Cui and O'Sullivan (2005) validated this algorithm analytically by simulating the response of a specimen of spheres with a face-centred-cubic packing and comparing the results with the expressions for the theoretical peak strength proposed by Rowe (1962) and Thornton (1979). Subsequently experimental validation was achieved by simulating physical triaxial compression tests on specimens of chrome steel ball bearings under vacuum confinement. Specimens of both uniformly sized spheres and spheres with a range of sizes were subject to monotonic and cyclic triaxial tests (refer to Cui et al. (2007) and O'Sullivan et al. (2008)).

Cui et al. (2007) compared the response of a test simulated using circumferential periodic boundaries and circumferential rigid boundaries. As illustrated in Figure 5.18, the overall response differed. Analysis of the internal structure of the material revealed that the difference in response can be attributed to the disruption of the network of contact forces in the specimens where the rigid walls are used. Referring to the contact force network plots given in Figure 5.19, where the periodic boundaries are used there is a continuous network of particle-particle contacts throughout the specimen, and this network is broken where the rigid wall boundaries are used.

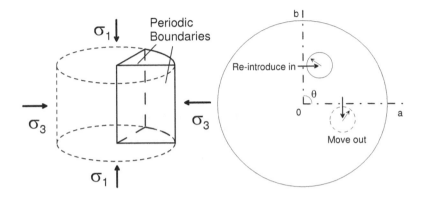

(a) Schematic diagram of periodic boundaries

(b) Illustration of the re-introduced particles

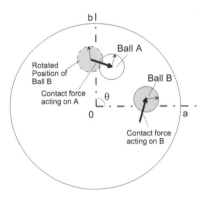

(c) Illustration of contact considerations

Figure 5.17: Schematic illustration of periodic boundaries Cui and O'Sullivan (2006)

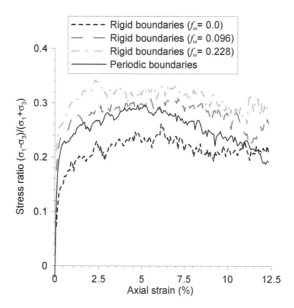

Figure 5.18: Sensitivity of macro-scale response to choice of vertical boundaries in axisymmetrical DEM simulations

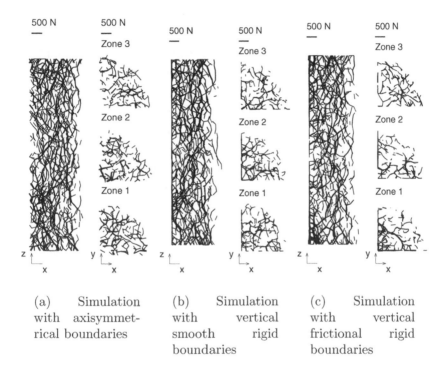

(a) Simulation with axisymmetrical boundaries

(b) Simulation with vertical smooth rigid boundaries

(c) Simulation with vertical frictional rigid boundaries

Figure 5.19: Sensitivity of contract force distribution to vertical boundary conditions.

5.6 Mixed Boundary Condition Environment

The most effective way to achieve a simulation result in reasonable time may be to use a mixed boundary environment, for example a combination of rigid walls and periodic boundaries in only one direction (Weatherley, 2009). Cheung (2010) explored this possibility by comparing the response of a specimen of parallel-bonded spheres in strain-controlled triaxial compression. The response observed in a simulation with rigid boundaries and an aspect ratio of 2:1 (height:diameter) was compared with the response of subvolumes that were extracted along the central axis as illustrated in Figure 5.20(a). The thickness of the slices considered ranged from 20% to 50% of the specimen height. In all cases the responses observed were close to the overall response, considering both the stress-strain response and the volumetric strain response. Zeghal and El Shamy (2004) used a mixture of periodic and rigid boundaries in their simulation of liquefaction.

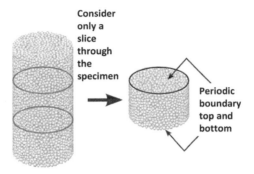

(a) Illustration of slice extraction

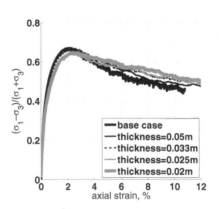

(b) Comparison of overall response

Figure 5.20: Comparison of triaxial test response using differing periodic slice thicknesses

Chapter 6

Fluid-Particle Coupled DEM: An Introduction

6.1 Introduction

The principle of effective stress as proposed by Terzaghi (1936), is one of the most fundamental concepts in soil mechanics. This principle states that when a soil element is subject to a given stress state (described by the three total principal stresses σ_1, σ_2 and σ_3) and when the soil voids are completely filled with water, it is the effective stress state, described by the effective principal stresses σ_1', σ_2' and σ_3' that governs the soil response. These effective principal stresses are calculated as the difference between the total stresses and the water pressure u, i.e. $\sigma_1' = \sigma_1 - u$, $\sigma_2' = \sigma_2 - u$, and $\sigma_3' = \sigma_3 - u$. A soil's response to a change in the stress state (compression, distortion, and a change of shearing resistance) is then exclusively due to a changes in the effective stresses. To study the response of a granular material to a change in effective stresses using DEM, it is valid to model dry conditions; then the applied total stresses will equal the effective stresses.

There are many applications where explicit consideration of the fluid-particle interaction merits consideration. For example, in some cases the variation in the total head in the fluid causes particle motion. This may be the mechanism underlying sand production in oil reservoir sandstones or internal erosion in dams and beneath flood embankments. Fluid flow can play a major role in triggering slope instabilities. Liquefaction (a phenomenon more commonly associated with earthquakes) is a phenomenon where a load is applied rapidly to a loose soil, resulting in an increase in the fluid pressure and with the resultant decrease in effective stress causing a reduction in shear strength. The consequences include flow slide failures and large permanent deformations.

Numerical models that simulate the response of systems that include two or more phases or interacting physical sub-systems are often required for engineering applications. When the independent solution of the response of one phase or system is impossible without simultaneous solution of the other systems or phases, the problem is said to be "coupled" (Zienkiewicz and Taylor, 2000a). The response of each phase in the system is then described using different equations and there are expressions to link or "couple" the phases together. For example, many geotechnical engineers will be familiar with Biot theory. Using Biot theory, a model can be created that considers both the deformations of the soil matrix and the flow of water in the pores. Biot theory considers the soil to be a continuum. Zienkiewicz and Taylor (2000a) draw a distinction between two main classes of coupled systems. In the first type of system, the coupling is at domain interfaces and achieved by the imposition of boundary conditions, while in the second type of coupling the physical domains overlap. Both of these approaches to coupling have been used in particulate DEM modelling.

Simulations with coupled particle and fluid motion appear to have had a bigger impact in process engineering applications, and Zhu et al. (2008) give overviews of various applications in this discipline that have benefitted from the use of coupled particle-fluid simulations. Curtis and van Wachem (2004) presented a review of particle-fluid coupling from a chemical engineering perspective, where simulating particle laden fluid flow is important. In chemical and process engineering applications, the particle packing densities are typically relatively low and turbulent fluid flow conditions often occur. However in geomechanics the particle packings are denser and it is usually assumed that the flow is laminar. A large number of algorithms for coupling particle motion and fluid flow have been presented outside of the geomechanics literature. Here consideration is restricted to some of the ways coupled-particle fluid systems have been implemented in geomechanics; in each case the particle interactions are simulated using conventional (soft sphere) particulate DEM (i.e. the algorithm proposed by Cundall and Strack (1979a)).

Given the significant influence of porewater pressures on fluid response, many geotechnical engineers want to know whether fluid can be modelled using DEM and how the particle-fluid coupling is achieved. This Chapter aims to answer this question by firstly introducing some of basic concepts of fluid flow in porous media and fluid-particle interaction. Then three approaches to simulate coupled fluid-particle systems that have been used in geomechanics are considered. These are (in order of increasing complexity) the assumption of a constant volume during undrained loading (Section 6.4), the use of Darcy's law (Section 6.5), and the numerical solution of the Navier-Stokes equation on a coarse grid (Section 6.6). Section 6.7 considers some additional alternative approaches to simulating coupled fluid-particle systems. The methods presented here are applicable to fully saturated flow. As already discussed in Chapter 3, if the soil is partially saturated, or unsaturated, additional forces will be imparted to the particles arising from the surface tension at the water-air interface.

6.2   Modelling Fluid Flow

It is useful to start by considering the general governing equations for fluid flow. Considering the general case of an infinitesimal element in a fluid with position vector $\mathbf{x}$, continuity considerations mean that the density and velocities are related as

$$\frac{\partial \rho}{\partial t} + \nabla \left(\rho \mathbf{v}^f \right) = 0 \qquad (6.1)$$

where ρ is the fluid density, and $\mathbf{v}^f$ is the fluid velocity. (Note that in tensorial notation $= \left(\rho v_i^f \right)_{,i}$ is equivalent to $\nabla \left(\rho \mathbf{v}^f \right)$). If a fluid is incompressible and has a constant density, this equation can be rewritten as

$$\nabla \mathbf{v}^f = 0 \tag{6.2}$$

Furthermore, for incompressible flow the differential equation for momentum is given by the Navier-Stokes equation:

$$\rho \mathbf{g} - \nabla u + \mu \nabla^2 \mathbf{v}^f = \rho \frac{\partial \mathbf{v}^f}{\partial t} \tag{6.3}$$

where g is a body force vector, u is pressure, and $\mathbf{v}^f$ is the fluid velocity vector. Knowing the initial and boundary conditions, this equation can be analytically solved for some simple problems. However more often it is solved numerically using, for example, the finite difference method or the finite element method. The application of numerical methods to solve the Navier-Stokes equation is generally called *computational fluid mechanics* or CFD.

Generally in geomechanics Equations 6.2 and 6.3 are not explicitly considered. Rather, it is assumed that Darcy's law is applicable. Darcy's law is an empirically derived one-dimensional expression that relates the gradient in total head in direction j to the fluid velocity in direction j as

$$v_j^f = -kh_{,j} = -ki_j \tag{6.4}$$

where k is the permeability, h is the total head and i_j is the hydraulic gradient in direction j. The term "head" is used to denote the energy per unit weight in the fluid and is the sum of the pressure head, the elevation (or potential) head and the velocity head. In engineering practice the permeability k is either determined experimentally or estimated from a knowledge of the particle size distribution for the soil. It represents the energy loss as the water passes along the rough, tortuous flow paths formed by interconnecting voids in the material. While the permeability is an empirical or phenomenological parameter, the Kozeny-Carmen equation, which relates permeability to void ratio, is derived by assuming that flow through soil is analogous to flow through an

212

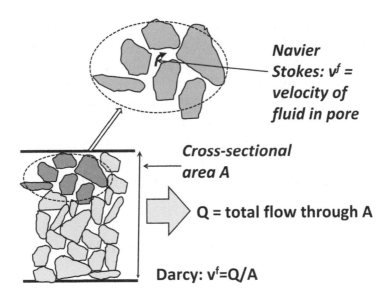

Figure 6.1: Difference between Darcy and Navier-Stokes fluid modelling approaches

assembly of capillary tubes (refer to Mitchell (1993) for the full derivation).

Referring to Figure 6.1, the velocity vectors considered in Equations 6.4 and 6.3 are not the same. In the Navier-Stokes equation the actual fluid velocity is considered. Darcy's law considers the *discharge velocity*, i.e. $\mathbf{v}^f$ is taken to be the measured flow rate, $\mathbf{Q}$ in $(\mathrm{m}^3/\mathrm{s})$, divided by the cross sectional area of the sand filled flow domain, A (m^2), i.e.

$$\mathbf{v}^f = \frac{\mathbf{Q}}{A} \tag{6.5}$$

However, the fluid flow will only take place through the voids (or pores), and not through the solid phase. Therefore the actual fluid velocity will be greater than this calculated value. A measure of the average velocity in the voids is given by

$$\mathbf{v}^f_s = \frac{\mathbf{v}^f}{n} \tag{6.6}$$

where n is porosity (strictly speaking the area porosity should be used, i.e. the area of a given cross section that is occupied by the void space, divided by the total area). Many authors refer to this average velocity as the *seepage velocity*. However, the *real velocity*, $\mathbf{v}_r^f$ of the fluid flow will vary from void to void depending on the actual void dimensions, and it is this real velocity that is considered in the Navier-Stokes equation.

For two-dimensional or three-dimensional flow, combining considerations of continuity with Equation 6.4, and assuming homogeneity, it can easily be shown that flow is given by

$$\begin{aligned} k_1\frac{\partial^2 h}{\partial x_1^2} + k_2\frac{\partial^2 h}{\partial x_2^2} &= 0 \quad (2D) \\ k_1\frac{\partial^2 h}{\partial x_1^2} + k_2\frac{\partial^2 h}{\partial x_2^2} + k_3\frac{\partial^2 h}{\partial x_3^2} &= 0 \quad (3D) \end{aligned} \tag{6.7}$$

where k_1, k_2 and k_3 are the velocities in the x_1-, x_2- and x_3- directions respectively. The contribution of the velocity head term to the total head is usually negligible, and as the elevation head is simply given by the elevation (x_3), Equation 6.7 can be expressed in terms of pressure, u:

$$\begin{aligned} k_1\frac{\partial^2 u}{\partial x_1^2} + k_2\frac{\partial^2 u}{\partial x_2^2} &= 0 \quad (2D) \\ k_1\frac{\partial^2 u}{\partial x_1^2} + k_2\frac{\partial^2 u}{\partial x_2^2} + k_3\frac{\partial^2 u}{\partial x_3^2} &= 0 \quad (3D) \end{aligned} \tag{6.8}$$

When the permeability is isotropic (i.e. $k_1 = k_2 = k_3$) Equations 6.7 and 6.8 reduce to the Laplace equation, i.e. $\nabla^2 u = 0$ and $\nabla^2 h = 0$.

Fluid flow is often classified as being either laminar or turbulent. Considering flow along a pipe, if laminar conditions exist, the fluid "particles" move in an orderly manner and maintain the same positions relative to the pipe bounding the flow. In contrast, if the flow is turbulent, there will be strong random high-frequency fluctuations in the magnitude and direction of the "particle" velocities. From the perspective of flow in granular materials, knowledge of the flow regime is important for two reasons. Firstly Darcy's law is applicable only to laminar flow. Secondly, as explained further in Section 6.3, one of the equations used to calculate drag force impacted on a particle (the Wen and Yu equation) depends on the flow regime. The type of flow regime can be identified by calculating the Reynolds number Re of the flow. The Reynolds number is the ratio of the inertial and viscous forces in the flow. Referring to Tsuji et al. (1993) the Reynolds number for flow through a particulate material can be calculated as

$$Re = \frac{n \rho_f d_p |\bar{\mathbf{v}}^p - \mathbf{v}^f|}{\mu_f} \tag{6.9}$$

where n is the porosity, ρ_f is the fluid density, μ_f is the fluid viscosity, d_p is the particle diameter, $\bar{\mathbf{v}}^p$ is the average particle velocity, and $\mathbf{v}^f$ is the fluid velocity.

Re≈1	Re≈100	Re≈800	
Darcy regime	Forcheimer regime	Transition regime	Turbulent regime
Creeping flow, no inertia influence	Laminar flow, increasing inertia influence	Inertia flow with increasing random, irregular flow	Flow entirely random and irregular
i=v/k	i=αv+βv²		i=α$_t$v+β$_t$v²

Figure 6.2: Regimes of flow in porous media after Trussell and Chang (1999)

As discussed by Cheung (2010), the review of flow through porous media by Trussell and Chang (1999) is useful when seeking to understand the implications of the Re values calculated. As illustrated in Figure 6.2, rather than simply classifying flow to be either laminar or turbulent, four flow regimes can be identified. For values of $Re < 1$ the flow is both laminar and "creeping", i.e. there is no significant inertial contribution. As Re increases the flow enters the Forchheimer regime, where it transitions between strictly steady laminar flow to a condition where inertial effects become increasingly important, as evidenced by the presence of a small contribution from $(\mathbf{v}^f)^2$ at the upper end of this regime. Then, as Re increases further, there is a transitional regime between full inertial flow to full statistical turbulence. Finally, above a Reynolds number of about 800 a regime of fully turbulent flow is encountered.

Equations 6.3 and 6.8 are both partial differential equations describing the variation in fluid pressure. The principal difference between these equations is that different scales are considered. The Navier-Stokes equation (Equation 6.3) directly considers the fluid flow, and can be applied to model the flow within the voids of a granular material. Considering Darcy's law, the permeability given by the values of k_1, k_2, k_3 used in Equation 6.8 is a macro-scale parameter, representing the frictional loss incurred as water flows through a complex network of connected voids, i.e. it is used to describe an average response for an assembly of particles.

Ideally when simulating a system of particles and fluids interacting, the fluid phase would be modelled by numerical solution of the Navier-Stokes equation. The motion of the particles would be determined using DEM, and some means to account for the fluid-particle interaction would be found. The challenge is that large numbers of grains are usually required in the particle DEM model, resulting in large numbers of voids with complex geometries and often very small throat widths. The solution scheme for the Navier-Stokes equation would need to use some type of grid or mesh with a very fine discretization that can accurately capture the geometry of the voids. While models with sub-void resolution can be created, they are complex and computationally very expensive. Consequently in geomechanics simplified approaches to simulate coupled systems are more commonly used.

6.3 Fluid-Particle Interaction

When submerged in a fluid, particles will interact with the fluid and the particle motion will be influenced by the presence of the fluid. Different types of forces act on the particle and these can be classified either as hydrostatic or hydrodynamic forces (Zhu et al. (2007) and Shafipour and Soroush (2008)). The hydrostatic force is the buoyancy force due to pressure gradient around the particle. The hydrodynamic forces include the drag force, the virtual mass force and the lift force. The virtual mass force is the force required to accelerate or move the fluid surrounding the particle. It is called a virtual mass as its effect is equivalent to adding a mass to each particle. Viscous effects may cause a delay in boundary layer development, and this is accounted for by including the Basset force. The lift forces are due to particle rotation, and research, including that by Morsi and Alexander (1972), has shown that the lift forces are much smaller than the drag forces. Zhu et al. (2007) give key references discussing each of these interaction forces. Consideration will be restricted here to the pressure gradient force and the drag force as these are the dominant interaction forces and have a measurable influence on the particle motion and fluid flow. The other interaction forces are not relevant for the coupled fluid-particle systems of interest in geomechanics.

The drag force is the dominant fluid-particle interaction mechanism and it depends on a drag coefficient, C_d, the particle-fluid relative velocity and the particle size. For a single isolated particle moving through a fluid the drag force $\mathbf{f}_d$ is given by:

$$\mathbf{f}_d = C_d \pi \rho_f d_p^2 \left| \mathbf{v}^f - \mathbf{v}^p \right| \frac{\mathbf{v}^f - \mathbf{v}^p}{8} \tag{6.10}$$

where C_d is a drag coefficient, ρ_f is the fluid density, d_p is the particle diameter, $\mathbf{v}^f$ is the fluid velocity and $\mathbf{v}^p$ is the particle velocity. (Here the sign convention used by Kafui et al. (2002), Van Wachem and Sasic (2008) and Zeghal and El Shamy (2008) is followed). This drag force cannot be directly applied for systems of particles, as the fluid flow regime will be affected by the other particles in the system. As explained by Zhu et al. (2007), the presence of other particles reduces the space for the fluid and this results in a steep gradient in the fluid velocity and an increased shear stress on the particle surface. The effect of the presence of other particles on the drag force is most often accounted for using

217

a corrective function that depends on the porosity.

Usually the drag force for a system of particles is calculated using either the Ergun equation or the Wen and Yu equation. In their implementation Tsuji et al. (1993) calculated the drag force as

$$\mathbf{f}_d = \beta \frac{\mathbf{v}^f - \mathbf{v}^p}{\rho_f} \tag{6.11}$$

where $\mathbf{v}^p$ is the velocity of the DEM particle, $\mathbf{v}^f$ is the fluid velocity and ρ_f is the fluid density. If the porosity is less than 0.8 then β is given by the Ergun equation (Ergun, 1952) as follows:

$$\beta = 150\mu \frac{(1-n)^2}{d_p^2 n^2} + 1.75 \frac{(1-n)\rho_f|\mathbf{v}^p - \mathbf{v}^f|}{n d_p} \tag{6.12}$$

If the porosity exceeds 0.8 then β is given by the Wen and Yu expression (Wen and Yu, 1966):

$$\beta = \frac{3}{4}C\frac{|\mathbf{v}^p - \mathbf{v}^f|\rho_f(1-n)}{d_p}n^{-2.7} \tag{6.13}$$

where C depends on the Reynolds number, i.e.

$$
\begin{aligned}
C &= 24\left(1 + 0.15Re_p^{0.687}\right)/Re_p \quad (Re_p < 1,000)\\
C &= 0.43 \qquad\qquad\qquad\qquad\quad (Re_p > 1,000)
\end{aligned}
\tag{6.14}
$$

Rather than using two different equations for the drag coefficient that depend on the fluid flow regime, Kafui et al. (2002) and Itasca (2008), adopted the empirical expression proposed by Di Felice (1994) that gives a single expression for the drag force for a range of flow regimes. There is a slight difference in the way the drag force is calculated. The expression used by Xu and Yu (1997) is given by

$$\mathbf{f}_d = n\frac{1}{8}C_d^{\text{DiF}}\rho_f\pi d_p^2|\mathbf{v}^f - \mathbf{v}^p|(\mathbf{v}^f - \mathbf{v}^p)n^{-\chi} \tag{6.15}$$

while Zhou et al. (2008), Kafui et al. (2002), Xu et al. (2000) and Zhu et al. (2007) use

$$\mathbf{f}_d = n\frac{1}{8}C_d^{\text{DiF}}\rho_f\pi d_p^2 n^2 |\mathbf{v}^f - \mathbf{v}^p|(\mathbf{v}^f - \mathbf{v}^p)n^{-(\chi+1)} \qquad (6.16)$$

The porosity function $n^{-\chi}$ is included to correct for the presence of other particles, where n is the porosity of the current cell. This function depends on the flow, i.e.:

$$\chi = 3.7 - 0.65 exp\left[-\frac{(1.5 - log_{10}Re^p)^2}{2}\right] \qquad (6.17)$$

The drag coefficient is given by

$$C_d^{\text{DiF}} = \left[0.63 + \frac{4.8}{\sqrt{Re^p}}\right]^2 \qquad (6.18)$$

where the Reynolds number for the particle Re^p is determined using Equation 6.9. Kafui et al. (2002) showed that this expression gave a close correlation to a drag force calculated using the Ergun equation at a porosity of 0.4 and good agreement with a drag force calculated using the Wen and Yu equation at a porosity of 0.8. Using the DiFelice expression there is a smooth variation in the drag force as a function of porosity.

The drag coefficient depends on the Reynolds number of the flow and the liquid properties. The empirical expressions proposed by Ergun (1952) for low-porosity materials were determined from experimental observation. As noted by Zhu et al. (2007), high-resolution numerical models that simulate the fluid flow at a sub-particle-scale resolution can also be used to determine the coefficients. Curtis and van Wachem (2004) note that the ability of the empirical drag coefficients to accurately model the interaction decrease with decreasing particle sphericity and they suggest reference to Chhabra et al. (1999) for a review of drag force coefficients for non-spherical particles. Zhu et al. (2007) propose an adjustment to the Wen and Yu equation to account for particle shape.

The hydrostatic force is given detailed consideration by Kafui et al. (2002). Citing the work of Anderson and Jackson (1967), they give an expression for the average stress tensor in the fluid to be

$$\xi_{ij}^f = -u\delta_{ij} + \tau_{ij}^f \tag{6.19}$$

where u is the fluid pressure, δ_{ij} is the identity tensor (as defined in Chapter 1) and $\boldsymbol{\tau}^f$ is the viscous stress tensor (sometimes called the deviatoric stress tensor). Kafui et al. (2002) give two expressions for the hydrostatic force imparted onto the particle as follows:

$$\mathbf{f}^{\text{hydrostatic}} = \frac{4}{3}\pi r_p^3 (-\nabla u + \nabla \boldsymbol{\tau}_f) \tag{6.20}$$

and

$$\mathbf{f}^{\text{hydrostatic}} = \frac{4}{3}\pi r_p^3 (\rho_f \mathbf{g} + \nabla \boldsymbol{\tau}_f) \tag{6.21}$$

As outlined by Kafui et al. (2002), when used in a numerical simulation the choice of equation depends on how the momentum equations are solved. If the pressure gradient force model is used then Equation 6.20 is applicable, and if the fluid density based bouyancy model is used then Equation 6.21 is adopted. Kafui et al. (2002) noted that Tsuji et al. (1993) neglected viscous stress and so their hydrostatic force is given by

$$\mathbf{f}^{\text{hydrostatic}} = \frac{4}{3}\pi r_p^3 (-\nabla u) \tag{6.22}$$

Zeghal and El Shamy (2004) also neglect viscous effects and use Equation 6.22.

The expressions to account for drag and bouyancy effects have been presented here in a way that can be directly incorporated in a DEM model. If the particle velocties are known, the porosity of the granular material is known and the velocity of the surrounding fluid is known, the influence of the fluid on the particle motion can be simulated by adding a drag force to the resultant forces acting on each DEM particle. The particle velocities can be determined using DEM, while some type of flow model is needed to determine the fluid velocities and pressures.

6.4 Simulation of Undrained Response Using a Constant-Volume Assumption

When a saturated soil is subjected to undrained loading, the bulk modulus of the pore fluid is assumed to be sufficiently large relative to the bulk modulus of the soil that the material deforms at a constant volume. DEM simulations where the sample volume is maintained constant as a deformation is imposed on the system can be used to model undrained material response. The response of the mixed particle-fluid system can then be simulated without explicit consideration of the fluid phase. In effect, the fluid and particle phases are decoupled. The mechanics of the fluid response is not explicitly considered, rather the pressures are inferred from the measured response of the particle system. This is the simplest way to model the response of the particle-fluid system and is therefore considered prior to discussing more complex coupled simulations. This approach is restricted to completely undrained response and is therefore really only applicable to simulate the ideal situations considered in laboratory element tests.

As outlined by Ng and Dobry (1994) for this type of DEM simulation a triaxial sample is initially compressed (consolidated) to an isotropic stress of $\sigma_{xx} = \sigma_{yy} = \sigma_{zz} = \sigma_0$. During the test, either a compressive or shear strain is applied. As the sample must deform with a constant volume, changes in the horizontal stresses in the sample ($\sigma_r = \sigma_{xx} = \sigma_{yy}$) will occur during shearing. Assuming the horizontal stress represents the confining pressure, the excess porewater pressure is then taken to be $\Delta u = \sigma_0 - \sigma_r$. As noted by Yimsiri and Soga (2010), a key assumption in this approach is that the soil skeleton, or network of contacting particles, is significantly more compressible than either the particles or the pore fluid. Examples of documented simulations using this approach include Ng and Dobry (1994), Sitharam et al.(2002, 2009) and Yimsiri and Soga (2010).

The easiest way to assess the success of this approach to simulating undrained soil response is to consider simulation of a mono-

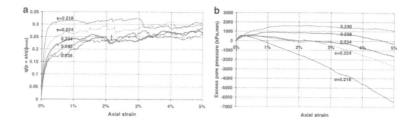

Figure 6.3: Results of DEM simulations of undrained biaxial compression tests by Shafipour and Soroush (2008) (a) Stress ratio (p'/q) versus axial strain (b) Excess pore water pressure versus axial strain.

tonic undrained test. Shafipour and Soroush (2008) simulated undrained biaxial tests using 4,000 disks adopting the constant-volume approach. A range of void ratios was considered and the results are included in Figure 6.3. Referring to Kramer (1996) or Mitchell and Soga (2005), the key characteristics one would expect to see in an undrained test on a sand are evident. The denser samples dilate, with a reduction in the excess porewater pressure giving an increase in the mean effective stress. For the looser samples, a phase transformation point, marking a transition from dilative to compressive response is observed, most notably for the specimen with a void ratio of 0.240. Drained and undrained tests on the same sample yielded equivalent effective stress responses. Sitharam et al. (2002) also presented results for monotonic undrained response on specimens of polydisperse spheres and again the loose sample compressed and generated positive excess porewater pressures, while the dense sample generated negative excess porewater pressures.

Notable contributions to understanding the criteria that trigger liquefaction during earthquakes have been made using undrained laboratory tests where (usually) the deviator stress is cycled and the number of cycles required to initiate liquefaction is recorded. Ng and Dobry (1994) used the constant-volume approach to apply shear strain cycles to a sample of spheres in a periodic cell. Qualitative agreement was observed with previously documented laboratory tests to examine liquefaction; there was a build up in excess porewater pressure and a decrease in stress as the load cycles progressed. As illustrated in Figure 6.4, the cyclic undrained triaxial compression tests by Sitharam et al. (2009) also gave a response that is similar to that observed in physical cyclic undrained laboratory experiments.

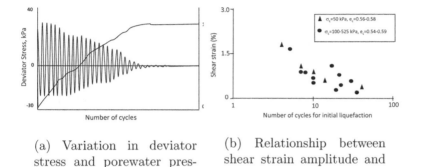

(a) Variation in deviator stress and porewater pressure for simulation with cyclic strain amplitude of 0.6%

(b) Relationship between shear strain amplitude and number of cycles to initial liquefaction

Figure 6.4: Results of DEM simulations by Sitharam et al. (2009)

6.5 Modelling of Fluid Phase using Darcy's Equation and Continuity Considerations

The simplest way to explicitly consider the fluid phase is to use Darcy's equation to model the fluid flow. The most basic implementation restricts the flow to be one-dimensional. Calvetti and Nova (2004) proposed this type of approach to account for seepage forces in slope stability analysis. In their method, an infinite slope with inclination α is considered. The seepage force $\mathbf{J}$ is assumed to act parallel to the slope, the force acting on a representative subvolume V in the slope is then given by

$$\mathbf{J} = V\gamma_w \mathbf{i} = V\gamma_w \sin\alpha \qquad (6.23)$$

where **i** is the hydraulic gradient and γ_w is the unit weight of water. The macro-scale seepage force is used to determine an equivalent, seepage-induced force on each particle ($\mathbf{J}_p$). This force is calculated so that

$$\mathbf{J}_p = C\gamma_w \sin\alpha \tag{6.24}$$

where the constant C is a parameter that is introduced so that $V = C\Sigma V_p$, where V_p is the particle volume and the summation is over all the particles in volume V. This condition is met if $C = \frac{1}{1-n}$, where n is the porosity. The water pressures act in all directions and will impose an uplift thrust on the particles that is given as $\mathbf{f}^{\text{uplift}} = \gamma_w V_p$. In a 2D analysis with disk particles (inhibited from rotating) Calvetti and Nova (2004) compared the critical water table levels obtained using this model with the analytical solution and obtained very good agreement.

Cheung (2010) proposed a simple model of axisymmetric flow through an assembly of particles to a central opening or well. In her model she discretized the domain using the radial boundary system illustrated in Figure 6.5 and assumed continuity of flow along each of the radially directed channels. A series of rings each centred on the origin of the system form circumferential boundaries to the fluid cells in the system. The porosity of each of these cells can be calculated. Then, if the discharge velocity at the centre of the assembly is known, the average fluid velocity for each cell in the system is given by

$$u_i = \frac{1}{n}\frac{R_{\text{cavity}}}{r_i}u_{\text{cavity}} \tag{6.25}$$

The average particle velocity in the cell and average particle diameter are known from the DEM simulation results. Knowing the fluid velocity for the cell, a drag force can then be calculated using the Ergun, Wen and Yu or DiFelice expressions given in Section 6.3 above.

Jensen and Preece (2001) and Shafipour and Soroush (2008) both developed coupled fluid-particle algorithms based on Darcy's

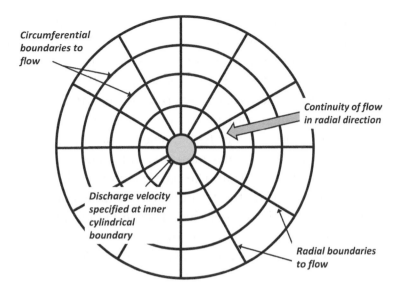

Figure 6.5: Discretization applied by Cheung (2010) where fluid velocities are calculated from continuity assumptions

flow model. In both cases the fluid phase is considered by creating a system of discrete cells that overlay the particles, with Jensen and Preece using a triangular mesh and Shafipour and Soroush using a Cartesian grid. In the implementation proposed by Shafipour and Soroush (2008) three steps are required to simulate the fluid flow. Firstly the change in pore volume due to particle motion is calculated. This will cause an increment in excess pore water pressure that is the product of the volumetric strain and the bulk modulus B of the fluid-particle system. For the fully saturated case

$$\frac{1}{B} = \frac{1-n}{B_p} + \frac{n}{B_w} \qquad (6.26)$$

where B_p is the bulk modulus of the particles and B_w is the bulk modulus of the water (fluid). Flow between cells is governed by Darcy's law. Realizing that the increment in pressure due to the flow of fluid from the cell corresponds to the volume of fluid entering the cell, a large system of linear equations can be formed to solve for the change in pressure due to water flow during the

225

current time increment. Shafipour and Soroush's implementation was applied to constant-volume simulations and changes to the hydraulic conductivity due to changes in void ratio were ignored.

While Shafipour and Soroush argue that an approach based on Darcy's law is preferable to the Navier-Stokes equation because of its simplicity, its implementation remains complex, and the addition of volumetric strains due to particle motion and volumetric strains due to fluid flow seems non-physical. The cell size must also be significantly larger than the particle size to allow a macro-scale permeability parameter to be used. These considerations suggest that grid-based schemes that use the Darcian flow models are unlikely to become a dominant means for simulating coupled fluid-particle systems in geomechanics. The simpler types of flow model such as those proposed by Calvetti and Nova (2004) and Cheung (2010), may more be useful as, while they are approximate models, they do allow efficient simulation of large boundary value problems involving fluid flow. An estimate of the errors associated with the simplifications in these models should ideally be established.

6.6   Solution of Averaged Navier-Stokes Equations

The most commonly implemented approach that solves the Navier-Stokes equation to determine the fluid motion is the coarse-grid approximation method proposed by Tsuji et al. (1993). In this approach the flow within each individual pore of the granular material is not modelled. Referring to Figure 6.6, the fluid phase is discretized at a scale that is typically five to ten times the average particle diameter. Zeghal and El Shamy (2008) propose that the ideal cell dimension should be large in comparison with the micro or particle-scale and small compared with macroscopic variations in the boundary value problem of interest. (The concept of scale in DEM is considered further in Chapters 9 and 10). The average velocity and pressure for each fluid cell are calculated. These velocities and pressures are then used to determine the drag and buoyancy forces acting on the particles. This approach does not simulate flow along individual pore pathways in the material. The average parameters in each cell determine the fluid flow. Anderson and Jackson (1967) first proposed the idea that the continuity and momentum equations for the fluid could be calculated using local average values. This method is within the second category of coupling introduced in Section 6.1 as the fluid flow domain overlaps the particles.

A stable 2D granular material will have a relatively high packing density, and the material will be dominated by closed voids. Discrete pathways allowing fluid flow will be largely absent. However, in the coarse-grid-averaged Navier-Stokes approach the fluid model considers the particles and voids to be merged into a continuous porous medium. The method can therefore be applied in 2D; the earliest demonstration of the efficacy of this approach was the 2D simulations of fluidized beds by Tsuji et al. (1993). Examples of the use of this approach include Kafui et al. (2002), El Shamy and Zeghal (2005) and Zhou et al. (2008) amongst many others. Kafui et al. (2002) and Zhu et al. (2007) both include discussions on the different implementations that have been used.

Two partial differential equations determine the response of the fluid: these are the averaged Navier-Stokes continuity and momentum equation. The averaged continuity equation is given by (Tsuji et al., 1993)

$$\frac{\partial n}{\partial t} + \nabla \left(n \mathbf{v}^f \right) = 0 \qquad (6.27)$$

where n is the local porosity at position $\mathbf{x}$, t is time and $\mathbf{v}^f$ is the fluid velocity vector at position $\mathbf{x}$ and time t. This implementation does not account for fluid compressibility; the continuity equation for compressible flow is (Kafui et al., 2002)

$$\frac{\partial (n\rho_f)}{\partial t} + \nabla \cdot \left(n\rho_f \mathbf{v}^f \right) = 0 \qquad (6.28)$$

Implementations of the averaged Navier-Stokes equations in geomechanics have tended to use Equation 6.28.

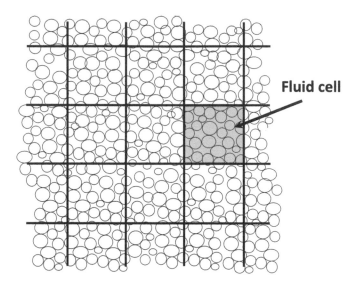

Figure 6.6: Schematic diagram of coarse-grid approach used to solve the averaged Navier-Stokes equations

Kafui et al. (2002) and Zhu et al. (2007) both give two formulations for the momentum equation; these equations were termed "Model A" and "Model B" by Gidaspow (1994). In model A it is assumed that the pressure drop is shared between the fluid and solid particle phases, it is termed the pressure gradient force model (PGF) by Kafui et al. (2002).

$$\frac{\partial \left(n\rho_f \mathbf{v}^f\right)}{\partial t} + \nabla \cdot \left(n\rho_f \mathbf{v^f v}^f\right) = -n\nabla u + \nabla \cdot (n\boldsymbol{\tau}_f) - \mathbf{f}_{fp}^A + n\rho_f \mathbf{g} \quad (6.29)$$

where $\boldsymbol{\tau}_f$ is the viscous or deviatoric stress tensor. Zeghal and El Shamy (2004) used this expression, neglecting viscous stresses, i.e. the term $\nabla \left(n\boldsymbol{\tau}_f\right)$ was excluded. In model B (called the FBD model by Kafui et al. (2002)) it is assumed that the pressure drop takes place in the fluid phase only and the momentum equation is given by

$$\frac{\partial \left(n\rho_f \mathbf{v}^f\right)}{\partial t} + \nabla \left(n\rho_f \mathbf{v^f v}^f\right) = -\nabla u + \nabla \left(n\boldsymbol{\tau}_f\right) - \mathbf{f}_{fp}^B + n\rho_f \mathbf{g} \quad (6.30)$$

The volumetric fluid-particle interaction forces $\mathbf{f}_{fp}^A$ and $\mathbf{f}_{fp}^B$ differ for the two models. For model A

$$\mathbf{f}_{fp}^A = \bar{N}_p n \mathbf{f}_d \qquad (6.31)$$

where $\bar{N}_p$ is the number of particles per unit volume and $\mathbf{f}_d$ is the drag force on a single particle (calculated using Equation 6.16 for example). The force $\mathbf{f}_{fp}^A$ is the volumetric average of the forces acting on the particle and in discrete form it can also be calculated for each cell as

$$\mathbf{f}_{fp}^A = \frac{\sum_{p=1}^{N_p} \mathbf{f}_d^p}{\Delta V} \qquad (6.32)$$

where $\mathbf{f}_d^p$ is the drag force acting on particle p within the fluid cell, N_p is the number of particles in the cell and ΔV is the volume of the fluid cell. Zeghal and El Shamy (2004) calculated an average drag force $\bar{\mathbf{f}}_d^p$ using the Ergun equation directly (Equation 6.12). However they used the average particle velocity, $\bar{\mathbf{v}}^p$, instead of the actual particle velocity ($\mathbf{v}^p$), along with the average fluid velocity $\bar{\mathbf{v}}^f$ instead of $\mathbf{v}^f$ and an average particle diameter, $\bar{d}_p$, giving a modified version of the Ergun equation

$$\bar{\mathbf{f}}_d^p = 150\mu \frac{(1-n)^2}{\bar{d}_p^{\,2} n} \left(\bar{\mathbf{v}}^f - \bar{\mathbf{v}}^p\right) + 1.75 \frac{(1-n)\rho_f |\bar{\mathbf{v}}^p - \bar{\mathbf{v}}^f|}{\bar{d}_p} \left(\bar{\mathbf{v}}^f - \bar{\mathbf{v}}^p\right) \qquad (6.33)$$

Zeghal and El Shamy calculated their representative particle diameter as $\bar{d}_p = 6/S_a$, where S_a is the average specific surface. In this approach $\bar{\mathbf{f}}_d^p$ is directly included in the momentum equation, i.e. $\bar{\mathbf{f}}_d^p = \mathbf{f}_{fp}^A$.

The interaction force for model B is given by

$$\mathbf{f}_{fp}^B = -(1-n)\nabla p + N^p n \mathbf{f}_d \qquad (6.34)$$

The model A formulation seems to be more commonly used, although this is difficult to assess exactly. From a geomechanics perspective, the limited number of published coupled particle-fluid studies using the averaged Navier-Stokes approach all seem to have used the a model A approach (refer to Zeghal and El Shamy (2004), Jeyisanker and Gunaratne (2009), and Suzuki et al. (2007)). Looking outside of the geomechanics literature, Tsuji et al. (1993), Kawaguchi et al. (1998), Van Wachem and Sasic (2008) and Xu and Yu (1997) also all used the model A equation. An example of the use of the model B approach is given in Zhou et al. (2008).

Some of applications of the averaged Navier-Stokes equations in geomechanics have neglected viscous effects (e.g. Zeghal and El Shamy (2004) and Suzuki et al. (2007)), as well as assuming an incompressible fluid. The momentum equation used is then given by

$$\rho_f \left(\frac{\partial \left(n\mathbf{v}^f \right)}{\partial t} + \nabla \left(n\rho_f \mathbf{v}^f \mathbf{v}^f \right) \right) = -n\nabla u - \mathbf{f}_{fp}^A + n\rho_f \mathbf{g} \quad (6.35)$$

Jeyisanker and Gunaratne (2009) accounted for viscous effects, but assumed the fluid was incompressible.

The drag force applied to each particle in the DEM simulation is added to the resultant force acting on the particles. Zeghal and El Shamy (2004) calculated the individual forces acting on the particles due to interaction with the fluid, $\mathbf{f}_{fp}$, from the average drag force and the pressure gradient:

$$\mathbf{f}_{fp} = \left(\frac{\bar{\mathbf{f}}_d}{1-n} - \nabla u \right) V_p \quad (6.36)$$

In other implementations where the drag force is calculated for each particle, the fluid-particle interaction force is given by

$$\mathbf{f}_{fp} = \left(\frac{\beta}{1-n} (\bar{\mathbf{v}}^f - \mathbf{v}^p) - \nabla u \right) V_p \quad (6.37)$$

where $\bar{\mathbf{v}}^f$ is the average fluid velocity within the current fluid cell and β is calculated from the Ergun equation.

Having established the concept of calculating the fluid-particle interaction in an averaged sense, and introduced the concept of solving the Navier-Stokes equations using a grid of relatively coarse cells, a solution strategy is needed. Tsuji et al. (1993), Zeghal

230

and El Shamy (2004), Xu and Yu (1997), Zhou et al. (2008), Van Wachem and Sasic (2008) and others all used the SIMPLE algorithm (Semi-Implicit Method for Pressure-Linked Equation) to solve for the fluid pressures and velocities. This method is detailed by Patankar (1980). As illustrated in Figure 6.7, in many of these implementations, the nodes where the pressure values are calculated are offset from the nodes at which the velocity values are calculated; this is called a "staggered cell" system. The porosities, and average particle velocities, and diameters are assigned to the centre of the cell, i.e. points indicated in Figure 6.7(a).

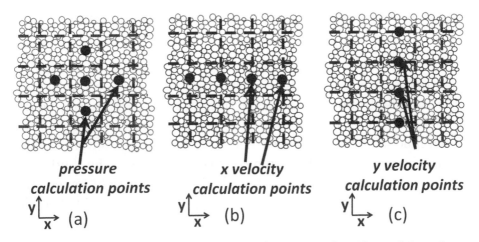

Figure 6.7: Schematic 2D diagram of staggered grid used in solution of the averaged Navier-Stokes equations (a) Pressure nodes (b) X−velocity nodes (c) Y−velocity nodes

A flow chart illustrating how the fluid model is coupled with a particle DEM code is given in Figure 6.8. The particle coordinates are calculated using the DEM model and this information is passed to the fluid model, where the fluid velocities and pressures are determined in an iterative sequence of calculations, knowing

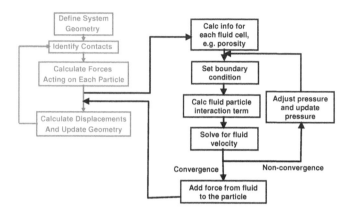

Figure 6.8: Flow chart to illustrate integration of particle DEM calculations and fluid phase calculations (Cheung, 2010)

the porosity in each cell. Knowing the pressures and velocities, the drag and buoyancy forces acting on each particle can be determined. The SIMPLE solver uses a series of iterations within each fluid time-step and the calculated values will converge to the numerical solution. Once convergence is achieved, the fluid pressure and velocities are used to calculate the particle forces, and these forces are passed back to the DEM code and added to the resultant force acting on each particle that is used to calculate the particle motion.

Two examples demonstrating the efficacy of the coarse-grid, averaged Navier-Stokes equations are given in Figures 6.9 and 6.10. Figure 6.9 illustrates the particle velocity vectors in a fluidized bed for different values of the inlet air velocities at the bottom of the assembly. This is a 2D simulation with disk particles. This figure was chosen for inclusion as Tsuji et al. (1993) are commonly credited with being the first authors to successfully demonstrate the use of the coarse-grid approach. The observed flow fields qualitatively agree with observations of equivalent experiments on samples of spheres. The results presented in Figure 6.10 are of more interest from a geotechnical perspective. This figure presents results of simulation of flow through an assembly of spheres with a gradation equivalent to Nevada Sand described by Zeghal and El Shamy (2008). The sample had dimensions of 42 mm × 42 mm × 84 mm and the particle sizes ranged between 0.06 and 0.4 mm. Samples with different initial void ratios were subject to different pressure gradients and the fluid velocities were observed. As illustrated in Figure 6.10, a linear relationship was observed between the applied hydraulic gradient and the measured discharge velocity (i.e. the velocity calculated using Equation 6.5). The calculated permeability increased with porosity and Zeghal and El Shamy (2008) found the permeabilities to be in good agreement with experimental data for the physical Nevada Sand.

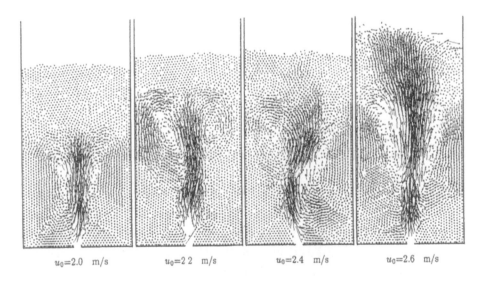

$u_0=2.0$ m/s $u_0=2\,2$ m/s $u_0=2.4$ m/s $u_0=2.6$ m/s

Figure 6.9: Illustration of fluid velocity vectors in a fluidized bed (Tsuji et al., 1993)

Examples of application of the coarse-grid averaged Navier-Stokes equations in geomechanics are given by El Shamy and Aydin (2008) (who simulated flow beneath a levee), Zeghal and El Shamy (2004 and 2008) (who simulated liquefaction in a column of

233

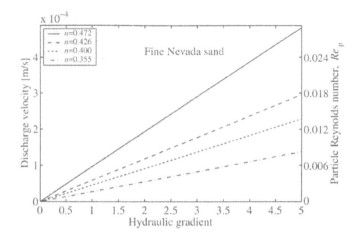

Figure 6.10: Variation in measured discharge velocity with applied hydraulic gradient for DEM simulations by Zeghal and El Shamy (2008)

soil subject to cycling at its base) and Jeyisanker and Gunaratne (2009) (who simulated water flow through a pavement layer).

6.7 Alternative Modelling Schemes

Alternative fluid-particle algorithms have been proposed and these differ in terms of their level of sophistication and accuracy. Some researchers such as Li and Holt (2002) have proposed that the pore network can be simulated as a network of pipes with the Hagen-Poiseille Law then being applied to calculate the "conductance" of each pipe. Quantitative correlations of this type of model with real soil response is likely to be difficult. Approaches that simulate the fluid flow using a discretization that is significantly smaller than the particles are more relevant. This approach is schematically contrasted with the coarse-grid approach in Figure 6.11. Sub-particle discretization uses the first of the two general coupling approaches introduced in Section 6.1, i.e. the coupling is at the fluid-particle interface and the two domains effectively do not

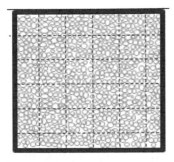
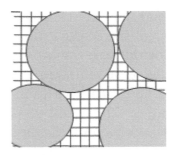

**Grid for Averaged
Navier-Stokes
Equations**
Fluid grid spacing
≈ 10 x D_{50}

**Sub particle
discretization
approaches**
Fluid grid spacing
<< D_{50}

Figure 6.11: Comparison of coarse grid used in averaged Navier-Stokes solution with a grid with sub-particle resolution (D_{50} is the mean particle diameter)

overlap.

One example of a sub-particle-scale approach is given by Cook et al. (2004), who used the Lattice Boltzman approach to solve the Navier-Stokes equation. In the Lattice Boltzman method the fluid is represented as packets of mass that move about a regular grid. The particles then overlap with the grid and a no-slip condition is enforced along the particle boundaries. Feng et al. (2007) proposed a more complex Lattice-Boltzman implementation capable of handling turbulent flow. Potapov et al. (2001) used smoothed particle hydrodynamics (SPH) to simulate the fluid phase with sub-particle discretization. SPH is a meshless method and the "particles" are interpolation points used in the evaluation of the Navier-Stokes equation. As in the Lattice-Boltzman method, a no-slip criterion is enforced at the particle-fluid boundary. Another example of a method that resolves the fluid flow at a sub-particle-scale was given by Mark and van Wachem (2008) who used a technique called the immersed boundary method to enforce the flow conditions at the surface of the particles. Zhu et al. (2007) contrasted the coarse-grid averaged Navier-Stokes approach with methods that use a sub-particle discretization. While they classified the coarse-grid approach as being computationally demanding, the approaches with sub-particle grid resolution are extremely demanding. Zhu et al. (2007) suggest that while the sub-particle discretization approaches are probably the most suitable for highly fundamental research into the mechanics of particulate materials, the coarse-grid approaches show more promise for application to industrial boundary value problems.

Chapter 7

Initial Geometry and Specimen/System Generation

A DEM simulation is a transient analysis where the response of the system at discrete points in time is predicted based upon the system state at slightly earlier times. Therefore, specification of the initial conditions is as important as specifying the boundary conditions. From an applied, geomechanics perspective, the response of a granular material is known to be highly dependent on its initial state (packing density and stress level), the anisotropy of the fabric (determined from the particle and the contact orientations), the anisotropy of the stress state, and the orientation of the principal stresses relative to the fabric. Just as experimentalists expend significant effort in preparing their samples for physical tests, DEM analysts need to give careful consideration as to how they construct their specimen. While Pöschel and Schwager (2005) (who consider particulate simulations from a general perspective) argue that as the initial conditions are used only once in each simulation, users should not need to expend significant effort in defining the initial conditions, this statement is not really valid from a geomechanics perspective. In fact Bagi (2005), looking at granular materials from a general perspective, identifies the production of the random arrangement of densely packed particles needed for many applications, to be a "challenging task." It is worth noting that significant effort has been expanded by mathematicians to develop algorithms to generate dense random assemblies of spheres (e.g. Sloane (1998) and Jodrey and Tory (1985)) and there is potential to adopt these algorithms in DEM specimen generation. This Chapter discusses some of the approaches used to generate initial particle configurations for DEM simulations. While consideration is restricted here to generation of specimens to fill simple geometries, e.g. cylinders for triaxial compression tests, the methods can also be applied to more complex boundary geometries that might be encountered in simulation of field- and industrial-scale problems.

Bagi (2005) includes a review of various approaches to generate random, dense specimens. The methods are divided into two categories. *Dynamic* methods are algorithms that include periods where the system is modified or adjusted and then DEM calculation cycles are invoked to either bring the system to a state of static equilibrium or to strain the sample to achieve a target stress level or packing density. In a *constructive* method the system is created without need for DEM calculation cycles. While it may often be easy, especially in preliminary simulations, to use uniform particle sizes in simulations, it is important to recognize that assemblies of uniform spheres and disks behave in a different manner to materials with a range of sizes (polydisperse materials). This is because uniform disks or spheres will tend to find a stable lattice packing and then there will be a preferential tendency to move along the lattice planes. This will result in a material response that differs significantly from natural geomaterials both at the scale of the overall material response and at the particle scale. Both the macro- and particle-scale responses are sensitive to the particle size distribution. In many cases the initial topology of the granular material packing (i.e. the material fabric) has a marked influence the mechanical response of the material, and it is good practice for analysts to assess the sensitivity of the system they are modelling to the approach used to generate the "virtual"

238

granular material.

7.1   Overall Initial Geometry of Assemblies of Granular Materials

In a DEM analysis a packing density and a stress level for the system under consideration cannot simply be specified as an input requirement. Rather a sample must be explicitly created and it must be in a state of equilibrium under the prescribed stress level. DEM analysts with laboratory experience will appreciate this, as the range of packing densities that can be considered in physical laboratory tests is similarly limited, and physical specimen preparation requires very careful consideration. Thus the first stage in a DEM simulation involves creating the initial geometry of the system, and in many cases, achieving a specified stress level via a preliminary simulation. In some cases this phase of the analysis can be more challenging than the subsequent simulation of the boundary value problem.

For most problems of geo-engineering interest, the physical system of particles will initially be in equilibrium under gravitational load. Almost all the particles will have at least one contact with another particle, i.e. the packing density will be relatively high. In contrast, low-density systems are often of interest in chemical or process engineering, e.g. when simulating fluidized beds. The concept of percolation merits consideration at this point, as these requirements mean that for geomechanics analyses systems that are "percolating" are sought. The term "percolation" comes from the mathematical discipline of network analysis (e.g. Grimmett (1999) or Watts (2004)). The percolation threshold marks the boundary of a situation where the network of contacts spans the entire system and can transmit the applied boundary stresses across the system (supercritical state). At packing densities lower than the percolation point, there may be clusters of contacting particles but these clusters are not connected and so an overall stress transmission cannot occur. In assemblies of granular materials there is a minimum number of contacts and a minimum packing density that must be attained before a link of contacts that can transmit stress from one side of the system to the other is formed. In geomechanics we are only interested in packing densities above this minimum point as non-percolating systems will not be deposited in nature (assuming we are not interested in materials in suspension). This percolation threshold is not uniquely defined; it will depend on the particle size distribution. In any case the specimen generation stage of a geomechanics DEM analysis must, at a minimum, create a configuration of particles whose density is sufficiently large such that the number of particles will exceed the percolation threshold. As clarified in Chapter 10, the terms "loose" and "dense" are often used in geomechanics to describe the current state of the particle assembly relative to the critical state line.

In nature or in the geotechnical laboratory, the particles are deposited by falling downwards under the action of gravity. The particles may fall through air (termed dry pluviation in the laboratory) or through water (wet pluviation). The method of deposition is known to affect the particle packing density and the geometrical arrangement of the particles (referred to as the fabric, as discussed further in Chapter 10). Intuitively in a DEM simulation, we might want to simply replicate this process, by generating particles at some height above the final analysis domain and then allowing them to fall downwards under a vertical body force. This process will involve significant particle movement and there will be a large number of collisions resulting in a varying contact configuration. Consequently, the computational cost of this stage in the analysis will be high, and is likely to take significantly more time than the simulation of the boundary value problem of interest. When simulating laboratory tests, for example, a number of virtual DEM samples will be required to demonstrate the statistical validity of the results or to complete a parametric study.

240

Typically then, rather than using the gravitational depositional approach, analysts use alternative means to generate relatively dense particle configurations and achieve the required stress state. The number of available methods is very large and the optimal method will depend on the analysts need. Here some of the more commonly used approaches are presented; many of these can easily be adapted or varied as necessary. While most of the methods discussed here consider the generation of "virtual" specimens of particles to be used in simulation of element tests, they can be modified to create the initial particle geometries for simulation of field-scale problems.

7.2 Random Generation of Particles

Provided they are not interested in a lattice or regular particle packing, most DEM analysts will employ random number generation at some point in their specimen generation. Most programming languages (such as C, C++ or Fortran) as well as mathematical software packages (including MATLAB and Excel) include a function for random number generation. Typically a random number generator is a complex function that takes the current time as a seed. Where this type of approach is used, repeated calls to the function will generate different numbers. However this is not always the case and when using this type of function care should be taken to ensure that repeated calls on the function will generate different numbers. A number of random numbers will be generated to describe each particle (depending on whether the simulation is 2 or 3 dimensional and what geometry is used). For example, where a spherical particle is used 4 numbers are needed for each particle: the values for the x, y and z coordinates of the particle centroid and the particle radius. Most random number generators output real values between 0 and 1. These can be scaled and translated to get a random distribution of particles and positions within the size limits and geometrical domain of interest in the analysis.

The idea of random particle generation is illustrated in 2D in Figure 7.1. Each particle position is determined by randomly gen-

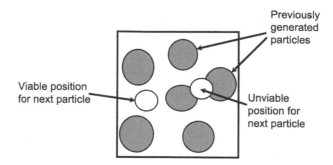

Figure 7.1: Schematic diagram of random number approach to generate specimens for DEM analyses.

erating the x and y coordinates and the particle size. Then, after generating each particle the contact detection phase of the DEM algorithm is invoked to compare the particle position with all previously generated particles in the system. If the particle overlaps with an existing particle this combination of centroidal position and size is considered non-viable. Then, while the particle could be abandoned and the process repeated to generate new position and a new size, it is better practice to retain the particle geometry and chose an alternative centroidal location at random. If both the particle size and the position are regenerated, there will be a bias in the particle size distribution, with a relatively large number of small particles being generated. Figure 7.2 illustrates the variation in void ratio as a function of the range of disk radii and the number of attempts to insert particles using the random number approach implemented in PFC2D. The void ratio is a measure of the packing density: the greater the void ratio, the smaller the packing density. As illustrated in Figure 7.2, the broader the range of particle sizes the greater the void ratio. This is a function of the random generation and will not be the case when the assembly is compressed, as then the smaller particles will fit inside the voids formed by the larger particles. It is also obvious that while increasing the number of attempts to insert particles will increase the packing density, there is a point where this has little effect, i.e. little difference was seen between the cases where there were

1 million attempts to insert a particle and 5 million attempts to insert a particle.

Using the random number generation approach, specific particle size distributions can be attained. In this case the target particle size distribution (PSD) is divided into a number of intervals and the radius values are randomly generated within each interval. This approach is not ideal; however, in real laboratory tests the PSD is usually obtained by sieving and the distribution between the discrete sieve size intervals is not accurately known. (More sophisticated PSD characterization tools do exist; however, their use is not yet widespread in geotechnics). There is also scope to combine the random number generator approach with a specified probability density function to get a more continuous PSD. Jiang et al. (2003) proposed a relatively simple equation to determine the number of particles required to match a specific distribution as follows:

$$N^{r_i} = \frac{P^{r_i}}{r_i^d P} N_p \tag{7.1}$$

where N^{r_i} is the total number of particles with radius r_i required, P^{r_i} is the percentage (by mass) of particles with radius r_i in the system, d is the dimension (2 for 2D, 3 for 3D) and N_p is the total number of particles in the system. The parameter P depends on the ratio between the weight percentage and the radius for each particle in the system, i.e.

$$P = \sum_{i=1}^{n_{\text{types}}} \frac{P^{r_i}}{r_i^d} \tag{7.2}$$

where n_{types} is the total number of particle sizes to be considered.

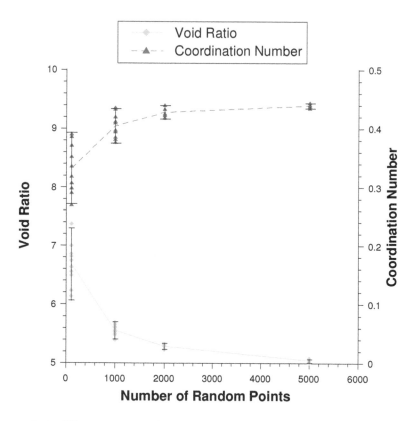

Figure 7.2: Illustration of variation in packing density (measured by considering void ratio and coordination number) with the number of attempts to insert particles when using the random number approach to generate specimens for DEM simulations (Cui and O'Sullivan, 2003)

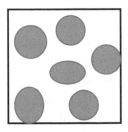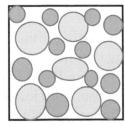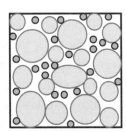

Step 1: Random generation of largest particles

Step 2: Random generation of intermediate particles

Step 3: Random generation of smallest particles

Figure 7.3: Illustration of approach used to generate a psd matching a specified size distribution

When a target PSD is to be achieved by discretizing the curve and generating the sizes within these discrete bins, it is best to start the generation considering the bin containing the largest particles, and then move successively through the bins in order of decreasing particle size, as illustrated in Figure 7.3. Cheung (2010) presents a discussion on the generation of a particle size distribution to match the physical particle size distribution observed for Castlegate Sandstone. Ferrez (2001) describes how, for his 3D simulations, he firstly placed some large particles (spheres) in a cylinder. Then, the remaining space was filled with medium-sized spheres, and subsequent iterations are used to densify the specimen, using increasingly smaller spheres to fill the voids.

An additional consideration when comparing the particle size distribution against laboratory data is that in the laboratory the particle size distribution is determined by considering the percentage of particles by volume that pass each standard sieve. In a DEM simulation it is tempting (and very easy) to look at the percentage of particles in terms of the number of particles that exist within each size interval. This will not give equivalent results: a large number of small particles may take up a very small fraction of the overall volume relative to a small number of very large particles. The median particle size as calculated by simply considering the particle size data will not be equivalent to the D_{50} calculated from the sieve analysis. D_{50} represents the particle size exceeded by the 50% of the particles by volume.

The mass of a particle with radius r is proportional to r^3. Therefore, a relatively large number of small particles may exist within a very small mass. From a DEM perspective these particles will add significantly to the computational cost of the simulations, as the simulation time will increase as the number of particles increases. Furthermore (and possibly more significantly) the critical time increment for stable analysis is proportional to the ratio $\sqrt{m/K}$ where m is the particle mass and K is the effective contact stiffness. These restrictions have led analysts to neglect the smallest fractions of the particle size distribution curve. For example in her simulations of Castlegate Sandstone, Cheung (2010) did not include the smallest 5% fraction by mass of the particles. This can be justified at the particle scale by assuming that these particles do not contribute to the strong force chains that transmit stress across the sample. In their simulations of rock mass, Potyondy and Cundall (2004) describe a procedure for removing the particles with no contacts (termed "floaters").

For disk- or sphere-shaped particles size is quantified simply by considering the particle radii or diameters. However, where the particles are non-spherical a variety of size measures can be considered. The size parameter that most closely approximates the size values obtained in sieving is the Feret minimum diameter. The Feret diameters are calculated by measuring the distance between two parallel lines that are tangent to the particle surface.

7.2.1 Radius expansion

The random number generation approach essentially generates a "cloud" of non-contacting particles. One approach that can be used to increase the packing density is to isotropically compress the sample by moving the side walls inwards (where rigid wall boundaries are used) or by applying a compressive strain in all

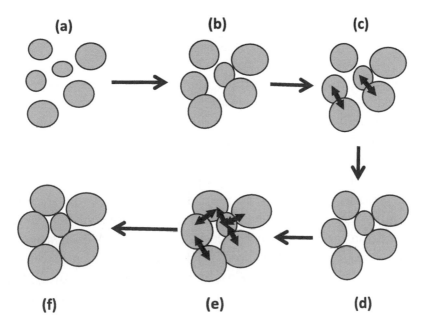

Figure 7.4: Schematic diagram of radius expansion process

directions where periodic boundaries are used. These approaches can be time-consuming. As discussed by Itasca (2004) a more convenient way to increase the specimen density is to gradually increase the particle sizes, by expanding the particles. In the case of disk or sphere particles this approach involves multiplying all the radii values for the particles in the system by a factor α where $\alpha > 1.0$. This concept can easily be adapted for more complex geometries. In any case it is best to increase the particles in a stepwise manner. For example, the following expression can be used to calculate the value of α:

$$\alpha = 1.0 + \frac{\beta}{n^\gamma} \qquad (7.3)$$

where $\beta < 1.0$, n is an integer indicating the step, which increases from a value of 1 until the expansion phase of the simulation is terminated, and γ is an integer (≥ 1). An appropriate value of β might be 0.2, while γ might be 1; a value of $\gamma > 1$ will give a greater rate of decrease of α during the expansion process. Other analytical expressions can be developed to control the radius expansion process, as long as the particle sizes increase by increasingly smaller amounts as the expansion process continues. The radius expansion process is illustrated in Figure 7.4. A each stage in the expansion process, once the particle sizes have been increased a series of DEM calculations should be carried out to bring the system into a state of equilibrium. These DEM calculations are required because when the particle sizes increase, particles that previously did not touch will now contact, and the overlap at the contact point may be significant. The resultant repulsive forces will "push" these particles to contact, and impart forces to other particles, causing accelerations. The DEM calculation cycles are thus needed to allow the disturbances to propagate through the system. The continuous movement and readjustment involved in the radius expansion approach led Bagi (2005) to classify it as a dynamic specimen generation approach. In the initial phases of the expansion the particles will be relatively far apart and the particles will be relatively small. As the process continues the particles will be both closer together and larger; thus, unless the expansion factor α is reduced using an expression such as that presented in Equation 7.3, there is potential to induce significant particle overlaps, resulting in very large accelerations. Accurate control of the packing density or the stress state also requires a small α value.

After the particle radii are expanded, a series of DEM calculation cycles should be invoked until the system comes into a state of equilibrium. A criterion is therefore required to assess whether the assembly is in equilibrium. One option is to consider the resultant force acting on each particle. This resultant force is the force that causes the particles to accelerate is sometimes called the "out-of-balance force." When this force is small, the particles are almost at rest. It can be difficult to chose an out-of-balance force value that determines equilibrium based directly on the force values. It is probably most appropriate to monitor the ratio of the resultant force to the particle mass and then specify the equilibrium point to have been achieved when the maximum ratio (considering all particles in the system) is smaller than a specified value. Alternatively or additionally the stress state and the total number of contacts in the system can be monitored and equilibrium be judged to be the point at which these parameters attain a constant value.

In the early stages of the radius expansion, in particular, there is a risk that a small number of particles may each experience one or more very large contact forces inducing significant accelerations and velocities. When this occurs these particles can experience a very large displacement in a single time step. This can cause problems, as a particle that was originally distant from a rigid wall boundary might move through the boundary and escape from the simulation space. The particle could also collide with another particle at a high velocity, generating another large overlap and causing very large velocities to propagate through the system. To avoid this, it may be appropriate to include damping in the system in these early stages. For example, the global damping approach discussed in Chapter 2 might be used. It is recommended, however, that this damping be reduced as the packing density increases, otherwise there is a risk of generating non-uniform stresses in the assembly. One way to ensure that the stress state is homogeneous is to compare the stress in a specified volume within the assembly with the stresses along the boundary. Where these are close there is a high probability that the stress state is homogeneous. If the difference between the internal stress, as measured over a significant volume of the specimen is large (i.e. more than 1 or 2%), it is not recommended to progress to the next stage of the expansion.

This radius expansion approach can either continue for a specific number of stages (e.g. until the parameter n in Equation 7.3 attains a specified value) or until a target packing density or stress state is achieved. The radius expansion procedure will generate an isotropic stress state in the specimen. It is then relatively straightforward to implement an algorithm to continue the process until the stress state is close to this value, i.e. to terminate the expansion when $|\sigma_{ii}^{\mathrm{target}} - \sigma_{ii}^{\mathrm{meas}}| \leq$ tol where $\sigma_{ii}^{\mathrm{target}}$ is the target stress, $\sigma_{ii}^{\mathrm{meas}}$ is the measured stress, and tol is a user-specified tolerance which might be 1% or so. The accuracy with which the target stress can be attained will depend on the α value. Typically it is best to approach the target stress for the simulation monotonically, and so the expansion should also be terminated with an error statement when $\sigma_{ii}^{\mathrm{target}} - \sigma_{ii}^{\mathrm{meas}} \leq 0$. Just as in the case of a physical sand, the sample will retain a memory of the stress history and if the target stress is exceeded during the specimen preparation process, the subsequent observed response will be that of an overconsolidated sample, which might not be desired. It should also be noted that during the radius expansion approach the increase in particle size gives an increase in particle mass and this has implications for the total energy of the system of the system.

Results for some of the radius expansion simulations of Summersgill (2009) are given in Figure 7.5; the specimen contains 441 disks enclosed within a set of 4 rigid walls forming a square. The results indicate that in the early phase of the expansion there is an increase in the stress and the coordination number (number of contacts per particle) just after the radii are expanded. As the particles move away from each other the stress drops off and the coordination number returns to zero. There is, however, a point where the packing density becomes sufficiently large that the par-

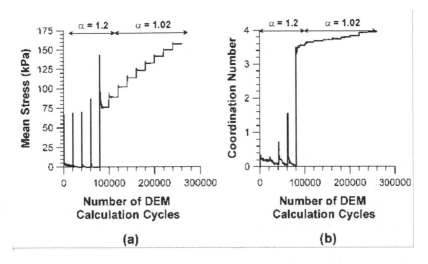

Figure 7.5: Variation in mean stress and coordination number during radius expansion process

ticles will remain in contact after the adjustment period. From this point onwards increases in the particle radii were controlled to give an almost monotonic increase in the stress state by reducing the value of α (Equation 7.3). This point is called the percolation threshold and beyond this point the system is percolating (refer to Section 7.1 above).

In their approach, Potyondy and Cundall (2004) calculate the α value by considering the current contact configurations and the required isotropic stress state. Assuming the (circular or spherical) particles are connected by linear contact springs if the desired increase the in mean stress, p, is given by Δp, then

$$\alpha = \frac{\lambda V \Delta p}{\sum_{b=1}^{N_b} \sum_{c=1}^{N_{c,b}} \tilde{R}_{c,b} K_{n,c} L_c} \tag{7.4}$$

where V is the volume of the region considered, N_b is the total number of particles, $N_{c,b}$ is the number of contacts associated with particle b, $\tilde{R}$ is the distance from the centroid of particle b to contact c, $K_{n,c}$ is the normal contact stiffness for contact c and L_c is the length of the branch vector for contact c. Where there is particle-to-particle contact, L_c is the sum of the radii of the

251

contacting particles; where a particle contacts a wall, L_c is the particle radius. The parameter λ equals the dimension of the system (i.e. 2 or 3), and the mean stress is given by $p = \frac{1}{\lambda}\sigma_{kk}$, i.e. in 2D, $p = \frac{1}{2}\sigma_{kk}$ or in 3D, $p = \frac{1}{3}\sigma_{kk}$. For details on the derivation of Equation 7.4, refer to Potyondy and Cundall (2004).

There are some approaches similar to the radius expansion process whose implementation in a DEM code would be relatively straightforward. One example is the "lily-pond" method proposed by Häggströ and Meester (1996) that was described by Bagi (2005). In this approach the starting point is a random configuration of points. Each point is expanded gradually as a disk/sphere until it touches another disk/sphere, when its expansion ceases. This algorithm would require modification were the analyst to desire the particle size distribution to remain constant. Again considering an initial sample of randomly generated particles, Han et al. (2005) proposed an alternative to radius expansion, where the initial packing is compressed to create space for addition of new particles. In this approach the particles adjacent to each particle are identified and the particle is moved to touch its closest neighbour in the direction of compression. Each disk is considered in a sequence; the process is iterative as there is an upper limit to the distance a particle can move through. Then new particles are inserted into the resultant voids that form, as described by Han et al. (2005). As the particles are compressed in a specified direction, care must be taken to ensure that a configuration of highly aligned force chains does not develop.

The combination of packing density and stress level, sometimes referred to as the "state" of the soil, significantly influences the mechanical behaviour of a granular material. It is not possible to simultaneously specify a packing density and a stress level, either in a physical experiment or in a DEM simulation. In laboratory tests on sand the void ratio can be controlled to some extent by using different methods to pour the particles into the confining mould (i.e. different pluviation techniques). There may be subsequent vibration or compaction of the sample. In a DEM simulation the packing control over the packing density can be achieved by varying the coefficient of friction for the particles. For exam-

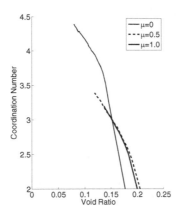

Figure 7.6: Illustration of variation in void ratio and coordination number values obtained as a function of inter-particle friction coefficient (Summersgill, 2009)

ple, Cundall (1988a) described the use of friction to control the final porosity of specimens constructed for periodic cell simulations. If the coefficient of friction is set to 0, a dense specimen will be achieved, while a friction coefficient of 1 will generate a loose packing. A further reduction in packing density can be achieved by also specifying an inter-particle cohesion. This sensitivity of the packing density to friction and cohesion can be understood by recognizing that the friction and cohesion provide a resistance to the particles sliding past each other and effectively add stability to local packing configurations. Using this approach, specimens with a specified void ratio can be created or a parametric study considering a range of specimens with differing void ratios can be completed. Figure 7.6 illustrates the variation in void ratio and coordination number as a function of inter-particle friction during the radius expansion process. The particles with the higher values of inter-particle friction have higher void ratios and lower coordination numbers.

7.2.2   Controlling the stress state

As noted above, even where the required stress state is isotropic, it is difficult to achieve exact control of the stresses solely using radius expansion. Thus at the end of every radius expansion procedure there will most likely be a sequence of DEM cycles carried out to bring the sample accurately to the required stress level. It is important that the target stress state be achieved monotonically, i.e. the stress should be gradually increased from a lower value to the target level. Even if the analyst wants to test an over-consolidated sample, they must first bring the system to the higher level of stress in a controlled manner, prior to reducing the stress to the value for the simulated test. If the sample is confined within a system of moveable rigid walls, the wall positions should be slowly moved at velocities that vary based on the current (measured) stress value. Consider the 2D case of a sample bounded laterally by two vertical planes, $x = x_{\min}$ and $x = x_{\max}$. To achieve a target stress level, $\sigma_{xx}^{\text{target}}$ the velocity of the wall defined $x = x_{\min}$ will be given by

$$v_x = \alpha_x \left| \sigma_{xx}^{\text{target}} - \sigma_{xx}^{\text{meas}} \right| \tag{7.5}$$

where α_x is a user-specified factor, v_x is the wall velocity and $\sigma_{xx}^{\text{meas}}$ is the measured stress. The wall defined by $x = x_{\max}$ should then move with velocity $-v_x$ The measured stress might be measured along the boundaries or within the sample. In a similar manner the strain rate applied in a periodic cell can be linked to the difference between the current and target stresses. Where flexible boundaries are used to achieve the initial stress condition it is most convenient to enclose the sample within a system of rigid boundaries and move the boundaries to achieve a prescribed stress condition prior to introducing these boundaries in the system.

It is important that the induced boundary velocities be sufficiently small to ensure that the sample is in a quasi-static stress state. Otherwise the stress equilibrium phase might be erroneously terminated based upon the measurement of a stress wave moving dynamically through the system, rather than consideration of the equilibrium stress state. This error can be avoided by comparing the average stresses acting on opposite rigid walls or by comparing the boundary stresses with the average stresses within the sample. Both stress levels should be met within a small tolerance, of 1% or less. If the target stress is overshot and the compressive stresses are too high, the stress will be reduced by relaxing the wall positions. However, in this case it may take multiple iterations to achieve the target stress level. An additional potential problem with rapid motion of the walls is that potentially a heterogeneous fabric might be induced (with larger packing density close to the edges of the sample in comparison with the remaining material) as highlighted by Jiang et al. (2003). Figure 7.7 illustrates the response observed when a sample of 460 disks is taken from an initial isotropic stress state of 112 kPa to an isotropic state of $\sigma_{xx} = \sigma_{yy} = 200$ kPa. Where the smaller α value of 0.1 is used a controlled convergence to the desired stress state is attained and the stress condition is maintained at the end of the compression. In the first simulation using the larger value of $\alpha = 2.0$, the stress control process stopped when both σ_{xx} and σ_{yy} were within 5% of the target stress level. The stress condition was not maintained and decreased to a lower value when the walls ceased moving. In the second simulation with $\alpha = 2$ the simulation oscillated for a long period and eventually when the wall movements ceased the stress came to a value close to the target. However the stress path followed to get to this point was complex, and the attainment of the final stress state was fortuitous rather than controlled. The appropriate α value will be problem-dependent, and can be found in a simple parametric study.

As in the case where radius expansion is used, the inter-particle coefficient of friction can be varied to control the packing density during the compression process. Barreto et al. (2008) described a series of periodic cell simulations where an isotropic compressive stress state was attained by moving the periodic boundaries, and hence the particles, inwards at a constant strain rate. A servo-controlled algorithm was used to control the strain rate to monotonically converge to the required confining stress. Typical results following isotropic compression to a confining stress of 200 kPa for various inter-particle friction values are presented in Figure 7.8. It

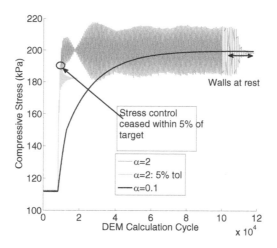

Figure 7.7: Illustration of sensitivity of response to servo-control rate selected

is clear from Figure 7.8 that, as in the case of radius expansion, low values of friction result in denser specimens. Care should be taken when using generation friction values (μ_{gen}) higher than the value that will be used during subsequent shearing (μ_{shear}). When the value of friction is reduced prior to shearing the soil structure will collapse if $\mu_{gen} > \mu_{shear}$, giving a reduction in void ratio. This was observed in several simulations where friction value was reduced from μ_{gen} to μ_{shear} and the specimen was allowed come into equilibrium at the same isotropic stress state of 200 kPa. The simulations (S1 and S2) illustrated in Figure 7.8 illustrate the degree of compression following a reduction in friction from $\mu_{gen} = 0.325$ to $\mu_{shear} = 0.3$ (S1) and $\mu_{gen} = 0.6$ to $\mu_{shear} = 0.3$ (S2).

As already noted, the stress state attained using particle expansion will be isotropic. Natural particulate assemblies rarely have an isotropic stress state. Many natural deposits have a K_0 stress state in situ. The term "K_0" is used to describe a stress state attained via a stress path where the vertical stress increases in a controlled or measurable manner, while the material is restricted from deforming laterally or horizontally. Typically, the stress conditions are then cross-isotropic, i.e. if the vertical nor-

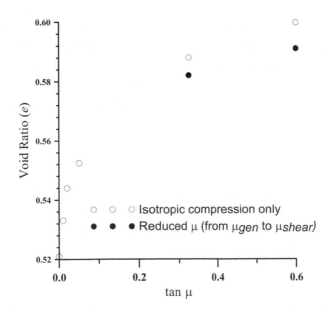

Figure 7.8: Illustration of range void ratios obtained using different inter-particle friction values for isotropic compression in a periodic cell (Barreto et al., 2008)

mal stress is $\sigma_v = \sigma_{33}$, the two horizontal normal stresses are equal ($\sigma_h = \sigma_{11} = \sigma_{22}$). The vertical and horizontal stresses are then related by $\sigma_h = K_0\sigma_v$, where (based upon empirical evidence) K_0 is a constant that can be related to the angle of shearing resistance for a given soil. To achieve this anisotropic stress state one approach that could be adopted is to progress with the radius expansion procedure until a packing density that is just slightly looser than the percolation point is reached. The termination criterion used in the radius expansion algorithm should in this case most likely be based on consideration of the number of contacts per particle in the system. Then, assuming the sample is confined within vertical and horizontal boundaries, it can be slowly compressed by moving the horizontal boundaries towards each other to achieve the target vertical stress, while maintaining the horizontal boundaries fixed.

Barreto et al. (2008) described an interesting study where they examined the sensitivity of the system to the stress path followed to achieve a particular stress state of q=150 kPa and p'=200 kPa. Note that in geomechanics the deviator stress is given by $q = \sigma_1 - \sigma_3$ and the mean effective stress is given by $p' = \frac{1}{3}(\sigma'_1 + \sigma'_2 + \sigma'_3)$, where σ_1, σ_2 and σ_3 and σ'_1, σ'_2 and σ'_3 are the principal stresses and the principal effective stresses respectively. As can be seen by reference to Figure 7.9, six different stress paths were considered. The sensitivity of the void ratio, coordination number and fabric anisotropy to the stress path followed was considered. (The terms "coordination number" and "fabric anisotropy" are defined in Chapter 10.) For each of the stress paths A - F considered, an identical initial specimen was used, only the sample subject to stress path F did not touch the "K_0" line prior to reaching the target stress state. While, stress paths A-E yielded final void ratios in the range 0.586-0.591 and coordination numbers between 4.24 and 4.27, sample F was measurably less dense, with a void ratio of 0.597 and a coordination number of 4.20. This example illustrates that the stress path followed influences the final state of the sample.

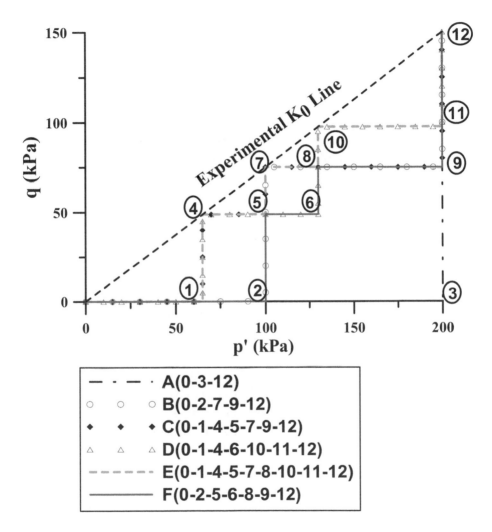

Figure 7.9: Stress paths followed to achieve a K_0 condition, Barreto et al. (2008)

7.2.3   Jiang's under-compaction method

One of the most important problems that merits consideration in geomechanics is liquefaction. Liquefaction occurs when loose, saturated deposits of sand are subject to rapid dynamic loading (e.g. earthquake shaking). The material responds to this rapid loading in an undrained manner, with an increase in porewater pressure causing a reduction in effective stress and consequently the shear strength reduces. This phenomenon has been responsible for significant damage to buildings and infrastructure leading to fatalities and having substantial financial implications (e.g. in Haiti in January 2010). Recognizing the need for DEM simulations to analyse liquefaction, Jiang et al. (2003) focussed on developing a specimen generation method that can form loose specimens. This method is similar to the experimental procedure for developing loose samples proposed by Ladd (1978). The general concept is illustrated in Figure 7.10. Rather than filling the entire volume at the beginning, the sample is built up layer by layer. At each stage in the development the sample is then compressed to achieve a target void ratio (e_1, e_2, ...) that is higher than the final required void ratio e_target. The void ratios should monotonically converge towards the target value, i.e. $e_1 > e_2 > ... > e_j.... > e_\text{target}$. Jiang et al. propose both linear and non-linear expressions (similar to the experimental expression of Ladd (1978)) that can be used to determine the "undercompaction" criteria, and hence target height for each layer. Recalling the discussion on percolation in Section 7.1, it is important to realize that where loose samples are sought, the target density must exceed the percolation threshold for them to be physically relevant.

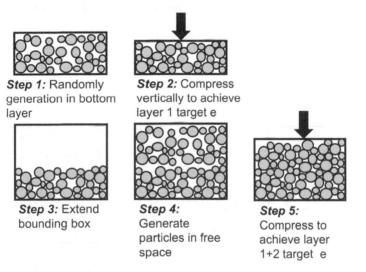

Figure 7.10: Illustration of undercompaction method proposed by Jiang et al. (2003)

7.3 Constructional Approaches

In the constructional approaches, the sample is created without any periods of DEM calculation. While these approaches can be very effective, and conceptually simple to understand, their implementation in a computer code is less straightforward and some coding effort must be applied to achieve the gains in specimen generation efficiency offered above the random generation approaches. Most of these constructive approaches have been implemented in two dimensions only and extrapolation to three dimensions is nontrivial. Here two representative constructional approaches are introduced.

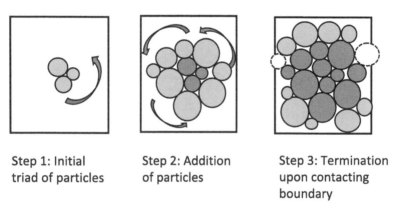

Step 1: Initial
triad of particles

Step 2: Addition
of particles

Step 3: Termination
upon contacting
boundary

Figure 7.11: Illustration of closed form of advancing front approach proposed by Feng et al. (2003)

The *advancing front* approach was proposed by Feng et al. (2003). In this approach the system of particles is incrementally constructed by starting from an initial small assembly of disks. The additional disks that are inserted into the system must contact the pre-existing disks, so that the specimen "grows" in size as the additional disks are added. Two implementations of this approach have been proposed: the closed form of the advancing front algorithm and the open form of the advancing front algorithm. In the closed form algorithm (illustrated in Figure 7.11) the specimen grows outwards from an initial triangular arrangement of three disks, while in the open form approach the specimen is constructed layer by layer within a specified geometry. Bagi (2005) expressed concern that this approach can generate samples with large voids close to the boundary.

Rather than starting from the inside of the specimen, the *inwards packing method* proposed by Bagi (2005) initially inserts disks along the boundary, and works inwards. As illustrated in Figure 7.12(a), in the first stage of the specimen generation a large disk is inserted in the upper left hand corner. Additional particles are then inserted, and each particle must touch both the boundary and a previously placed sphere, working downwards along the left boundary. The process continues until the lower left corner is reached. A large particle is inserted here, and the previously placed particle is adjusted to accommodate this insertion. The process continues in an anticlockwise manner around the boundaries until a closed loop of particles is formed (Stages 2 and 3). The final stage is to identify the innermost (smallest) closed loop formed by connecting the centroids of contacting particles; this is the *initial front*. Then the process continues and particles are inserted one by one along this front so that they just touch previously placed particles, considering an "active particle" and the previous and next particles that touch it. The determination of a valid disk insertion point will depend on the local geometry and Bagi (2005) outlines the various scenarios and the most appropriate approach to adopt.

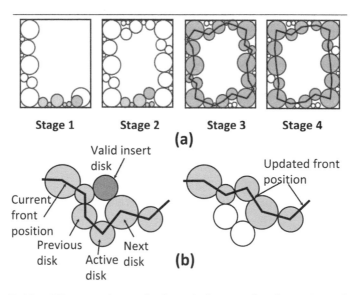

Figure 7.12: Illustration of closed form of advancing front approach proposed by Feng et al. (2003)

7.4   Triangulation-Based Approaches

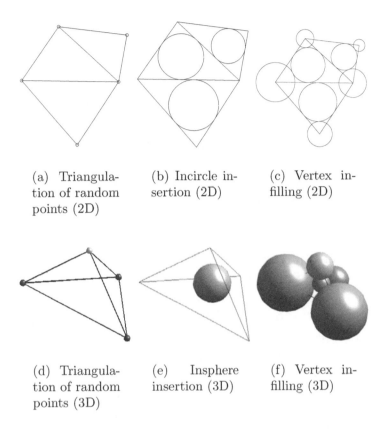

(a) Triangula-
tion of random
points (2D)

(b) Incircle in-
sertion (2D)

(c) Vertex in-
filling (2D)

(d) Triangula-
tion of random
points (3D)

(e)   Insphere
insertion (3D)

(f) Vertex in-
filling (3D)

Figure 7.13: Schematic diagram of specimen generation steps: Triangulation approach (Cui and O'Sullivan, 2003)

Cui and O'Sullivan (2003) proposed a triangulation-based scheme, where a triangular grid or mesh is created by applying Delaunay triangulation to a random cloud of points (Figure 7.13(a) and (d)) then disks (spheres) are inserted as incircles (inspheres) in the triangles (tetrahedra). Each incircle or insphere is entirely contained within its triangle or tetrahedron (Figure 7.13(b) and (e)); consequently a constructional method for specimen generation can be developed. As there will be large voids close to the triangle vertices, the final stage is to insert a circle or sphere at each vertex so that it just touches the circle/sphere that is closest to that vertex. The packing density can be further controlled by controlling the geometry of the triangles. Cui and O'Sullivan (2003) used the mesh refinement capabilities that are available within the Triangle program developed by Shewchuk (2002) for 2D work while the refinement in the Geompack ++ program developed by Joe (2003) was used. Figure 7.14(a) illustrates an unrefined 2D mesh, this can be compared with a refined mesh (where the minimum angle is restricted to be > 30°) by reference to Figure 7.14(b), details of the refinement process are given by Shewchuk (1996). Jerier et al. (2009) extended this method for the 3D case. Their algorithm is more complex than the approach proposed by Cui and O'Sullivan (2003), and involves placing spheres along the edges of each tetrahedron, then at the tetrahedra vertices, then on the tetrahedra faces, and then within the tetrahedron; for a more detailed description refer to Jerier et al. (2009). Weatherley (2009) proposed a triangulation-based specimen generation scheme in which a number of seed particles are inserted at random locations in the system so that they do not overlap. Then the 4 adjacent particles (in 3D) are triangulated and the centroid of the tetrahedron is calculated. The radius of the particle that just touches one of the 4 particles (subject to a tolerance) is calculated and if this radius is within the particle size range specified, it is inserted.

Bagi (2005) described the Stienen model proposed by Stoyan (1973), which has some similarities with the triangulation-based approaches. In this approach, random points are generated in space, as in Cui and O'Sullivan's triangulation approach. Then for each point, its nearest neighbour is identified and the disk with a radius equal to half the distance to the nearest neighbour is inserted. Bagi proposed that this process essentially generates the incircles of the Voronoi tessellation of the random points. In the approach proposed by Labra and Oñate (2008), the particles are randomly generated using the approach proposed in Section 7.2. Then the system is triangulated and each particle is connected to a number of other particles via the edges of the triangles or tetrahedra. The distance between the particle centroids along the triangle edges d^e is given by

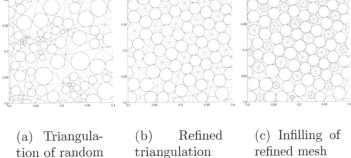

(a) Triangulation of random points without refinement

(b) Refined triangulation of random points

(c) Infilling of refined mesh

Figure 7.14: Disks generated using an initial coarse mesh (10 points) with subsequent refinement and vertex infilling (2D)

$$d^e = |\mathbf{x}^a - \mathbf{x}^b| - \left(r^a - r^b\right) \tag{7.6}$$

where the centroids of the particles with a common edge are given by $\mathbf{x}^a$ and $\mathbf{x}^b$ and the radii are r^a and r^b. The positions and radii of the particles are then adjusted iteratively to minimize the product DD^T, where $\mathbf{D}$ is given by

$$D = \sum_{e=1}^{N_e} d^e \tag{7.7}$$

where N_e is the total number of triangle edges in the system.

While effective in generating dense assemblies of particle without needing to run DEM calculation cycles, the triangulation approaches have a notable disadvantage of lacking control of the particle size distribution within the specified range. They are however, particularly useful when seeking to generate dense packings within awkward geometries, as may be encountered in process engineering applications.

7.5   Gravitation and Sedimentation Approaches

(a) Gravitational approach proposed by Feng et al. (2003)

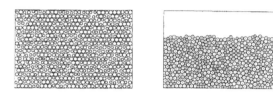

(b) Gravitational approach proposed by Thomas (1997)

Figure 7.15: Illustration of gravitational specimen generation approaches

In nature sand is typically deposited as particles fall downwards under gravitational action. They may fall out of a liquid suspension or be transported by wind to the site. In the laboratory sand specimens for testing are formed by pluviation, where the particles fall gently downwards from a container, through water or air. While ideally in a DEM simulation we would like to replicate this process, e.g. have our particles fall downwards through a funnel or hopper and settle down with a random mixture of different size disks in a container placed beneath the chute, as illustrated in Figure 7.15(a), this is expensive, there are very large deformations requiring significant continuous updating of the contact lists (refer to Chapter 3). It is therefore often easier to create a "cloud" of close but non-contacting particles within our system, and apply gravity to these particles. For example, Thomas (1997) developed a two-dimensional rectangular mesh and used a random number generator to select a different sized particle for placement in each cell of the mesh, and a similar approach was proposed by Abbiready and Clayton (2010). Then a vertical body force (i.e. gravity) was applied and the particles were allowed to settle into a rigid box in a DEM simulation; refer to Figure 7.15.

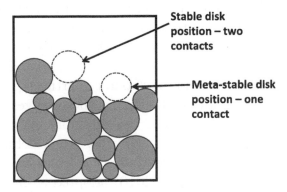

Figure 7.16: Illustration of valid stable disk position and invalid metastable disk position

A very interesting approach to simulate sedimentation was proposed by Marketos and Bolton (2010) who simulated the dry pluviation approach used to create samples in the laboratory by initially generating an assembly of non-contacting spheres using the random number generation approach described in Section 7.2 above. Having created the initial "cloud" of particles, they then moved the particles downwards in sequence until the particle contacted a particle that was already at rest. At this point the particle might not be in a state of stable equilibrium and it was allowed to slide along the surface of the already at rest particles until it found a stable equilibrium position where it came to rest. Bagi (1993) also used a similar this sedimentation-type approach in 2D. The concept of stability for a 2D simulation is illustrated in Figure 7.16, a valid stable inserted particle will require at least two contacts. As noted by Remond et al. (2008) where sedimentation is applied to non-spherical particles the particles must be rotated as well as translated to find a suitable stable point.

7.6 Bonding of Specimens

Where DEM is used to simulate the response of a cemented sand, a sandstone or other rock mass, the cementation must be inserted at some point in the sample creation process. In their description of the formation of a ponded particle model to simulate the response of rock mass, Potyondy and Cundall (2004) firstly randomly generated balls in space, then the ball radii were expanded, an isotropic stress state was achieved and the "floater" particles (i.e. those with no contacts) were removed. Following these stages, the parallel bonds were installed throughout the assembly between all particles considered to be "near" to each other (their definition of near was to have a separation of less than 10^{-6} times the mean radius of the two particles). Cheung (2010) adopted a slightly different approach. She directly matched a physical particle size distribution curve, and as she neglected the smallest 10% (by mass) of the particles in the physical measurement, elimination of floaters was not considered. She also considered the SEM images of the physical material obtained by Gutierrez (2007), and observed that only about 50% of the contacts were bonded. In her model she brought her sample to a required stress state and then bonded 50% of the contacts transmitting a compressive contact force, and randomly assigned a parallel bond α value of between 0.1 and 1.0 to each of these contacts (see Section 3.8.1). Wang and Leung (2008) used very small particles to represent the cement phase in their model of cemented sand, and they adopted a procedure where the larger sand particles were first generated and the sample was brought to an isotropic stress state. Then the small cement particles were introduced and "pulled" towards the particle contacts, at which point the parallel bonds were introduced.

270

7.7 Experimental Generation of DEM Packing Configurations

Just as it may be possible to use images of real sand particles to determine the particle morphologies used in DEM simulations (Section 4.7), there is also potential to use experimental data to create realistic packings for use in DEM simulations. Scanning electron microscopy (SEM) is often used to qualitatively examine the fabric of clays and sands. For example Gutierrez (2007) assessed the particle scale distribution of cementation in cemented sands using SEM. Optical microscopy can also be used, for example Cresswell and Powrie (2004), present images of thin sections through Reigate Sand gained by resin-impregnating the material, and cutting thin sections. As noted in Chapter 10, the analytical approaches used to quantify fabric from the results of DEM simulations can be applied to the images gained from SEM and optical microscopy to assess the material anisotropy. However these images are two-dimensional slices through a three-dimensional material. While it is technologically feasible to apply image analysis techniques to recreate an assembly of particles for input into a 2D DEM analysis, the validity of this approach should be carefully considered, as the real material is three-dimensional and the packing geometry arises due to a complex three-dimensional system of particle interactions.

Obtaining three-dimensional images of the material microstructure is possible, though slightly more difficult than for the 2D case. One example of the use of optical microscopy or two-dimensional imaging to gain quasi three-dimensional information on soil structure is the work of Kuo and Frost (1996) and Frost and Jang (2000). In their research they impregnated their samples with an epoxy resin, then they cut out subvolumes with horizontal and vertical orientations and made assessments of the void ratio (as measured in 2D) along these sections.

The potential to use micro-computed tomography to gain information on 3D soil microstructure is becoming more feasible with advances in the technology and increases in computational power. Desrues et al. (2006) and Viggiani and Hall (2008) give introductions into micro-computed tomography and its application in geomechanics, while Ketcham and Carlson (2001) give a more general introduction to the use of micro-computed tomography on geological specimens. Micro-computed tomography uses the information on the way in which materials attenuate x-ray images to gain a tree dimensional description of the sample. The data are stored as voxels (three-dimensional pixels) and each voxel is assigned a number to represent the attenuation of x-rays by the material at that point in the sample. To date, a lot of the work in soil mechanics has considered the application of micro-computed tomography to look at "meso-scale" variations in density. In general, when using micro-computed tomography the resolution required to accurately resolve the particle positions and orientations as well as the location of the inter-particle contacts can only be achieved on very small samples. To enable reconstruction of a soil sample fabric for a μCT scan a high-resolution dataset is required, with a voxel size that is significantly smaller than the particle diameters. Once the data has been obtained, the subsequent processing involves two main stages. Firstly the image must be thresholded; this is the process of associating each voxel with either void space or solid particles. The information gained at this point can be used to assess the void ratio. The segmentation process required to identify the individual particles is slightly more challenging, as discussed by Fonseca et al. (2009).

Once the particles are segmented, one of two approaches can be use to generate particles for DEM simulation. In the first case polyhedral particles could be reconstructed using a mesh generation or triangulation-based approach. Alternatively the μCT data can be used to create clumps or clusters of spheres. Algorithms to reconstruct irregular particle geometries as assemblies or clusters of disks or spheres from images of real particles are considered in Section 4.2.

7.8 Assessing Success of Specimen Generation Approaches

While a variety of specimen generation methods for particulate materials have been developed, there is no uniform agreement on the optimum specimen generation approach; however the success of the approach can be quantitatively assessed. Cui and O'Sullivan (2003), Bagi (2005) and Jiang et al. (2003) all quantified the efficacy of their specimen generation algorithms considering the packing density (void ratio) of the system of particles generated to be an appropriate measure of the effectiveness of a particle generation algorithm. The computational cost of the specimen generation method is an important measure of its success. The topology of the packing, or the material fabric must also be considered. Methods to quantify the material fabric are presented in Chapter 10. It is important to recognize the difference between an inherent fabric that describes the initial arrangement of the particles and a stress-induced fabric that arises due to changes in the stress state.

Consideration should also be given to the range of particle sizes and the shape of the particle size distribution curve. Typically in geomechanics the size distribution is quantified either by using the coefficient of uniformity (C_u) or the coefficient of curvature (C_z); these measures are calculated as

$$C_u = \frac{D_{60}}{D_{10}} \tag{7.8}$$

and

$$C_z = \frac{D_{30}^2}{D_{60}D_{10}} \tag{7.9}$$

where 60% of the particles (by volume) are smaller than the size D_{60}, 30% of the particles are smaller than the size D_{30} and 10% of the particles are smaller than the size D_{10}.

Jiang et al. (2003) highlighted the need to assess the homogeneity of the packing density generated. They quantified the homogeneity by measuring the accurately the void ratio within sub-volumes, and they chose horizontal strips (as illustrated in Figure

7.17) and vertical strips. If the void ratio in layer i is given by e_i, they measured the extent of the homogeneity in a given direction by calculating the variance S:

$$S = \frac{1}{N_{\text{layer}} - 1} \sum_{i=1}^{N_{\text{layer}}} (e - e_i)^2 \qquad (7.10)$$

where N_{layer} is the number of layers and e is the overall void ratio. As highlighted by Jiang et al. (2003), for this measure to function it is important that the layers be sufficiently thick, and a minimum layer thickness of $2.25d_{50}$, where d_{50} is the median particle diameter, was found necessary to register a variance of less than 5% on a homogeneous sample.

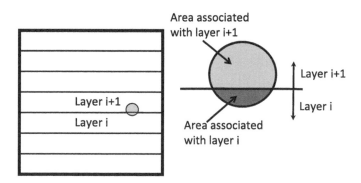

Figure 7.17: Illustration of subdivision approach used by Jiang et al. (2003) to analyse sample homogeneity

7.9 Concluding Comments on Specimen Generation

Many of the specimen generation approaches described here were proposed to minimize the time spent creating the virtual granular systems for use in simulations. A lot of the algorithms considered have been implemented in 2D and extension to 3D is non-trivial. It is important to recognize that with improvements in computer hardware and software, the computational costs of DEM simulations will decrease. The simulation of a more realistic physical pluviation process within reasonable time frames will become more viable and the need to use the approaches considered in this Chapter will be eliminated. Where pluviation is used, an initial, inherent anisotropy will be induced in the system. In fundamental research into granular materials there will always be scenarios where an isotropic fabric is required and in these cases the radius expansion approach is likely to be the best option to create specimens. The mechanical response of a granular material will be sensitive to the topology of the particle packing, as discussed further in Chapter 10. As noted at the beginning of this Chapter, analysts interested in simulating the response of a particular physical system should give careful consideration to the procedure used to generate the initial packing and develop an understanding of the sensitivity of their model to the specimen generation procedure.

Chapter 8

Postprocessing: Graphical Interpretation of DEM Simulations

8.1   Introduction

The enormous wealth of data that is available for output from a DEM simulation presents many alternatives for detailed observation of the particulate system. Management and interpretation of these data is in itself a challenge. Prior to starting a simulation, a DEM analyst therefore must consider not only details relating to the DEM simulation itself but also what data to monitor, how to extract these data in a manageable form, and how then to examine it. Chapter 9 explains how to interpret DEM data within a continuum mechanics framework and outlines approaches to obtaining stress and strain (both average parameters for the system and local values). In Chapter 10 methods to characterize the packing of the system of particles are considered. As argued by Rapaport (2004), DEM simulations (or in Rapaport's case molecular dynamics) simulations are the computational equivalent of an optical microscope; in three dimensions it may be more correct to consider them analogous to micro-computed tomography. DEM simulations offer an advantage over both these physical microscale observation tools because as well as allowing the geometry to be observed, data on the contact forces and internal stresses are also available. This Chapter considers selected approaches used to look inside our particulate systems.

Creating plots of the system of particles or the contacts linking them can generate interesting and colourful images and attractive animations. These images are not simply pretty pictures. In the first instance, visual observation of the system as it evolves during the simulation is essential as a means to assess whether there has been a gross error in the analysis. Some software includes the option for graphical representation of the system on screen as the simulation runs. However, typically if DEM simulations are run with this graphics option enabled, there will be a significant increase in the simulation run time. For large simulations in particular, it is more appropriate to generate images of the system by intermittent output of data and subsequent plotting in a postprocessing tool. Even when using a commercial DEM software package, use of a more general programming language/software for postprocessing is advantageous. This is true in particular for research applications where the analyst may want flexibility in the visualization approach to better understand the mechanics of the system.

The variety of ways one can graphically interpret the results of DEM simulations is limited only by the curiosity and imagination of the analyst. The objective of this Chapter is to present some of the more common approaches that have been used, mainly to point a novice user of DEM in the right direction. The discussion is illustrated by relatively simple simulations completed by the author or students working with the author. Most of the plots presented here were generated using MATLAB; however, many other software packages for visualization exist; for example the open-source code Paraview is used by some DEM analysts (Ahrens et al., 2005). While emphasis is placed here on the generation of figures (static graphics), these approaches can easily be adapted to generate animations. Observation of animations may provide useful insight into the evolution of the system. To get an impression of how graphical interpretation of DEM data can provide insight into the mechanics of the material response, reference to Kuhn (1999) is recommended. Kuhn used a variety of visualization and analysis approaches to study the particle mechanics of a dense 2D granular material and only some of his ideas are directly referred to here.

8.2 Data Generation

In order to generate the data required to create diagrams of the system evolution, digital "snapshots" of the system must be taken at selected points during the simulation. These snapshots are not images themselves, rather they contain the information needed to create the images. The snapshot process should generate distinct data files containing information about the state of the particles and the contacts. Referring to Figures 1.7 and 1.8, during the calculation loop the DEM code will be recording and updating information for each particle and contact. Typically in geomechanics simulations the number of contacts will greatly exceed the number of particles and so the data files saving the contact state are larger than those saving information on the particles. Furthermore, while the number of particles will normally remain constant, the number of contacts will evolve as previously contacting particles move away from each other, and new contacts form, and so the size of the contact data file will also change.

The minimum "snapshot" information for the particles includes the particle reference number, the current particle locations and cumulative rotations. Additional particle information might include the individual particle stresses (calculated using the approach presented in Section 9.4.2), the number of contacts the particle participates in, or the particle velocity. The contact information should include the contact location, the orientation of the contact normal, the magnitude of the contact normal and tangential forces, and the reference numbers for the particles involved in the contact. More sophisticated analysis can also be carried out within the DEM code to generate data for visual inspection. For example Kuhn (2006) proposes calculation of parameters including the energy dissipation at each sliding contact, the rate of change in contact force and the inter-particle movements at the contacts. The necessary snapshot files are then either taken (i.e. output) at specified intervals during the simulation, or when a trigger criterion is met. This information will be saved to disk, often in the form of a simple ASCII text data file. The data also can be saved in binary format; however, Rapaport (2004) notes that while this may reduce the storage requirements, there is potential to constrain future processing of the data as these files may not be readily transportable between different kinds of computer and operating systems. The snapshot files can also be used to calculate local strains and to analyse the fabric of the system (refer to Chapters 9 and 10).

8.3 Particle Plots

The results of a DEM simulation of a 2D biaxial compression test on a sample made up of 2,376 unbonded disks enclosed within a stress-controlled membrane boundary are used here to illustrate some of the ways the particle data can be presented to gain insight into the mechanics of the material response. This sample contains a relatively small number of particles and was selected for consideration simply because individual particles can be identified in the plots generated for inclusion. The disk sizes were between 0.48 mm and 0.72 mm, the particle density was 2.650×10^3 kg/m^3. Linear springs with a stiffness of 1×10^5 kN/m were used to model the inter-particle contacts and the inter-particle friction was 0.5. The top and bottom platens were modelled as smooth rigid linear boundaries. The overall stress-strain response is given in Figure 8.1. The two points selected to take "snapshots" of the system are indicated in Figure 8.1.

The plots illustrated in Figure 8.2 show how the internal deformation in the sample can be observed by creating a regular grid within the sample. The grid was generated by colouring horizontal and vertical layers of disks according to their initial position, and the deformation pattern can be appreciated by comparing the initial disk positions and the disk positions at an axial strain of 12.0%. An alternative means to examine the deformations is to

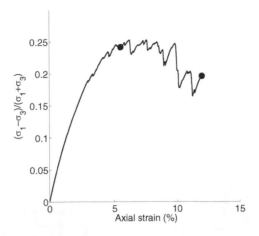

Figure 8.1: Plot of overall stress-strain response for selected sample simulation

illustrate both the original grid configuration and the deformed configuration on a single plot, as in Figure 8.3. Schöpfer et al. (2007) demonstrated the use of bands of disks shaded in this way to analyze the failure mechanisms involved in large-scale geological faulting processes; this approach could be applied to other boundary value problems (e.g. deformation around a pile during installation).

The exact inclination and extent of any shear band or localization that may have formed in the sample is not evident. In Figure 8.4 the deformed specimen at axial strains of 5.5% and 12.0% is again considered. Now each disk is shaded according to the magnitude of the net rotation it has experienced since the start of shearing; the zones of localized displacement are more much obvious in the resultant plots. Figure 8.5 is included to show the potential benefit of considering the rotation direction. The simulation considered a biaxial compression of 12,512 disks and the simulation parameters are detailed in O'Sullivan et al. (2003) (see also Figure 11.2). Following the visualization approach adopted by Iwashita and Oda (2000), the mean particle rotation magnitude was calculated. Then only particles whose rotations that

 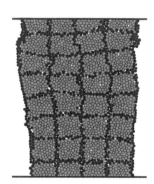

(a) Initial Configuration   (b) Deformed sample at an
                                    axial strain of 12.0%

Figure 8.2: Plot of deformation mechanism for biaxial simulation of 2,376 unbonded disks, confining pressure applied using a stress-controlled membrane.

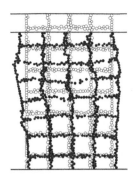

Figure 8.3: Plot of deformed grid of tracked particles at an axial strain of 12.0% imposed on original grid geometry for biaxial simulation of 2,376 unbonded disks, with confining pressure applied using a stress-controlled membrane.

were greater in magnitude than these mean values are considered. As illustrated in Figure 8.5, the particles whose rotation exceeds the mean counterclockwise rotation are presented as solid black circles, while the particles whose rotation exceeds the mean clockwise rotation are plotted as hollow circles. A magnified image of the central zone where the two localizations intersect is given in Figure 8.5(b). In both cases the shear band locations are defined by the locations of the zones of maximum rotation. The disks on the dominant localization tend to rotate counterclockwise. An example of authors who have considered rotations in some detail is Bardet and Proubet (1991) who proposed that the gradients in the rotations are useful in characterizing the deformation of the material inside shear bands.

While visualization of 3D simulation data is complicated by the opaque 3D nature of the particles; as before, plots of the overall system geometry from the exterior are obviously useful to understand how the specimen has captured the physical system. An example 3D plot is given in Figure 8.6; in this case the sensitivity of the deformed specimen geometry at the end of triaxial

Figure 8.4: Plot of deformed sample at (a) 5.5% and (b) 12% for biaxial simulation of 2,376 unbonded disks, confined with stress-controlled membrane. Shading illustrates magnitude of disk rotation in radians

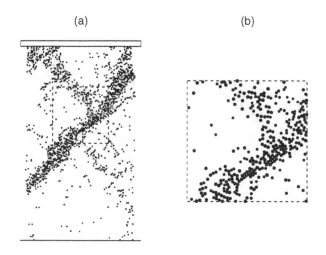

Figure 8.5: Plot of rotation directions for a specimen of 12,512 disks subject to biaxial compression, with an axial strain of 3%, only rotations exceeding the mean value are considered

compression to the friction along the top and bottom boundaries is considered. The axisymmetric boundary condition discussed in Section 5.5 was used and so only a quarter of the sample is shown. While the influence of the boundaries on the overall specimen deformation is evident, it is more difficult to identify the local, particle-scale deformations.

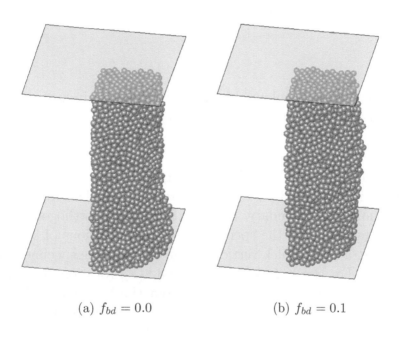

(a) $f_{bd} = 0.0$ (b) $f_{bd} = 0.1$

Figure 8.6: Sensitivity of deformed specimen shape ($\varepsilon_a = 15.3\%$) to particle-boundary friction coefficient (Cui et al., 2007)

Just as in the 2D case, bands of particles can be coloured according to their initial positions to give an impression of their movement (Figure 8.7). For 3D simulations, rather than plotting a 3D plot it can be more useful to take slices through the system and plot 2D projections of the particles. Cheung and O'Sullivan (2008) described a triaxial compression test on a sample of 12,622 spheres

Figure 8.7: Use of a coloured band of particles to indicate deformation in a 3D simulation of the direct shear test

where cementation was modelled using the parallel bond model presented in Section 3.8.1 and a 3D stress-controlled membrane formed the lateral boundary. Figure 8.8 shows two orthogonal cuts through the centre of the specimen, and the rotations of spheres intersecting a 2 mm thick vertical plane through the specimen are considered. For the YZ view in Figure 8.8(a), rotation about the X-axis is considered, while rotation about the Y-axis is XZ view in Figure 8.8(b). Each sphere is represented as a filled disk whose radius equals the disk radius and the disk shading indicates the magnitude of the rotation (in radians) of the sphere. By presenting the data in this way, the presence of a localization through the centre of the sample is easily identified.

8.4   Displacement and Velocity Vectors

Figure 8.9(a) and (b) illustrate the total (or cumulative) displacement of each disk in the simulation considered in Figures 8.2 to 8.4, up to axial strains of 5.5% and 12% respectively. In both

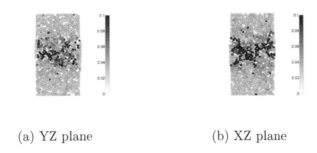

(a) YZ plane   (b) XZ plane

Figure 8.8: Particle rotations in radians at an axial strain of 4.5% for a (3D) triaxial test on bonded specimens with a membrane lateral boundary condition (Cheung and O'Sullivan, 2008)

cases the displacement vectors are scaled by a factor of 2 to allow visualization of the local deformation mechanisms. As in the case of the cumulative particle rotations illustrated in Figure 8.4, the deformation is largely homogeneous at an axial strain of 5.5%. At the larger axial strain value (Figure 8.9(b)) it is clear that the localizations, evident from the rotation plot (Figure 8.4(b)) bound regions within which the particles move together, almost like a single deformable body.

Rather than plotting the cumulative displacements, more insight into the mechanics of the system can be gained either by considering the displacement increments between two selected points or the instantaneous velocities. For example Cundall et al. (1982) used plots of the velocity vectors to identify discontinuities in the deformation mechanism. Figure 8.10 is a plot of the velocity vectors at the points considered in Figure 8.9. The velocity vectors are scaled, with different scaling factors being applied at the two axial strain values. At an axial strain of 5.5% the velocity vectors provide a better indication of the sample failure mechanism than the displacement vectors. The velocity vectors illustrate the complexity of the internal deformation pattern; in particular, where the two localizations intersect, the particles are almost moving in a circular trajectory. At an axial strain of 12% it is clear that deformation along the localization dipping to the left dominates.

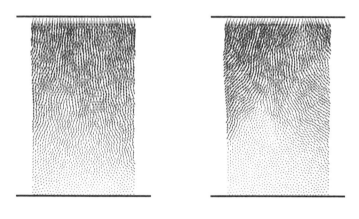

(a) $\epsilon_a=5.5\%$ (b) $\epsilon_a=12\%$

Figure 8.9: Plot of particle displacement vectors for biaxial simulation of 2,376 unbonded disks, confined with a stress-controlled membrane

Circular, almost vortex-like deformation patterns are again evident.

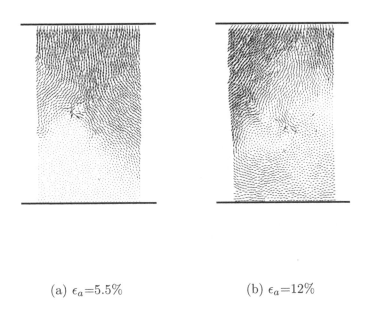

(a) ϵ_a=5.5%   (b) ϵ_a=12%

Figure 8.10: Plot of particle velocity vectors for biaxial simulation of 2,376 unbonded disks, confined with a stress-controlled membrane

Kuhn (1999) presented the results of a biaxial test simulation on about 4000 disk particles in a periodic cell and analysed the particle velocity vectors in some detail. In his analysis, rather than observing the particle velocities directly, Kuhn considered the velocities of the particles relative to the "background velocity" that is calculated from the global strain rate, and assuming a unform deformation within the sample, i.e. for particle k

$$\mathbf{v}^{\text{rel},k} = \frac{\mathbf{v}^k - \mathbf{L}\mathbf{x}^k}{D_{50}|\mathbf{L}|} \tag{8.1}$$

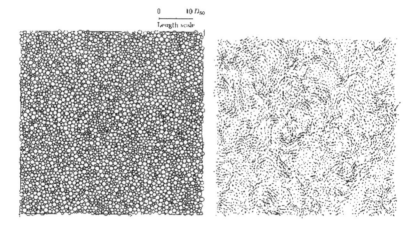

(a) Initial disk configuration (b) Normalized relative velocity vectors ($\mathbf{v}^{\mathrm{rel},k}$) at axial strain of 0.6%

Figure 8.11: Illustration of "vortex" nature of disk velocity vectors observed in the 2D simulations by Kuhn (1999)

where $\mathbf{v}^{\mathrm{rel},k}$ is the relative velocity for particle k, $\mathbf{v}^k$ is the velocity vector for particle k and $\mathbf{L}$ is the global velocity gradient. The difference between the particle velocity and the background velocity was divided by the product of the median particle diameter and the magnitude of the velocity gradient ($D_{50}|\mathbf{L}|$) so that the results could be presented in dimensionless format. As illustrated in Figure 8.11, even though the sample was subject to a simple biaxial strain field, Kuhn observed complex circulation patterns with a vortex-like geometry. Kuhn noted that similar observations of circulation cells or vortex structures had been made by researchers analyzing 2D DEM data including Murakami et al. (1997), while Zhu et al. (2008) referred to the observations of Williams and Rege (1997). Additional approaches that can be applied to analyse the distribution of particle velocities within the system (including analysis of the relative velocities of contacting particles) are presented in Kuhn (1999).

When considering the particle velocities, it is useful to note that many researchers use the term "fluctuation" to refer to the difference between the actual particle velocities and the background velocity (i.e. $\mathbf{v}^{\mathrm{rel},k}$ in Equation 8.4). Campbell (2006) defines a granular temperature T_G that is analogous to a thermodynamic temperature and which can be calculated from the velocity fluctuations as follows:

$$T_G = \left\langle \left(v_x^{\mathrm{rel},k}\right)^2 \right\rangle + \left\langle \left(v_y^{\mathrm{rel},k}\right)^2 \right\rangle + \left\langle \left(v_z^{\mathrm{rel},k}\right)^2 \right\rangle \tag{8.2}$$

where $v_x^{\mathrm{rel},k}$ is now termed the velocity fluctuation and the brackets $\langle$ and $\rangle$ refer to the arithmetic mean of the parameter over the total number of particles considered. Cleary (2007) gives a modified version of this expression in 2D to include the particle rotations, as follows:

$$T_G^{2D} = \left\langle \left(v_x^{\mathrm{rel},k}\right)^2 \right\rangle + \left\langle \left(v_y^{\mathrm{rel},k}\right)^2 \right\rangle + \left\langle (r\omega)^2 \right\rangle \tag{8.3}$$

where ω is the particle rotational velocity and r is the particle radius.

Plots of displacement vectors can also be generated for 3D simulations, and in this case orthogonal plots of the displacement

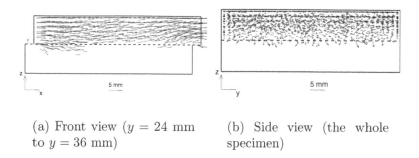

(a) Front view ($y = 24$ mm to $y = 36$ mm)

(b) Side view (the whole specimen)

Figure 8.12: Incremental displacement vectors for the global shear strain interval from $\gamma=0\%$ to $\gamma=15.3\%$ (Cui and O'Sullivan, 2006)

vectors can be very useful. For example, Figure 8.12 illustrates the displacements of particles in a simulation of a direct shear test (described by Cui and O'Sullivan (2006)). In this simulation the macro-scale shearing was in the $x-z$ plane, i.e. the top half of the shear box was moved in the x direction, and the macro-scale dilation during shear was in the z direction. Referring to Figure 8.12(b) it is clear that the magnitude of the particle displacements in the y-direction, i.e. orthogonal to the global direction of shearing, is finite. The vector plots are not necessarily the most useful means to gain quantitative insight into the material response. For this simulation, a plot of the particle displacements in the directions of each of the coordinate axes against the original vertical position of the particle (Figure 8.13) more clearly illustrates the finite magnitude of the displacements orthogonal to the zone of shearing than the displacement vector data in Figure 8.12.

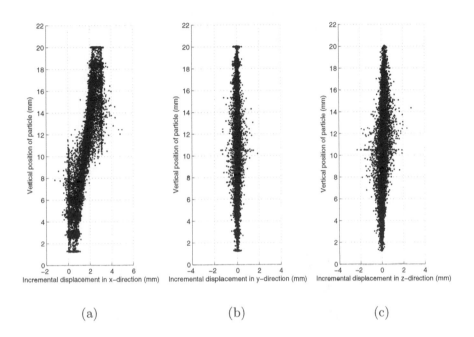

(a) (b) (c)

Figure 8.13: Incremental displacements of particles as a function of depth for the global shear strain interval from $\gamma=0\%$ to $\gamma=15.3\%$ in the direct shear test simulation described by Cui and O'Sullivan (2006)

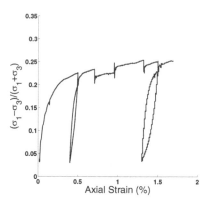

(a) Overall response

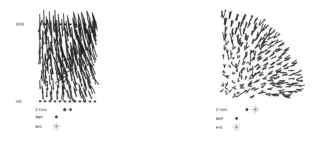

(b) Cycle 2 - vertical view   (c) Cycle 2 - horizontal view

Figure 8.14: Particle trajectories during the load-unload cycles for a specimen of spheres

As well as examining incremental or cumulative displacements of particles, it can be useful to consider the displacement trajectories of individual particles. For example, Figure 8.14(a) illustrates the macro-scale response observed in an axisymmetric DEM simulation of a triaxial compression tests on a sample of 1173 spheres. Two load reversals were included in the simulation; the simulation details are given by O'Sullivan and Cui (2009 a,b). The trajectories of a subset of the particles in horizontal and vertical views is given in Figure 8.14(b) and (c). The energy dissipation evident from the hysteresis in the stress-strain plot indicates a plastic response and there is a net movement of many of the particles during the load reversal.

8.5   Contact Force Network

The fact that a DEM simulation can provide information on the particle interactions and contact forces is central to its use in advancing our understanding of particulate material response. To visualize the contact forces, the convention generally adopted is to illustrate contact forces by drawing a line between the centroids of contacting particles whose thickness is proportional to the magnitude of the force. As illustrated in Figure 8.15, the resulting image is one of a highly complex web or network, which is referred to here as the contact force network. Cundall and Strack (1978) explain that this approach to modelling contact forces was initially adopted as it was used to present data obtained in physical experiments on photoelastic disks, therefore allowing comparison of DEM simulation results with physical test data.

Figure 8.15 illustrates the evolution of the contact force network for the sample of 2,376 disks considered in Figure 8.1. At all three values of axial strain considered ($\epsilon_{axial} = 0\%$, $\epsilon_{axial} = 5.5\%$ and $\epsilon_{axial} = 12.0\%$), the contact force network is highly complex, heterogeneous and difficult to describe quantitatively. As the sample is strained and the deviator stress increases, the network becomes noticeably anisotropic, as the larger contact forces tend to align with the direction of the major principal stress. It is also

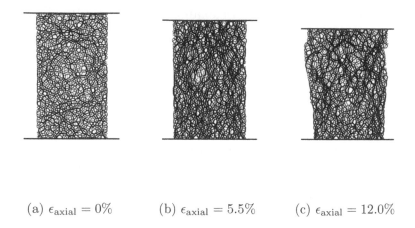

(a) $\epsilon_{\text{axial}} = 0\%$ (b) $\epsilon_{\text{axial}} = 5.5\%$ (c) $\epsilon_{\text{axial}} = 12.0\%$

Figure 8.15: Evolution of contact forces during biaxial test simulation of 2,376 disks

evident that the topology evolves as previously contacting particles disengage and new contacts form. Comparing Figure 8.15(b) and (c) in detail, it is possible to observe chains of contacts within the system bending and eventually buckling as the specimen deforms. Figure 8.16 is a plot of the mean particle stresses at the strain values considered in Figure 8.15. In this case each particle is coloured according to the average stress acting on it (the approach used to calculate the particle stresses is described in Section 9.4). The heterogeneity in the contact force network is reflected in the heterogeneity of the particle stresses. It is clear that some of the particles carry little or no stress, and these particles are sometimes called "rattlers." Kuhn (1999) adopts a slightly different terminology: he recognizes that as the system is deformed particles will "disengage" from the force network and he refers to this process as one where those particles become "dormant." Note that many DEM analysts use their simulation data to generate a probability density function to characterize the range of contact forces in their systems.

Visualization of the contact force network in three dimensions

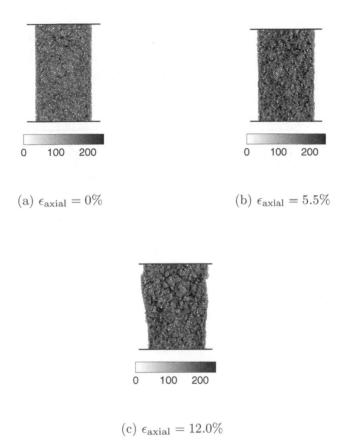

(a) $\epsilon_{\text{axial}} = 0\%$ (b) $\epsilon_{\text{axial}} = 5.5\%$

(c) $\epsilon_{\text{axial}} = 12.0\%$

Figure 8.16: Evolution of mean particle stresses during the biaxial test simulation of 2,376 disks

is more challenging. However, some insight can be gained by considering "slices" through the system and looking at the projection of the forces onto selected planes. It is also often useful to restrict consideration to the larger forces only. Figure 8.17 illustrates a plot of the contact force distribution in a direct shear box simulation. For clarity, to generate the plot only the central third of the sample was considered and only forces that exceed the mean contact force plus one standard deviation are illustrated. A diagonal distribution of contact forces across the shear box has also been observed in two-dimensional simulations (e.g. Masson and Martinez (2001)).

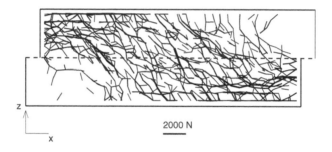

Figure 8.17: Contact force vectors at a global shear strain (γ) of 15.3% (Cui and O'Sullivan, 2006)

To create dimensionless contact force plots, Kuhn (2006) proposes that the line thickness used to represent contact k should be proportional to $\frac{|\mathbf{f}^k|}{pD_{50}}$ where $|\mathbf{f}^k|$ is the contact force, D_{50} is the median particle diameter and p is the mean stress. Kuhn also includes expressions so that the magnitude of the contribution of a given contact force to the average stress tensor can be considered.

Even for simple, 2D systems of disks, the contact force network is very complex. Quantitative analysis of this system and extraction of data to characterize the force network is non-trivial. The force network is normally highly redundant or statically indeterminate (using the language of structural engineering). As noted since the earliest DEM analyses (e.g. Cundall et al. (1982)), "force chains" can be traced through the sample with nodes located at

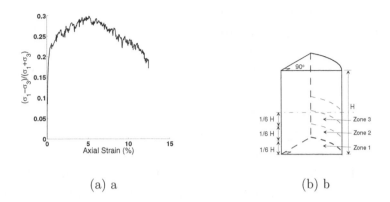

(a) a   (b) b

Figure 8.18: (a) Macro-scale response of sample considered in Figure 8.19(b) Zones used to generate horizontal (XY) views of contact force network

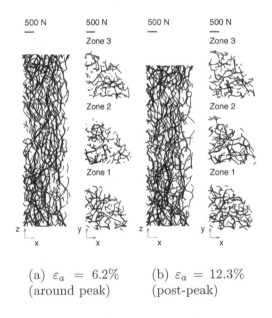

(a) $\varepsilon_a = 6.2\%$ (b) $\varepsilon_a = 12.3\%$
(around peak)   (post-peak)

Figure 8.19: Network of contact forces in three zones for monodisperse specimen (only forces exceeding the mean force plus one standard deviation are illustrated)

299

the particle centroids and a link exists between any two nodes if the particles are contacting and transmitting a force. Almost infinite numbers of possibilities exist to trace force chains or paths through the material. As illustrated in Figure 8.15, and confirmed in countless other DEM simulations the contact force network is both heterogeneous and anisotropic.

Considering firstly the heterogeneity of the contact force network it is clear that some contacts transmit much more force than others. Looking just at the larger contact forces and considering Figures 8.15(b) and (c), paths can be traced down through the sample along the thicker lines, representing the larger contact force. These thicker lines are referred to as "strong force chains" or "load columns" within the material. Behringer et al. (2008) provide a nice definition of a strong force chain as being a "filamentary" structure comprising the contacts that carry a disproportionately large % of the contact force. Tordesillas and Muthuswamy (2009) describe force chains to be "quasi-linear, chain-like particle groups through which above average contact forces are transmitted." Analysts may want a more specific criterion to select strong force chains and a reasonable definition is given by Pöschel and Schwager (2005) who consider a particle b in contact with at least two other particles, including particles a and c. If the contacts with particles a and c both exceed a threshold force value and if the centres of the three particles are almost along a straight line then particles a, b and c are all members of a strong force chain. The three particles can be considered on a straight line if the obtuse angle between the two line segments ab and bc exceeds a specified value. Pöschel and Schwager (2005) suggest a threshold angle of 140°.

As noted above, the contact force networks in Figure 8.15(b) and (c) are anisotropic. Initially, when the stress state is almost isotropic ($\frac{\sigma_1}{\sigma_2} \approx 1.1$), there is no apparent preferential orientation for the larger contact forces. However as the sample is compressed the stress state becomes more anisotropic at $\epsilon_{axial} = 5.5\%$, $\frac{\sigma_1}{\sigma_2} \approx 1.8$ and at $\epsilon_{axial} = 12.0\%$, $\frac{\sigma_1}{\sigma_2} \approx 1.65$, and this results in the anisotropic force network evident in the figure. The larger contact forces are clearly aligned with the direction of the major principal stress, which acts vertically, and the strong force chains are clearly transmitting the stress across the sample. Approaches for quantifying the degree of the anisotropy of the contact force network are considered in Chapter 10.

The examples given of contact force networks in this Chapter are restricted to circular or spherical particles. Where non-spherical particles are used the contact normal orientation and the branch vector will no longer be coincident. Where the particles are non-convex, there may be multiple contact points between two contacting particles. In these cases, more detailed consideration must be given as to how to plot the contact force network. It would seem reasonable, in the first instance, to plot one line between the centroids of contacting particles and make the line thickness proportional to the magnitude of the resultant force acting between the two particles.

Observations of the contact force network and its evolution in a granular material have fundamentally changed the way failure is considered in a granular material. The strength of soil is frictional, i.e. the shear strength of the material increases as the confining stress increases. While this has in the past been related to the friction at particle contacts, an alternative hypothesis is to consider the failure to be directly related to the buckling or collapse of the strong force chains. These force chains are laterally supported by a "weak network" of contacts oriented orthogonal to the major principal stress direction. Dean (2005) proposed that the majority of the plastic work in the system is done in those particles supporting the strong force chains. While these ideas have been proposed by a number of researchers, some of the most interesting work in this area is described by Tordesillas and her colleagues (e.g. Tordesillas (2007 and 2009)). Tordesillas (2009) identifies four stages in the progression towards buckling of a force chain. There is an initial pre-buckling stage prior to attainment of a critical buckling load, then there is a phase of elastic buckling. Then, the system evolves into an initial plastic buckling phase where either the contacts in the force chain column or in the lateral supports will remain elastic. At the fully plastic stage the contacts in the lateral support and the force chain column have reached their plastic threshold.

These findings, amongst other research indicate that the buckling of the strong force chains is the governing mechanism behind the formation of shear bands.

Experimental evidence on real sand specimens seems to confirm the hypothesis that the material response is dominated by the strong force chain kinematics. Oda and Kazama (1998) identified "columnar-like" structures when they studied the microstructure of shear bands in thin sections formed by resin-impregnated sand samples that had been subject to plane strain compression. From their 3D micro-computed tomography (μCT) images Hasan and Alshibli (2010) could identify arch-like structures of contacting particles within a shear band that are indicative of strong force chains buckling. By applying two-dimensional digital image correlation to images of a sand subject to a plane strain compression test, Rechemacher et al. (2010) observed patterns of variation in strain along a shear band, which they attributed to force chain formation and collapse.

While displacement mechanisms can be observed in digital images of sands, no information can be gained on the forces and their magnitudes. Experimental observation of force chains has, however, been achieved using photoelasticity. In photoelastic experiments transparent birefringent materials are considered. These are materials whose optical properties vary as a function of the stress conditions. As discussed by Behringer et al. (2008), when images of photoelastic materials are observed using polarizers, the magnitude of the contract forces can be deduced. Experiments using photo elastic disks provided key insight into the micromechanics of granular material response prior to the development of DEM. Some of the early notable contributions in this area include the work of Dantu (1957, 1968), de Josselin de Jong and Verrujit (1969) and Drescher and de Josselin de Jong (1972). As an example, results from the photoelastic experiments of Behringer et al. (2008) are illustrated in Figure 8.20. In their experiments Berhinger and his colleagues compressed an assembly of photoelastic disks in the biaxial tester illustrated in Figure 8.20(a). No clear pattern in the contact force network could be deduced from the initial image of the assembly (Figure 8.20(b)(i)). However, during shearing an anisotropic stress condition developed and chains of contacting particles transmitting high stresses are visible in the sample (Figure 8.20(b)(ii)). Three-dimensional verification of the existence of strong force chains is also possible; for example Mueth et al. (1997) describe a carbon-paper-based approach that can measure the contact forces along the boundary of a granular material and the force distributions obtained for samples of ballotini are in agreement with observations in 3D DEM simulations on spheres.

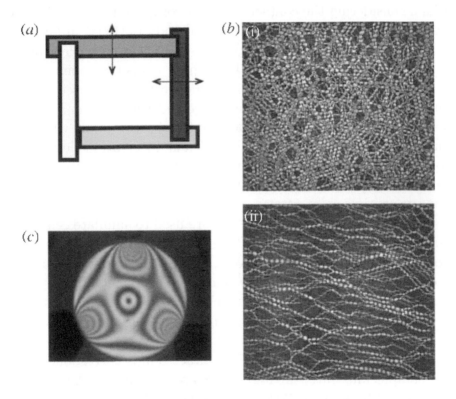

Figure 8.20: Illustration of results of photoelastic experiments described by Behringer et al. (2008) (a) Diagram of biaxial testing apparatus used (b) Photo-elastic images of granular samples that have been (i) isotropically compressed and (ii) subject to pure shear. (c) Photoelastic image of a single disc illustrating the distribution of stress within a particle.

The growth of the internet and global social connectivity, amongst other issues, has fuelled research into networks and network topologies and a very accessible introduction to networks is provided by Watts (2004). It seems likely that some of these developments can be exploited to better understand our contact force network. Adopting the language of network analysis, in a granular material each node will be a particle and the connections are the force transmitting contacts between the particles. As discussed further in Chapter 10 some of the related concepts used in network analysis are of interest in granular mechanics. When considering the contact force network, the idea of "percolation" threshold might be useful. As already outlined in Chapter 7, generally for networks percolation is associated with a phase transition (Grimmett, 1999). This phase transition point marks the transition between a connected network and a disconnected network (Watts, 2004). Specifically considering granular materials, the percolation threshold appears to be the point when the number of contacts per particle sufficient to create a network of contact forces capable of transmitting stress through the sample (Summersgill, 2009). Some authors refer to this point as the jamming transition point.

Where contact or parallel bonds are used to simulate cemented materials or rock mass (e.g. Potyondy and Cundall (2004)) the bond breakages can be monitored and considered analogous to the evolution of micro-cracks within the material. One way to visualize this is to adopt the approach used to generate the contact force plots, but rather than plotting every contact in the system, only the intact bonds are plotted, as illustrated in Figure 8.21 where two orthogonal views of the intact bonds corresponding to the sample illustrated in Figure 8.8 are presented. Authors including Fakhimi et al. (2002) have related the locations of bond breakages to crack locations deduced from acoustic emission in the laboratory.

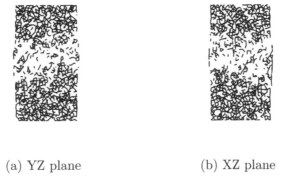

(a) YZ plane   (b) XZ plane

Figure 8.21: Intact bonds at an axial strain of 4.5% for a (3D) tri-axial test on bonded specimens with a membrane lateral boundary condition

Chapter 9

DEM Interpretation: A Continuum Perspective

9.1   Motivation for and Background to Homogenization

Much of our understanding of soil response has been developed within a continuum mechanics framework. Continuum-based analyses tools are dominant in geotechnical engineering practice. Use of a continuum mechanics framework requires knowledge of the stresses and strains in the material. In a slightly provocative tone, Cundall et al. (1982) suggest that for a granular material these stresses and strains are "fictitious" parameters. The deformation and strength responses that distinguish granular materials from other materials arise from their particulate nature and special, highly complex constitutive models are needed to apply continuum mechanics to granular materials. Continuum mechanics cannot capture many important mechanisms that operate at the particle scale. Geomechanics cannot exclusively adopt either a continuum or a discrete approach, leading Muir Wood (2007) to state that there is a "particulate-continuum duality" in geomechanics. From a geomechanics perspective, DEM micro-scale analyses serve little purpose, and will have little impact on research or practice, if the particulate measurements are not interpreted or translated into continuum mechanics terminology. The need to relate discrete and continuum parameters has been recognized since the early work of Cundall and Strack (1978) who presented formulae for the average stress tensor, the average moment tensor and the average displacement gradient tensor within an assembly of disks. In this Chapter approaches that have been proposed to "translate" the results of DEM simulations into the language of continuum mechanics are discussed, considering firstly stress and then strain.

9.2 Representative Volume Element and Scale

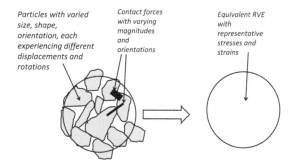

Figure 9.1: Illustration of concept of homogenization

A key inherent assumption in continuum mechanics is that, if a specific point in the material is considered, the material itself and the stresses and strains in an infinitesimal region around that material point are assumed to be uniform. As shown in Chapter 8 the contact force distributions and particle deformations are highly non-uniform or heterogeneous. Dean (2005) quoted an observation by Lambe and Whitman (1979) that "when we talk about stresses at a point in a soil .... we often must envisage a rather large point." Continuum mechanics does not explicitly accommodate sudden transitions in geometry and material properties (e.g. a particle void interface) which will result in highly non-uniform stress and strain fields at the particle scale. There are many materials in which the material properties are locally discontinuous when sufficiently small scales are considered and continuum-based approaches that can accommodate such inhomogeneities exist, as discussed by Nemat-Nasser and Hori (1999).

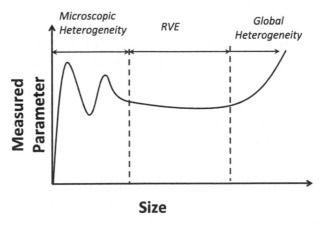

Figure 9.2: Variation in measured parameter with sample size (developed with reference to Masad and Muhunthan (2000))

Recognizing the inherent inhomogeneity in a granular material, when considering the material to be a continuum, a scale that is significantly larger than the particles themselves must be considered. It is useful to introduce the concept of a "representative volume element" or RVE as described by Nemat-Nasser and Hori (1999), amongst others. As illustrated in Figure 9.1 the microstructure within the RVE may be highly complex considering both the particles themselves and the contact force orientations and magnitudes. The RVE for a material is defined to be a volume that can be considered to be statistically representative of the material under consideration. It must be sufficiently large that an increase in size will not change the measured parameter (Figure 9.2). This RVE must therefore include a (potentially large) number of particles.

Nemat-Nasser (1999) introduced the general concept of local

and global heterogeneity, while Masad and Muhunthan (2000) distinguish microscopic heterogeneity and macroscopic heterogeneity (Figure 9.2). For example, there will be local heterogeneities in the sample as a consequence of the particle packing and the development of a localized shear band will introduce heterogeneities at a slightly larger scale. In addition, obviously, in a physical soil, whether it is naturally deposited or a reconstituted sample used in the laboratory, there will be depositional heterogeneities. Geological processes will introduce larger scale heterogeneities (e.g. bedding) in naturally deposited sands. Up until the time of writing most DEM analysts have been concerned with local, or microscopic, heterogeneity.

The appropriate diameter of the RVE (D_{RVE}) will most likely be a function of the particle diameter, i.e. if d_{50} is the mean particle diameter, then $D_{RVE}/d_{50} \gg 1$. In DEM analyses for soil mechanics the particle sizes may vary from about 100 μm to hundreds of millimetres, and so the RVE sizes can vary significantly. In a general discussion on material heterogeneities Nemat-Nasser (1999) cited the observation of Hill (1956). Hill's guidance on the selection on the size of the RVE for a heterogeneous material is that the ratio should be given by $D_{RVE}/d_{50} > 10^3$. In DEM simulations in geomechanics to date limited number of particles have been used (refer to Chapter 12) and the typical D_{RVE}/d_{50} ratios that have been considered are much smaller than this suggested limit.

The concept that there are multiple scales at which a material can be considered is closely associated with the transition from discrete to continuum mechanics, and the terms "macro-scale," "meso-scale," and "micro-scale" are often used. There is little ambiguity in particulate geomechanics on the definition of a micro-scale. This is taken to be a scale where individual particle responses are measurable, i.e. the contact forces and particle displacements can be distinguished. However if particle damage or crushing is important (refer to Section 4.3), the relevant micro-scale may be substantially smaller than the particle diameter. Recalling the discussion on the stress distributions around the contacts in Chapter 3, at this sub-particle micro-scale stress and strain may be relevant parameters. There are therefore two continua: a sub-particle continuum and a macro-scale continuum, with the discrete representation (contact forces and displacements) being relevant at the particle micro- and meso-scales.

The macro continuum length scale that is considered in continuum soil mechanics can be defined to be a scale where the individual particle properties cannot be distinguished, i.e. it will likely correspond with the size of the RVE. When considering the concept of scale for soil, some useful ideas are found in Cambou (1999), who presented an introduction to the concepts involved in homogenization, including a definition of scale. Cambou proposed that the macro-scale corresponds to the discretization of a boundary value problem in a continuum analysis, i.e. the dimension of the smallest element of a finite element mesh. The macro-scale is often considered to correspond to the size of a laboratory element test. Some authors have considered the existence of a "meso-scale." For example, Dean (2005) suggests that this intermediate scale is defined by "patterns" that can be identified by considering the material topology (particles, contact forces and void space). Other considerations determine the definition of a meso-scale; for example it could be considered to the size of the fluid cell adopted in coarse-grid coupled particle-fluid DEM simulations (Chapter 6).

There are times when, while the use of an RVE may not be valid, continuum terminology is still relevant. For example, when a shear band develops, the width of the localization may be significantly smaller than the ideal RVE dimension, yet an assessment of strain within the shear band may be sought. In the process of upscaling, important details on mechanisms and deformation patterns may be erased. Therefore this Chapter will consider how to calculate local stresses and strains as well as representative averaged stress and strain values. The use of a RVE is equally applicable when quantifying the packing or geometrical arrangement of particles in the material using the methods discussed in Chapter 10.

9.3 Homogenization

Having selected the reference volume, *averaging* procedures are needed to calculate the representative stresses and strains from the discrete forces and displacements calculated by DEM. The methods to translate particulate mechanics into continuum mechanics are often called *homogenization techniques*. This transition from discrete to continuum mechanics is equivalent to an upscaling from the micro- to the macro-scale and homogenization techniques are associated with "multi-scale" modelling. This homogenization is often achieved using a volume averaging approach. As outlined by Hori (1999) the use of volume averaging to relate micro- and macro-scale properties comes within the "mean field theory" in physics, and an alternative approach based on homogenization theory uses perturbation analysis, i.e. a singular perturbation is applied to the governing differential equations.

Zhu et al. (2007) classify the homogenization or averaging techniques used in DEM simulations to be either volume, time-volume, or weighted time-volume averaging methods. The volume averaging methods are applicable to quasi-static systems, where inertia effects are ignored. These are the methods that are most commonly used in geomechanics and are considered in this Chapter. Readers interested in applying DEM to rapid flows are advised to consult Zhu et al. (2007), who provide an overview of the relevant theories and cite useful references. Luding et al. (2001) proposed that in general when averaging parameters in granular materials an average or representative value for the quantity or parameter should first be assigned to each particle. Then, considering the volume or subvolume of the material of interest (referred to as the *measurement volume* here), the average value for the parameter Q is given by

$$Q = \langle Q \rangle = \frac{1}{V} \sum_{p \in V} w_v^p V^p Q^p \qquad (9.1)$$

where Q^p is the representative value of the parameter for particle p, V^p is the volume of particle p, and w_v^p is the weighting assigned to particle p. Luding et al. (2001) suggest that the parameter w^p

should equal the proportion of the total volume of particle p that exists within the measurement volume. Using this approach, the volume of solid particles considered in calculation of the void ratio (Equation 10.1) is given by

$$V_s = \sum_{p \in V} w_v^p V^p \tag{9.2}$$

As outlined in Section 9.4, calculation of stress from DEM simulations is relatively straightforward, there is, however, more variation in the approaches used to calculate strain (Section 9.5).

9.4 Stress

Various approaches have been used to derive expressions for the continuum parameter stress for a discrete system comprising contacting particles. Three approaches for stress calculation are considered here: these are calculating stresses by integration of forces along a boundary (Section 9.4.1), consideration of the stresses on individual particles (Section 9.4.2) and the summation of the product of the contact forces and the branch vectors within the granular material (Section 9.4.3).

9.4.1 Stress from boundaries

As noted by Weatherley (2009), in a DEM simulation we can trace measurable quantities, i.e. quantities that can be measured in conventional physical laboratory tests. In the first instance these measurable quantities can then be compared with physical test results for the purpose of DEM model validation or calibration. The most obvious of these measurable quantities are the forces acting along boundaries.

In a conventional triaxial apparatus (schematically illustrated in Figure 5.7), the vertical deviator stress $(\sigma_1 - \sigma_3)$ transmitted through the upper and lower horizontal boundaries is measured by a load cell located above the top platen or beneath the base. The load cell gives a measure of the deviator stress and the applied cell pressure is σ_3. In a DEM simulation the average stress on the upper and lower rigid walls gives the vertical stress (i.e. σ_1). The lateral stress (typically σ_3) is either applied through the stress-controlled membrane or calculated by considering the average stress acting on rigid wall, vertical lateral boundaries.

In any DEM simulation the average stresses along a rigid wall boundary are calculated by summing the contact forces along the boundary, and dividing by the surface area of the rigid boundary (3D) or rigid boundary length (2D). Using a similar approach the distribution/variation of stresses along the boundary can be determined by dividing it into subareas.

From an alternative perspective, referring to Bagi (1999b), for a closed continuous domain, with volume V and boundary area S, Gauss's integral theorem gives the relation between the stresses within the material (σ_{ij}), and the applied boundary force as

$$\int_V \sigma_{ij} dV = \oint_S x_i t_j dS \tag{9.3}$$

where the boundary force or boundary traction t_j is applied at position x_i along the surface of the volume considered. This force is normally not explicitly applied, it is more often a contact force that develops between the particle and a rigid wall boundary. The stresses acting normal to the boundary are related to the boundary traction by $\sigma_{ij} n_i = t_j$, where n_i is the normal vector directed outwards from the surface S.

The average stress within the material ($\bar{\sigma}_{ij}$) is then given by

$$\bar{\sigma}_{ij} = \frac{1}{V} \int_V \sigma_{ij} dV \tag{9.4}$$

$$\bar{\sigma}_{ij} = \frac{1}{V} \oint_S x_i t_j dS \tag{9.5}$$

In continuum analyses, the boundary forces t_j may be specified as a function of x_i, allowing integration. However, in a DEM analysis the forces act at discrete points, e.g. at the particle boundary-contact points (or the particle centroids where a numerical membrane is used, as discussed in Section 5.4). Rather than using integration, to interpret DEM simulation data, Equation 9.5 must then be reformulated in terms of a summation over a set of discrete forces applied at distinct points. Replacing the surface integration by a summation yields:

$$\bar{\sigma}_{ij} = \frac{1}{V} \sum_{k=1}^{N_{BF}} x_i^k t_j^k \qquad (9.6)$$

where the force t_j^k is applied at location x_i^k and a total of N_{BF} forces act along the boundary.

9.4.2 Local stresses: Calculation from particle stresses

Potyondy and Cundall (2004) and Li et al. (2009), amongst others, present detailed derivation of the calculation of the average stress within a RVE ($\bar{\sigma}_{ij}$) from the individual particle average, or representative stresses, σ_{ij}^p. As stress can exist only within the particles (i.e. the voids don't transmit force or stress), the product of the average stress and the volume V equals the sum of the stresses acting on the N_p particles within the volume weighted by the particle volumes V_p i.e.

$$\bar{\sigma}_{ij} V = \sum_{p=1}^{N_p} \sigma_{ij}^p V_p \qquad (9.7)$$

As the particles used in DEM are rigid, the idea of a particle stress is somewhat contradictory, and so strictly σ_{ij}^p is a notional or representative stress. An expression for σ_{ij}^p can be derived by firstly considering the stress equilibrium equation as follows:

$$\sigma_{ij,i}^p - \rho g_j = -\rho a_j \qquad (9.8)$$

where ρ is the particle density, g_j is a body force (e.g. gravity) and a_j is the particle acceleration vector. Assuming a quasi-static response (i.e. negligible accelerations) and neglecting consideration of the body force, the equilibrium equation reduces to

$$\sigma_{ij,i}^p = 0 \qquad (9.9)$$

Now imagine that a stress domain exists within the particle and that the stress at any point x within the particle volume is given by σ_{ij}^x. The equilibrium equation also holds at this point, i.e.

$$\sigma_{ij,i}^x = 0 \tag{9.10}$$

Equation 9.10 can be expanded by application of Gauss's divergence theorem. This theorem considers a vector function, e.g. F_{ij}, and relates the volume integral of the divergence, $\nabla \cdot \mathbf{F}$, of $\mathbf{F}$ over a volume V to the surface integral of $\mathbf{F}$ over the boundary, S, of V (provided certain smoothness conditions are met). In the general case it is given by:

$$\int_V \nabla \cdot \mathbf{F} dV = \int_V (F_{ij})_{,i} \, dV = \oint_S F_{ij} n_i dS \tag{9.11}$$

where $\int_V ...dV$ represents integration over the volume of the domain and $\oint_S ...dS$ represents an integration over the surface of the domain, and n_i is a unit vector normal to the surface and pointing away from the domain.

Some manipulation of terms is required to allow application of Gauss's theorem to Equation 9.9. The objective is to develop an expression for the particle stress σ_{ij}^p in terms of the contact forces f_i^c and their locations, x_i^c, these data are known from the DEM simulation results. To get to this point an expression for the stress tensor that includes the position vector $\mathbf{x}$ is sought. Applying the product rule for differentiation to the product $x_i \sigma_{kj}^x$ gives

$$\left(x_i \sigma_{kj}^x\right)_{,k} = x_i \sigma_{kj,k}^x + x_{i,k} \sigma_{kj}^x \tag{9.12}$$

Referring to Chapter 1, the Kronecker delta δ_{ij} equals $x_{i,j}$, where $\mathbf{x}$ is the position vector. From equilibrium considerations (Equation 9.9) $\sigma_{kj,k}^x = 0$. Therefore Equation 9.12 is equivalent to

$$\left(x_i \sigma_{kj}^x\right)_{,k} = x_{i,k} \sigma_{kj}^x \tag{9.13}$$

Referring to Equation 1.12

$$x_{i,k} \sigma_{kj}^x = \delta_{ik} \sigma_{kj}^x = \sigma_{ij}^x \tag{9.14}$$

so

$$\left(x_i \sigma_{kj}^x\right)_{,k} = \sigma_{ij}^x \tag{9.15}$$

The average, or representative, particle stress is related to the point stress σ_{kj}^x by

$$\sigma_{ij}^p = \frac{1}{V^p} \int_{V^p} \sigma_{ij}^x dV = \frac{1}{V^p} \int_{V^p} \left(x_i \sigma_{kj}^x\right)_{,k} dV \tag{9.16}$$

Applying Gauss's theorem gives

$$\sigma_{ij}^p = \frac{1}{V^p} \int_{V^p} \left(x_i \sigma_{kj}^x\right)_{,k} dV = \frac{1}{V^p} \oint_{S^p} x_i \sigma_{kj}^x n_k dS \tag{9.17}$$

where $\oint_{S^p} ...dS$ indicates integration over the particle surface. Along the particle surfaces the product of the stresses σ_{kj}^x and the outward-facing normal vector n_k will equal the applied forces or tractions acting on the particle surfaces, denoted here as t_j, so

$$\bar{\sigma}_{ij}^p = \frac{1}{V^p} \oint_{S^p} x_i t_j dS \tag{9.18}$$

In a DEM simulation these particle surface tractions will be the discrete contact forces, f_j^c acting at contact points x_i^c along the surface of the particle. So the integration can be replaced by a summation over the total number of contacts involving particle p, $N^{c,p}$

$$\sigma_{ij}^p = \frac{1}{V^p} \sum_{c=1}^{N^{c,p}} x_i^c f_j^c \tag{9.19}$$

Equation 9.19 can be directly used to calculate the average stress tensor for individual particles. As noted by Nemat-Nasser (1999) when considering a heterogeneous material comprising an agglomerate of crystals or particles, it is common to assume the stress and deformation fields within each grain to be uniform. Referring back to the discussion on particle crushing in Chapter 4, these definitions of stress do not account for the variation in stress that will exist within the particles themselves in the real, physical situation. These stress distributions within individual grains are highly non-uniform, as illustrated already in Figure 8.20. Reference to Russell et al. (2009) may also be useful to those considering the stress inhomogeneities in actual particles.

Figure 9.3 illustrates the stresses within an assembly of 460 disks. This is a relatively small specimen and its size was chosen so that individual particles can be observed. Two stress states are considered: in the first case the average stresses are given by $\sigma_1 = \sigma_3 = 120$ kPa, while in the second case $\sigma_1/\sigma_3 = 2.6$ and $\sigma_3 = 120$ kPa. In Figure 9.3(a) and (b) the distribution of inter particle-contact forces is given. The heterogeneity in the contact force network is reflected in the range of mean stresses the particles experience, (Figure 9.3(c) and (d)). The particles experiencing the highest contact forces tend to experience higher stresses (as illustrated in Figure 9.3(f) in particular). Note, however, as observed by Summersgill (2009), the calculated mean stresses depend on the particle volume. Using this approach small particles with relatively low contact forces can have higher mean stresses than larger particles participating in the strong force chains. In reality the stress distribution within the particles is non-uniform, and so the larger particles transmitting larger forces might experience much higher stresses (in particular close to the contact point) than the maximum stresses induced within smaller particles experiencing smaller contact forces.

Equation 9.19 can also be used to develop an expression for the average stress within a specified subvolume or measurement volume in the material. As outlined by Potyondy and Cundall (2004), each contact location can be expressed as

$$x_i^c = x_i^p + |x_i^c - x_i^p| \, n_i^{c,p} \tag{9.20}$$

where x_i^p is the position of the particle centroid, and $n_i^{c,p}$ is the unit vector directed from the particle centroid to the contact point, c. As the sum of the contact forces acting on a particle in equilibrium must equal 0

$$\sum_{c=1}^{N_{c,p}} f_j^c = 0 \tag{9.21}$$

For particles in equilibrium $\sum_{c=1}^{N_{c,p}} x_i^p f_j^c = x_i^p \sum_{c=1}^{N_{c,p}} f_j^c = 0$ and Equation 9.19 can be expressed as

$$\sigma_{ij}^p = \frac{1}{V_p} \sum_{c=1}^{N^{c,p}} |x_i^c - x_i^p| \, n_i^{c,p} f_j^c \qquad (9.22)$$

The average stress within an arbitrary subvolume, V can be determined by summing the contributions from each particle within that subvolume. Care must be taken to account for the intersection of particles with the measurement volume boundary, i.e. $\sum_{p=1}^{N_p} V_p$ is not the actual amount of the volume that is occupied by solid particles (V_s). (Latzel et al. (2000) includes a discussion on the sensitivity of the calculated stresses to the accuracy with which the volume of the particles that intersects the measurement subvolume boundary are considered in the calculations.) To rectify this, Potyondy and Cundall (2004) propose applying an adjustment to the volume by considering the porosity so that

$$V \approx \frac{\sum_{p=1}^{N_p} V_p}{1 - n} \qquad (9.23)$$

where n is the porosity of the measurement volume considered. In making this adjustment, it is assumed that the geometrical distribution of the particles within the measurement volume is statistically uniform and so the volume associated with each particle is $\frac{V_p}{1-n}$. Then the overall average stress within a selected measurement volume or measurement region containing N_p particles is given by

$$\bar{\sigma}_{ij} = \frac{(1-n)}{\sum_{p=1}^{N_p} V_p} \sum_{p=1}^{N_p} \sigma_{ij}^p V_p \qquad (9.24)$$

or

$$\bar{\sigma}_{ij} = \frac{1}{V} \sum_{p=1}^{N_p} \sigma_{ij}^p V_p \qquad (9.25)$$

where V is the measurement volume under consideration. Equation 9.24 can be expanded to

$$\bar{\sigma}_{ij} = \frac{(1-n)}{\sum_{p=1}^{N_p} V_p} \sum_{p=1}^{N_p} \left(\sum_{c=1}^{N_{c,p}} |x_i^c - x_i^p| \, n_i^{c,p} f_j^c \right) \tag{9.26}$$

This formulation does not include contributions from moment terms.

Equation 9.26 can be further developed to eliminate the double summation, i.e. summing over the contacts involving each particle and then summing over each particle. If each contact is shared between two particles, then each contact is considered twice, and the double summation in Equation 9.26 can be re-expressed as

$$\sum_{p=1}^{N_p} \left(\sum_{c=1}^{N_{c,p}} |x_i^c - x_i^p| \, n_i^{c,p} f_j^c \right) = \\ \sum_{c=1}^{N_c} \left(|x_i^c - x_i^{pa}| \, n_i^{c,pa} f_j^{ca} + \left| x_i^c - x_i^{pb} \right| n_i^{c,pb} f_j^{cb} \right) \tag{9.27}$$

where there are N_c contacts in the measurement region and each contact involves two particles, a and b, with centroidal coordinates x_i^{pa} and x_i^{pb} respectively. The forces exerted on each particle at the contact act in equal and opposite directions so that $f_j^{cb} = -f_j^{ca}$. Referring to Figure 9.4, by the rules of vector addition, subtracting the vector extending from the centroid of particle b to the contact point c from the vector from the centroid of particle a to the contact point c gives the vector between the two particle centroids (a to b). This is the branch vector (l_i^c) for contact c and it connects a and b, i.e.

$$l_i^c = |x_i^c - x_i^{pa}| \, n_i^{c,pa} - \left| x_i^c - x_i^{pb} \right| n_i^{c,pb} \tag{9.28}$$

Therefore

$$\sum_{c=1}^{N_c} |x_i^c - x_i^{pa}| \, n_i^{c,pa} f_j^{ca} - \left| x_i^c - x_i^{pb} \right| n_i^{c,pb} f_j^{cb} \\ = \sum_{c=1}^{N_c} l_i^c f_j^c \tag{9.29}$$

and so Equation 9.26 can be re-expressed as

$$\bar{\sigma}_{ij} = \frac{1-n}{\sum_{p=1}^{N_p} V_p} \sum_{c=1}^{N_c} l_i^c f_j^c \tag{9.30}$$

(a) Contact force network $\sigma_1 = \sigma_2 = 120kPa$

(b) Contact force network $\sigma_1/\sigma_2 = 2.6$

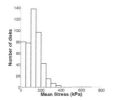

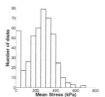

(c) Histogram of mean particle stresses $\sigma_1 = \sigma_2 = 120kPa$

(d) Histogram of mean particle stresses $\sigma_1/\sigma_2 = 2.6$

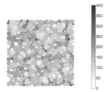

(e) Disks coloured according to mean particle stresses, $\sigma_1 = \sigma_2 = 120kPa$

(f) Disks coloured according to mean particle stresses, $\sigma_1/\sigma_2 = 2.6$

Figure 9.3: Distribution of particle stresses within an assembly of disk particles; isotropic and anisotropic stress distributions

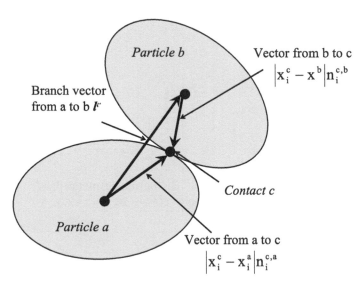

Figure 9.4: Illustration of particle contacts and branch vector

Thus, working from a consideration of the average stresses acting on individual particles, two alternative expressions for the average stress within an arbitrary volume in a granular material can be developed (Equations 9.24 and 9.30). Both expressions relate the stress to the magnitude of the contact forces and the locations of the particles and contacts.

9.4.3 Local stresses: Calculation from contact forces

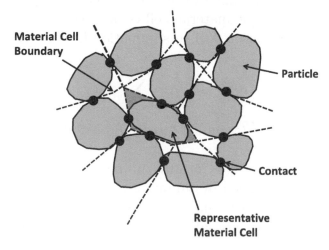

Figure 9.5: Schematic diagram of material cell system proposed by Bagi (1996)

Rather than working from a perspective where the average particle stresses are used to calculate the average stress in a subdomain of the system, Bagi (1996) showed how the average stresses can be directly calculated from the contact forces. Bagi developed her expression by considering the forces acting on the boundary of a subdomain within the system. Bagi selected her subdomains or graph cells so that their boundaries pass through the contact points. In the limiting case the smallest subdomains will be the material cells illustrated in Figure 9.5. This material cell system is essentially a space-filling tessellation; each cell is centred around a single particle in the system and the domains are contiguous. The material cell system is an example of a graph representation of the particle network and the concept of a particle graph is discussed in more detail in Chapter 10.

Similar to the case above where the average stress within a particle was considered (Equation 9.18), the average stress within a material cell, $\bar{\sigma}_{ij}^{MC}$, is given by

$$\bar{\sigma}_{ij}^{MC} = \frac{1}{V^{MC}} \oint_{S_{MC}} t_i x_j dS \qquad (9.31)$$

where V^{MC} is the volume of the material cell and $\oint_{S_{MC}} ...dS$ is a surface integral for the material cell considered.

Expressing Equation 9.31 in summation format gives

$$\bar{\sigma}_{ij}^{MC} = \frac{1}{V^{MC}} \sum_{c=1}^{N^{f,MC}} f_i^c x_j^c \qquad (9.32)$$

where a total of $N^{f,MC}$ contact forces (f_i^c) act along the boundary at points x_j^c.

A representative stress for any sub-region of the granular material can be determined by summing together the stresses in the material cells within the chosen "measurement region." The contribution from each material cell is weighted by its volume:

$$\bar{\sigma}_{ij}^{MR} = \frac{1}{V^{MR}} \sum_{MC=1}^{N^{MC,MR}} \bar{\sigma}_{ij}^{MC} V^{MC} \qquad (9.33)$$

where $\bar{\sigma}_{ij}^{MR}$ is the average stress within the measurement region, $N^{MC,MR}$ is the number of material cells within the measurement region and V^{MR} is the volume of the measurement region ($V^{MR} = \sum_{MC=1}^{N^{MC,MR}} V^{MC}$). Then combining Equations 9.32 and 9.33 gives

$$\bar{\sigma}_{ij}^{MR} = \frac{1}{V^{MR}} \sum_{MC=1}^{N_{MC,MR}} \sum_{c=1}^{N^{f,MC}} f_i^c x_j^c \qquad (9.34)$$

The position vector $\mathbf{x}^c$ of the contact force can be represented as $x_j^c = x_j^p + \left| x_j^c - x_j^p \right| n_j^{c,p}$, where x_j^p are the centroidal coordinates of the particle on which the stress acts and $n_j^{c,p}$ is the unit vector directed from x_j^p to x_j^c. Equilibrium considerations mean $\sum_{c=1}^{N_{f,SD}} f_i^c = 0$ (and so $\sum_{c=1}^{N_{f,SD}} f_i^c x_j^p = 0$), then

$$\bar{\sigma}_{ij}^{MR} = \frac{1}{V^{MR}} \sum_{MC=1}^{N_{MC,MR}} \sum_{c=1}^{N^{f,MC}} f_i^c \left| x_j^c - x_j^p \right| n_j^{c,p} \qquad (9.35)$$

and particle p is at the centre of material cell MC. Each contact force f_i^c is represented by two material cell boundary forces acting in equal and opposite in directions, and so Equation 9.35 can be rewritten as a summation over the contact points in the measurement region, i.e.:

$$\bar{\sigma}_{ij}^{MR} = \frac{1}{V^{MR}} \sum_{c=1}^{N^{c,MR}} \left(f_i^c \left| x_j^c - x_j^a \right| n_j^{c,a} - f_i^c \left| x_j^c - x_j^b \right| n_i^{c,b} \right) \qquad (9.36)$$

where x_j^a and x_j^b are the centroidal coordinates of the two particles participating in contact c. From considerations of vector addition

$$\left| x_j^c - x_j^a \right| n_j^{c,a} - \left| x_j^c - x_j^b \right| n_j^{c,b} = x_j^c - x_j^a - x_j^c + x_j^b = l_j^{ba} \qquad (9.37)$$

where $l_j^{ba} = l_j^c$ is the branch vector for contact c directed from particle a to particle b. Then

$$\bar{\sigma}_{ij}^{MR} = \frac{1}{V^{MR}} \sum_{c=1}^{N_{c,MR}} f_i^c l_j^c \qquad (9.38)$$

Comparing Equations 9.38 and 9.30, it is clear that while they adopted alternative perspectives, both Bagi (1996) and Potyondy and Cundall (2004) have developed equivalent expressions for the calculation of (continuum) stresses from DEM simulation data. Note that while the derivations presented here follow the steps outlined by Bagi (1996) and Potyondy and Cundall (2004), alternative earlier, derivations were given by Christoffersen et al. (1981) and Rothenburg and Bathurst (1989) amongst others. The more recent expression developed by Li et al. (2009) includes a term to account for particle rotation in the stress tensor.

9.4.4 Stresses: Additional considerations

Both Equations 9.38 and 9.30 give an expression for the stress within an arbitrary volume of a granular material in terms of the contact forces and the branch vectors. The stresses may either be calculated within the DEM programme or in a postprocessing stage using information output in the snapshot files that contain details on the particles and the contact forces (as described in Chapter 8).

Expanding the tensorial expressions for the 2D stress tensor gives

$$
\begin{pmatrix} \sigma_{xx} & \sigma_{xy} \\ \sigma_{yx} & \sigma_{yy} \end{pmatrix} = \frac{1}{V} \begin{pmatrix} \sum_{c=1}^{N_{c,V}} f_x^c l_x^c & \sum_{c=1}^{N_{c,V}} f_x^c l_y^c \\ \sum_{c=1}^{N_{c,V}} f_y^c l_x^c & \sum_{c=1}^{N_{c,V}} f_y^c l_y^c \end{pmatrix} \tag{9.39}
$$

while the 3D stress tensor is given by

$$
\begin{pmatrix} \sigma_{xx} & \sigma_{xy} & \sigma_{xz} \\ \sigma_{yx} & \sigma_{yy} & \sigma_{yz} \\ \sigma_{zx} & \sigma_{zy} & \sigma_{zz} \end{pmatrix} = \frac{1}{V} \begin{pmatrix} \sum_{c=1}^{N_{c,V}} f_x^c l_x^c & \sum_{c=1}^{N_{c,V}} f_x^c l_y^c & \sum_{c=1}^{N_{c,V}} f_x^c l_z^c \\ \sum_{c=1}^{N_{c,V}} f_y^c l_x^c & \sum_{c=1}^{N_{c,V}} f_y^c l_y^c & \sum_{c=1}^{N_{c,V}} f_y^c l_z^c \\ \sum_{c=1}^{N_{c,V}} f_z^c l_x^c & \sum_{c=1}^{N_{c,V}} f_z^c l_y^c & \sum_{c=1}^{N_{c,V}} f_z^c l_z^c \end{pmatrix} \tag{9.40}
$$

where $N_{c,V}$ is the total number of contacts in the volume V considered, (f_x^c, f_y^c, f_z^c) is the force vector for contact c and (l_x^c, l_y^c, l_z^c) is the branch vector for contact c. The typical convention for stress used in geomechanics takes compressive stresses to be positive. Referring to Figure 9.6, the contact between two particles a and b if we define the contact force to be the force applied by particle a on particle b and the branch vector to be directed from the centre of particle a to particle b then the calculated stresses will be positive in compression.

In geomechanics the stress state is normally characterized by the principal stresses and their orientations, rather than using the stress tensor directly. The principal stresses are given by the eigenvalues of the matrices represented in Equations 9.39 and 9.40, while the eigenvectors give the orientations. Alternatively a Mohr's circle construction or the characteristic equation can be used; these approaches are outlined in general undergraduate mechanics texts, e.g. Gere and Timoshenko (1991). In three dimensions the intermediate principal stress (σ_2) must be considered in addition to the major (maximum) and minor (minimum) principal stresses (σ_1 and σ_3 respectively). The extent of the stress anisotropy in a three-dimensional stress state is often quantified by considering the principal stress ratio, b, defined as

$$b = \frac{\sigma_2 - \sigma_3}{\sigma_1 - \sigma_3} \qquad (9.41)$$

Thornton (2000) and Barreto (2010) both considered the influence of the intermediate principal stress on the material response. Sand particles deposited in nature and in reconstituted samples in the laboratory are typically deposited under gravity and so the particle long axes will be preferentially oriented horizontally. Therefore, when quantifying the direction loading, the orientation of the major principal stress to the vertical is considered. The stress states are often quantified using the invariants of the stress tensor; the expressions for the invariants are defined in continuum mechanics texts such as Shames and Cozzarelli (1997).

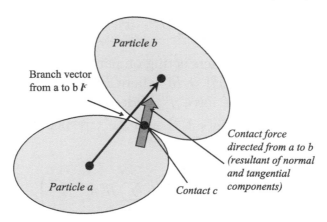

Figure 9.6: Convention for stress in geomechanics

The volume selected for calculating the representative or average stress obviously depends on the application and the information sought. As an example, Figure 9.7(a) illustrates a typical control volume, which takes the form of a measurement sphere that is used to calculate internal stresses and control the vertical stress in the DEM simulations of the direct shear test described by Cui and O'Sullivan (2006). Figure 9.7(b) illustrates the configuration of a rectangular box whose depth was varied to define different volumes of internal stress calculation. Figure 9.7(c) then illustrates how the calculated average shear stresses vary with the size of the volume used for stress homogenization.

As well as studying the heterogeneities within the domains simulated, the expressions for the stress tensor can be applied and manipulated to study fundamental mechanics. The papers by Thornton (2000) and Thornton and Antony (2000) describe a particularly interesting approach to the calculation of stress. These authors partitioned the stress tensor and decoupled the contribution to the average stress due to the normal (σ_{ij}^N) and tangential components of the contact force (σ_{ij}^T), i.e.

$$\sigma_{ij} = \sigma_{ij}^N + \sigma_{ij}^T \tag{9.42}$$

Recall that the representative particle stress is given by

$$\sigma_{ij}^p = \frac{1}{V_p} \sum_{c=1}^{N^{c,p}} |x_i^c - x_i^p| \, n_i^{c,p} f_j^c \tag{9.43}$$

where there are $N^{c,p}$ contacts acting on particle p. For spherical or circular particles, $|x_i^c - x_i^p| = R^p$ where R^p is the particle radius. If the resultant contact force f_j^c is decomposed into its normal and tangential components $f_j^{c,N}$ and $f_j^{c,T}$, respectively, then the particle stress is given by σ_{ij}^p:

$$\sigma_{ij}^p = \frac{1}{V_p} \left(\sum_{c=1}^{N^{c,p}} R^p n_i^{c,p} f_j^{c,N} + \sum_{c=1}^{N^{c,p}} R^p n_i^{c,p} f_j^{c,T} \right) \tag{9.44}$$

and the average value is taken to be the volume weighted sum of the contributions from the individual particles, divided by the total volume. Thornton and his colleagues applied this equation to data from a periodic cell and considered all contacts in the system, so it is valid to simply divide by the total volume of the cell, V. Equation 9.42 can then be expressed as

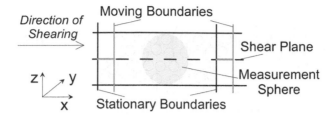

(a) Spherical volume used for control of vertical stress simulation (Cui and O'Sullivan, 2006)

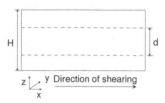

(b) Rectangular box used to explore stress non uniformities

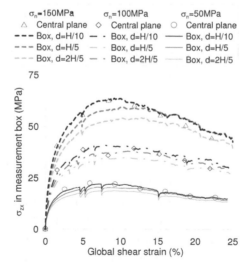

(c) Shear stress variation with internal volume considered

Figure 9.7: Measurement of internal stresses in direct shear box simulations (Cui and O'Sullivan, 2006)

329

$$\bar{\sigma}_{ij} = \frac{1}{V} \left[\sum_{p=1}^{N_p} \left(\sum_{c=1}^{N^{c,p}} R^p n_i^{c,p} f_j^{c,N} \right) + \sum_{p=1}^{N_p} \left(\sum_{c=1}^{N^{c,p}} R^p n_i^{c,p} f_j^{c,T} \right) \right] \quad (9.45)$$

where there are N_p particles within the volume V. Thornton (2000) used this decomposition effectively to describe the nature of stress transmission in granular materials. In a simulation of triaxial compression, it was shown that $\sigma_{ij}^N \gg \sigma_{ij}^T$, so the bulk of the deviatoric stress is transmitted by the normal component of the contact force. This reflects the fact that the material response is dominated by contact forces that are oriented in the direction of the major principal stress acting on the sample.

The discussion to date has considered quasi-static/static situations where the particles are assumed to be approximately in static equilibrium. Luding et al. (2001) define a dynamic component of the stress tensor to be given by

$$\sigma_{ij}^d = \frac{1}{V} \sum_{p \in V} w_v^p V^p \rho^p v_i^p v_j^p \quad (9.46)$$

Luding et al. noted that there are two components to this stress tensor, the stress due to the fluctuations of the velocities about the mean and the stress due to the mean mass transport in the overall direction of strain or flow.

9.5 Strain

Just as was the case for stress, a representative strain for the material in an element test simulation can be calculated by considering the boundary positions. For example, in a triaxial test simulation the position of the top and bottom boundaries can be used to calculate the overall axial strain, $\epsilon_a = \frac{\Delta_H}{H_0}$ where Δ_H is the axial compression and the original height is given by H_0. As highlighted by Marketos and Bolton (2010), just as in the laboratory, the local kinematics of particle motion close to the boundary may not be indicative of the overall response, and this should be considered when interpreting these strain values.

A DEM simulation gives detailed information on the displacement trajectories of all the particles in the system. Using this information, a representative, or average, strain for the overall system of particles or a subdomain of the system can be determined. In many applications the particle deformation field is not homogeneous and therefore calculating of local strains within the problem domain is of interest. Methods to calculate both average and local strains are introduced here.

9.5.1 Overview of calculation of strain from a continuum mechanics perspective

In continuum mechanics, the strains at a point are calculated from the local deformation or displacements, specifically the spatial gradients of these deformations. When a material experiences strain there is a change in geometry from an original configuration or "state" to a deformed configuration or state, and points within the material will be displaced. Considering a particle with a position vector in the original configuration given by $\mathbf{x}^0$, the same particle will have a position vector given by $\mathbf{x}^D$ in the deformed configuration. If the original geometry is taken as the reference configuration, then the incremental displacement is taken to be $\mathbf{u} = \mathbf{x}^D - \mathbf{x}^0$; however, if the deformed geometry is taken as the reference configuration, then the incremental displacement is given by $\mathbf{U} = \mathbf{x}^0 - \mathbf{x}^D$. In geomechanics applications the strains are usually calculated relative to the original configuration.

For many engineering applications, it is assumed the strains are small, and Cauchy's infinitesimal strain tensor is applicable (Fung, 1977):

$$e_{ij} = \frac{1}{2}(u_{j,i} + u_{i,j}) \tag{9.47}$$

where e_{ij} is the strain tensor, u_i is the displacement tensor and $u_{i,j}$ represents the partial derivatives (i.e. $u_{x,y} = \frac{\partial u_x}{\partial y}$, etc.). This definition is only appropriate where the partial derivatives, $u_{i,j}$, are sufficiently small that the squares and products of these terms are negligible.

In "finite strain" problems the products can no longer be neglected, and the definition of the strain tensor depends on whether the deformation measure is related to the reference (original) configuration or the current configuration (Zienkiewicz and Taylor, 2000b). The definition of the finite strain tensor depends on whether the deformation measure is related to the reference (original) configuration or the current configuration (Zienkiewicz and Taylor (2000b)). When the deformation measure is related to the original configuration, the Green strain tensor, E_{ij}, is applicable:

$$E_{ij} = \frac{1}{2}(u_{j,i} + u_{i,j} + u_{k,i}u_{k,j}) \qquad (9.48)$$

Whichever definition is used to calculate the strain, once the deformation gradients $u_{i,j}$ are known the strain can easily be calculated. This section focusses on how these gradients can be calculated. In a DEM simulation the displacement gradients can be calculated by taking "snapshots" of the system, i.e. by outputting all the particle coordinates at specific points in the simulation. The particle locations at the beginning of the chosen interval define the reference location and the calculated strains can be mapped to these points for visualization purposes. The displacement increments will be calculated by subtracting the initial particle coordinates from the particle coordinates in the deformed system. Alternatively a single "snapshot" can be taken and the particle velocities can be used to calculate the current strain rate.

An analytical expression (i.e. an equation) is needed to describe the incremental displacements, $\mathbf{u}$, so that the partial derivatives $u_{i,j}$ can be calculated. If the velocities are considered the partial derivatives of the deformation rate tensor, $\dot{u}_{i,j}$, are sought. The displacement and velocity fields are highly heterogeneous, and various approaches have been proposed to determine the displacement gradients. These can be broadly divided into best fit approaches, spatial discretization approaches, and local non-linear wavelet based approaches. The spatial discretization approaches are sometimes called equivalent continuum approaches (e.g. Bagi (2006)). Each of these approaches is considered below. Reviews of various strain calculation approaches are given by Bagi (2006) (comparison of 2D implementations) and Duran et al. (2010) (comparison of 3D implementations).

Rotations and strain

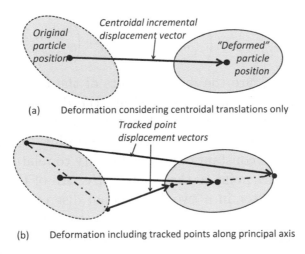

(a)   Deformation considering centroidal translations only

(b)   Deformation including tracked points along principal axis

Figure 9.8: Particle rotation and strain

The particle centroidal displacements describe only the translational motion of the particles. The particle rotations will also contribute to the material displacements at a micro-scale. As already highlighted in Chapter 8, rotations within shear bands can be significant, and where local strains within a shear band are needed, it would seem appropriate to capture the deformation due to these rotations in the calculated strain value. A schematic diagram illustrating the effects of particle rotation on the calculated strain values is illustrated in Figure 9.8(a). When a particle experiences a finite amount of rotation, the displacements of points on the edge of the particle may differ substantially from the centroidal displacements. Consequently, the displacement gradient values calculated considering the centroidal coordinates alone do not capture the actual strains that the assembly of particles is experiencing. Various approaches have been proposed for accounting for rotation. In one approach, proposed by O'Sullivan et al. (2003), the displacements of two points located at a distance from the particle centroid are tracked. For example these points might be located along the particle's principal axes as indicated in Figure 9.8(b).

Considering the two-dimensional case, for each particle the co-ordinates of the two "tracked" points (j=1,2) are as follows:

$$
\begin{pmatrix} u_x^j \\ u_y^j \end{pmatrix} = \begin{pmatrix} u_x^0 \\ u_y^0 \end{pmatrix} + \alpha(-1)^j r \begin{pmatrix} \sin\omega \\ \cos\omega \end{pmatrix}
\tag{9.49}
$$

where u_x^j and u_y^j are the displacements of the tracked points, in the x- and y-directions, u_x^0 and u_x^0 are the displacements of the particle centroid, ω is the particle rotation accumulated over the increment considered, r is the particle size (radius for circular particles), and α is a constant of proportionality to relate the position of the monitoring point to the particle size. All of these variables are easy to monitor in a two-dimensional discrete element analysis. For the three-dimensional case, the coordinates of the two "tracked" points (j=1,2) can be defined to be at the intersection of the principle axis of inertia of the particle with the particle boundary. The time integration approaches for three-dimensional analysis generally keep track of the orientation of the principal axes of inertia. Therefore, as with the two-dimensional case, no additional calculations are required within the body of the discrete element code.

Wang et al. (2007) also explicitly account for rotations in their meshless calculation approach. However, rather than tracking the motion of reference points located on the particle, they consider a background grid of reference points, whose motion depends both on the particle rotations and translations (this is considered further in Section 9.5.4 below).

9.5.2 Best fit approaches

The idea in the best fit approaches is find the equation or curve that most accurately describes the observed particle displacements within the region of the material considered. Once this best fitting curve is found, it can be differentiated to give the deformation gradient. Itasca (2008) and Potyondy and Cundall (2004) fitted a linear expression (i.e. a first-order polynomial) to the velocity data to get the deformation gradient rates ($\dot{u}_{i,j}$). The method can equally be applied to the incremental displacements to calculate actual strains (e.g. Marketos and Bolton (2010)).

To calculate $\dot{u}_{i,j}$, the velocity of each particle relative to the centroid of the selected monitoring region is firstly determined. If the velocity of a particle is denoted $\dot{u}_i^p$ and the position vector is denoted x_i^p, then the mean velocity, $\bar{\dot{u}}_i$, and mean position $\bar{x}_i$ within the region considered are given by

$$\bar{\dot{u}}_i = \frac{\sum\limits_{N_p} \dot{u}_i^p}{N_p}$$
$$\bar{x}_i = \frac{\sum\limits_{N_p} x_i^p}{N_p} \tag{9.50}$$

where N_p is the number of particles within the measurement region and $\sum\limits_{N_p}$ indicates a summation over N_p particles. The velocities and positions of every particle relative to these mean values are given by

$$\dot{u}_i^{p,\mathrm{rel}} = \dot{u}_i^p - \bar{\dot{u}}_i$$
$$x_i^{p,\mathrm{rel}} = x_i^p - \bar{x}_i \tag{9.51}$$

If every grain moved with a uniform deformation rate gradient the following would hold:

$$\dot{u}_i^{p,\mathrm{rel}} = \bar{a}_{ij} x_j^{p,\mathrm{rel}} \tag{9.52}$$

where $\bar{a}_{ij}$ is the deformation rate gradient. For an arbitrary deformation, the objective is to seek $\bar{a}_{ij}$ to minimize the following:

$$\sum\limits_{N_p}(\dot{u}_i^{p,\mathrm{rel}} - \bar{a}_{ij} x_j^{p,\mathrm{rel}})(\dot{u}_i^{p,\mathrm{rel}} - \bar{a}_{ij} x_j^{p,\mathrm{rel}}) \longrightarrow min \tag{9.53}$$

where N_p is the number of particles in the domain, and $|\dot{u}_i^{p,\mathrm{rel}} - \bar{a}_{ij} x_j^{p,\mathrm{rel}}|$ is the magnitude of the difference between the actual particle displacement and the approximate value determined using the best fit expression. The minimization problem is solved using a least squares approach, yielding a system of simultaneous equations for the gradients in direction i. For a 2D system the resultant equations are:

$$
\begin{pmatrix}
\sum_{N_p} x_1^{p,\text{rel}} x_1^{p,\text{rel}} & \sum_{N_p} x_1^{p,\text{rel}} x_2^{p,\text{rel}} \\
\sum_{N_p} x_2^{p,\text{rel}} x_1^{p,\text{rel}} & \sum_{N_p} x_2^{p,\text{rel}} x_2^{p,\text{rel}}
\end{pmatrix}
\begin{pmatrix}
\bar{\dot{a}}_{i1} \\
\bar{\dot{a}}_{i2}
\end{pmatrix}
=
\begin{pmatrix}
\sum_{N_p} \dot{u}_i^{p,\text{rel}} x_1^{p,\text{rel}} \\
\sum_{N_p} \dot{u}_i^{p,\text{rel}} x_2^{p,\text{rel}}
\end{pmatrix}
\tag{9.54}
$$

Cundall and Strack (1979b) noted that the strain rate approach based on instantaneous velocities will only be meaningful if the assembly is deforming steadily.

To get the average deformation gradients $\bar{a}_{ij}$, instead of the average deformation rate gradients, $\bar{\dot{a}}_{ij}$, the incremental relative displacements should be considered. The incremental displacement of a particle over the time increment t_1 to t_2 will be given by $\Delta u_i^p = x_i^{p,t_2} - x_i^{p,t_1}$, where the particle positions at times t_1 and t_2 are x_i^{p,t_1} and x_i^{p,t_2} respectively. Then the mean incremental displacement $(\overline{\Delta u_i})$ and relative incremental displacements $\Delta u_i^{p,\text{rel}}$ are given by

$$
\overline{\Delta u_i} = \frac{\sum_{N_p} \Delta u_i^p}{N_p}
$$
$$
\Delta u_i^{p,\text{rel}} = \Delta u_i^p - \overline{\Delta u_i}
\tag{9.55}
$$

The deformation gradients (again for direction i and in 2D) are found by solving the following system of equations:

$$
\begin{pmatrix}
\sum_{N_p} x_1^{p,\text{rel}} x_1^{p,\text{rel}} & \sum_{N_p} x_1^{p,\text{rel}} x_2^{p,\text{rel}} \\
\sum_{N_p} x_2^{p,\text{rel}} x_1^{p,\text{rel}} & \sum_{N_p} x_2^{p,\text{rel}} x_2^{p,\text{rel}}
\end{pmatrix}
\begin{pmatrix}
\bar{a}_{i1} \\
\bar{a}_{i2}
\end{pmatrix}
=
\begin{pmatrix}
\sum_{N_p} \Delta u_i^{p,\text{rel}} x_1^{p,\text{rel}} \\
\sum_{N_p} \Delta u_i^{p,\text{rel}} x_2^{p,\text{rel}}
\end{pmatrix}
\tag{9.56}
$$

Liao et al. (1997) proposed an alternative best-fit-type approach, where they considered only the contact displacements. Restricting consideration only to the contact displacements will, however not give a good estimate of the overall deformation experienced by the material. Both Bagi and Bojtar (2001) and Cambou et al. (2000) found this method yielded average strain values that underestimated the globally applied strain. Whether the information gained using a "best fit" approach will be representative depends on the heterogeneity of the strain field in the material, i.e. if the average strain over a relatively large volume is calculated, shear bands or localizations may not be picked up.

9.5.3 Spatial discretization approaches

The best fit approaches are applicable when the strain within a RVE is sought. If the local variation of strains at a micro- or meso-scale is of interest, other approaches are preferable. In the "spatial discretization" type approaches (also termed equivalent continuum approaches by Bagi), a graph, or tessellation, connecting the particles is created. While various graph topologies have been used, the edges of each cell in the tessellation are given by lines connecting particle centroids. The incremental displacement gradient is calculated by considering the relative incremental displacement along each edge of the graph. Normally the variation in the displacement values between adjacent nodes is assumed to be linear. Strain values are then assigned to each cell in the graph; these can then be volume-averaged to give a representative value for the RVE, where required.

Triangulation-based approaches

Referring to Figure 9.9(a), Thomas (1997) and Dedecker et al. (2000), proposed approaches based on a simple Delaunay triangulation of the particle centroids (Delaunay triangulation is described in Chapter 1). Over a given increment of deformation, the displacement of each triangle vertex triangles is given by the particle incremental displacements (Figure 9.9(b)). In each triangular element, a linear variation in displacement is assumed. These triangles are directly analogous to the constant strain triangles that can be used in finite element analysis. Referring to Figure 9.10, if the particle coordinates are plotted in the xy plane, we can create a 3D diagram where the displacements are plotted against the vertical axis. A 3D planar triangular surface can be plotted through

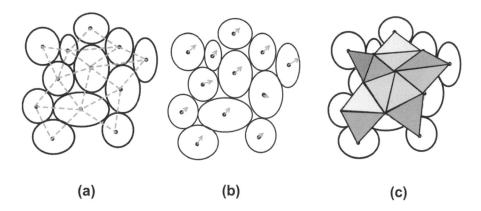

(a)   **(b)**   **(c)**

Figure 9.9: Calculating strain in a DEM simulation

the displacement points and the slopes of this surface in the x and y directions give the displacement gradients.

In this local, linear approach, the gradients are calculated by recognizing that at any point (x, y) within the triangle the displacement vector (u_x, u_y), can be expressed as

$$\begin{aligned} u_x &= \beta_1 + \bar{a}_{11}x + \bar{a}_{12}y \\ u_y &= \beta_2 + \bar{a}_{21}x + \bar{a}_{22}y \end{aligned} \tag{9.57}$$

The coefficients of these two linear equations $\bar{a}_{11}$, $\bar{a}_{12}$, $\bar{a}_{21}$, and $\bar{a}_{22}$ give the displacement gradients, i.e. $\bar{a}_{11} = \bar{u}_{x,x}$, $\bar{a}_{12} = \bar{u}_{x,y}$, $\bar{a}_{21} = \bar{u}_{y,x}$, and $\bar{a}_{22} = \bar{u}_{y,y}$. The terms β_1 and β_2 are constants. By substituting the nodal coordinates ((x^a, y^a), (x^b, y^b), and (x^c, y^c)) and nodal displacements into Equation 9.57, a system of linear equations is created that can easily be solved to get the displacement gradients, i.e.

$$\begin{aligned} \beta_1 + \bar{a}_{11}x^a + \bar{a}_{12}y^a &= u_x^a \\ \beta_1 + \bar{a}_{11}x^b + \bar{a}_{12}y^b &= u_x^b \\ \beta_1 + \bar{a}_{11}x^c + \bar{a}_{12}y^c &= u_x^c \end{aligned} \tag{9.58}$$

The equivalent process is used to get the values of $\bar{u}_{y,x}$, and $\bar{u}_{y,y}$.

O'Sullivan (2002) extended this approach to three dimensions, using a three-dimensional Delaunay triangulation and calculating the displacement gradient by reference to the approach used in

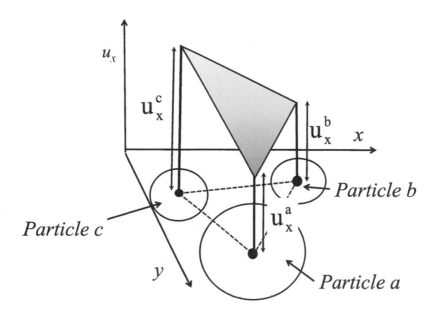

Figure 9.10: Schematic diagram to illustrate calculation of displacement gradients in x and y directions for 2D analyses

constant strain tetrahedra in FEM. The use of tetrahedral elements in three-dimensional finite element analysis is described by Zienkiewicz and Taylor (2000a). Referring to Figure 9.11, the four nodes of the tetrahedra are the particle centroids, or, when rotations are considered, the measured points. Using linear interpolation, the displacement at any point with coordinates (x, y, z) in the tetrahedron can be expressed as

$$
\begin{aligned}
u_x &= \beta_1 + \bar{a}_{11}x + \bar{a}_{12}y + \bar{a}_{13}z \\
u_y &= \beta_2 + \bar{a}_{21}x + \bar{a}_{22}y + \bar{a}_{23}z \\
u_z &= \beta_3 + \bar{a}_{31}x + \bar{a}_{32}y + \bar{a}_{33}z
\end{aligned}
\tag{9.59}
$$

where u_x, u_y and u_z represent the incremental displacements in

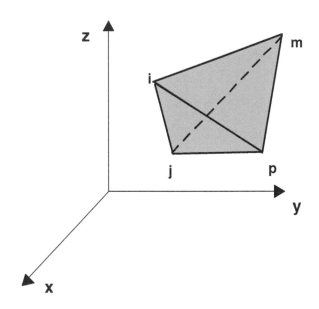

Figure 9.11: Diagram of tetrahedral element

the x, y and z directions respectively, the $\bar{a}_{ij}$ values are the averaged displacement gradients for the tetrahedron, and the β values are constants. Considering the x-component of displacement the $\bar{a}_{ij}$ values can be obtained by solving the following system of equations (which is equivalent to the two-dimensional system given in Equation 9.58):

$$
\begin{pmatrix}
1 & x^a & y^a & z^a \\
1 & x^b & y^b & z^b \\
1 & x^c & y^c & z^c \\
1 & x^d & y^d & z^d
\end{pmatrix}
\begin{pmatrix}
\beta_1 \\
\bar{a}_{11} \\
\bar{a}_{12} \\
\bar{a}_{13}
\end{pmatrix}
=
\begin{pmatrix}
u_x^a \\
u_x^b \\
u_x^c \\
u_x^d
\end{pmatrix}
\tag{9.60}
$$

where (x^a, y^a, z^a) are the coordinates of point a, and u_x^a is the x-component of the incremental displacement of point i. The remaining displacement gradient values are obtained in a similar fashion.

When implementing this approach, it is important to use a consistent order of numbering. Referring to Figure 9.11, for point d the other nodes should be numbered in an anticlockwise order as viewed from d (this is the convention adopted by Zienkiewicz and Taylor (2000a)). For the purposes of contouring, the strain values are taken to be the point values at the tetrahedron centroids.

When the strains are calculated using this triangulation-based approach, the calculation process can also be carried out in a slightly different way by recognizing the similarity with the finite element constant strain triangle (the principle is, however, the same). As already noted, the triangles used to calculate the strain are equivalent to finite elements. In the finite element method a continuous material is discretized into small elements and the displacement at specific points (nodes) in each element is sought. Then the displacement at any point $\mathbf{x}$ in the element, $\mathbf{u}$, can be approximated by interpolating between the nodal displacements $\mathbf{a}^{\text{node}}$ using the expression

$$
\mathbf{u} \approx \mathbf{N}\mathbf{a}^e
\tag{9.61}
$$

where $\mathbf{a}^e$ is a matrix that lists all the nodal displacements for that element and $\mathbf{N}$ is an interpolation function or shape function whose value varies with position **x**. The simplest type of triangular element is the constant strain triangle, and in this case each triangular element has three nodes that are positioned at the triangle vertices. The displacement gradient at any point can then be calculated as

$$u_{i,j} \approx (Na_i^e)_{,j} \tag{9.62}$$

As the nodal displacements are constants,

$$u_{i,j} \approx N_{,j} a_i^e \tag{9.63}$$

The discussion on shape functions is included to illustrate that if we can find a suitable interpolation function that is multiplied by the nodal displacements to calculate the displacements within the material, the strains can then be calculated by taking the product of the derivatives of the interpolation function and the nodal displacements. This concept is used in the local, non-linear approach described in Section 9.5.4 below.

As argued by Thomas (1997) when linear, triangulation-based, approaches are applied to problems where localizations emerge, it can be difficult to define the location of the shear band because plotting strain contours is complicated by the inter-element variation in strain values. O'Sullivan (2002) demonstrated that these limitations were amplified in 3D and discussed the use of smoothing operators to overcome this problem. The simplest type of smoothing is to assign a strain value to each node, which is the average of the strains in each triangle that includes that node, weighted by their area. This approach can give some improvements to the observed values.

Representative results for the linear triangulation approach were obtained by applying the method to a DEM simulation of a biaxial compression test on a specimen of 2,377 unbonded disks. Full details of the simulation parameters are given by Cheung and O'Sullivan (2008). The specimen configuration is illustrated in Figure 9.12(a) and the overall stress-strain response is given in Figure 9.12(b). The position of the localization that develops in the sample post-peak can be appreciated by considering the plot of the cumulative particle rotations up to an axial strain of 10% as illustrated in Figure 9.13(a). The tessellation obtained upon applying Delaunay triangulation to the initial particle positions is illustrated in Figure 9.13(b). This triangular mesh was used as the basis for calculating the strains. The local strains are then plotted on the deformed specimen configuration (at an overall axial strain of 10%). As illustrated in Figure 9.14(a), while there is significant heterogeneity in the local strain values, the dilation along the localizations is clear, and the position of the localization coincides with the zones of peak rotation in Figure 9.13(a). Figure 9.14(b) is included here to quantitatively demonstrate the heterogeneity in the response, while the overall axial strain is 10% the vertical strains (ϵ_{yy}) along the localizations greatly exceed this mean value. It is also interesting to note that there are zones of both local extension and compression evident in the sample.

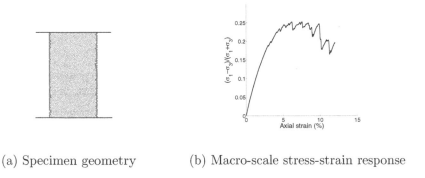

(a) Specimen geometry   (b) Macro-scale stress-strain response

Figure 9.12: Macro-scale response for specimen considered in linear triangulation example

Alternative spatial discretization approaches

As outlined by Duran et al. (2010) Gauss's integral theorem can be applied to relate the average deformation gradient within a system to the displacements along the boundaries,

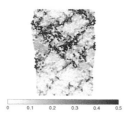

(a) Cumulative particle rotations (axial strain 10%)   (b) Triangular mesh used in strain calculations

Figure 9.13: Cumulative rotations and mesh used to calculate strain for linear triangulation example

$$\overline{u}_{i,j} = \frac{1}{V} \int_V u_{i,j} dV = \frac{1}{V} \oint_S u_i n_j dS \qquad (9.64)$$

where $\oint_S ...dS$ indicates a surface integral along the surface of the domain and n_j is vector normal to the surface. While direct implementation of this approach is non-trivial, a number of approaches use this concept, by looking at the orientation of the edges to cells in a tessellation and the relative displacement of the two nodes forming each edges. Consideration is restricted here to the methods proposed by Bagi (2006). However, other authors including Kruyt and Rothenburg (1996), Kuhn (1999) and Li et al. (2009) have proposed similar algorithms.

Bagi proposed a graph topology to calculate strain that is a dual to the material cell system outlined in Figure 9.5 above. This graph is called the "space cell system." As illustrated in Figure 9.15, the nodes of this space cell system correspond to the particle centroids and the edges are formed by connecting the centres of particles whose material cells have a common edge, i.e. the edges are the branch vectors. Each cell is a triangle in 2D or a tetrahedron in 3D. The duality of the material cell and space cell systems are similar to the duality of the Voronoi and Delaunay tessellations. However, in contrast to the material cell edges, which pass

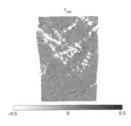

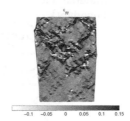

(a) Volumetric strains (axial strain 10%)

(b) Local axial strains (overall axial strain 10%)

Figure 9.14: Local volumetric and direct (ϵ_{yy}) strains calculated using linear triangulation-based approach.

through the particle contacts and do not overlap with the solid particles, if a Delaunay triangulation were created based on the particle centroids, the resultant Voronoi diagram edges would intersect the particles. Then Bagi (1996) showed that the average deformation gradient, ($\bar{a}_{ij}$), within a cell in the space cell system (with volume V_{SC}) is then given by

$$\bar{a}_{ij} = \frac{1}{V_{SC}} \sum_{k=1}^{N_{e,SC}} d_i^k \Delta u_j^k \qquad (9.65)$$

where the summation is made over the number of edges ($N_{e,SC}$) in the considered space cells and is carried out in a consistent manner (i.e. consistently clockwise or counterclockwise). Δu_j^k is the relative incremental displacement of the two grain centers of the edge k, i.e. referring to Figure 9.15, for the edge connecting the centre of particle a and the centre of particle b, $\Delta u_j^{ab} = \Delta u_j^b - \Delta u_j^a$. The vector d_i^c is the complementary area vector. The complementary area vector is a vector that characterizes the local geometry of the neighborhood of the kth edge, and the direction of a representative complementary area vector is illustrated in Figure 9.15(b), its magnitude will be one third the length of the line segment cd. The sum of the products of the branch vectors l_i that form the

345

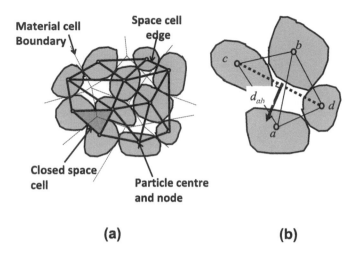

(a)   **(b)**

Figure 9.15: Schematic diagram of space cell system proposed by Bagi (1996)

space cell edges and the vectors d_i for every edge in a given domain equals three times the volume (3D) or twice the area (2D) of the domain, i.e. in the 2D case the area of the domain, A, is given by

$$A = \frac{1}{2} \sum_{k=1}^{N_{e,SD}} l_i^k d_i^k \tag{9.66}$$

where $N_{e,SD}$ is the total number of edges in the subdomain.

In Bagi's space cell system and the triangulation approach proposed above, the particles whose vertices are connected along one edge of the cell may or may not contact. Kuhn (1999) and Kruyt and Rothenburg (1996) only considered edges along which the particles are contacting, generating the tessellation illustrated in Figure 9.16. Expressions for deformation gradient/strain have also been proposed for this tessellation by these authors.

As noted by Li and Li (2009), where the spatial discretization approach is used to calculate the average strain for individual cells in selected volume, the average displacement gradient in the domain is calculated as a volume-weighted average:

$$\bar{a}_{ij} = \frac{1}{V} \sum_{N_{SD}} V^M a_{ij}^M \tag{9.67}$$

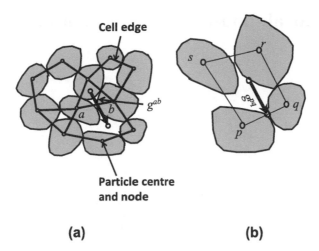

(a) **(b)**

Figure 9.16: Schematic diagram of tessellation system proposed by Kruyt and Rothenburg (1996)

where V^M is the volume of cell M, a_{ij}^M is the displacement gradient in cell M, and $\sum\limits_{N_{SD}}$ indicates a summation over all the cells in the volume.

The constant strain triangle approach outlined above is easier to implement then the edge-based methods of Bagi (2006) and Kruyt and Rothenburg (1996). However, these edge-based methods present an interesting alternative, as they allow a strain to be associated with sets of particles that define the void edges.

9.5.4   Local, non-linear interpolation approach

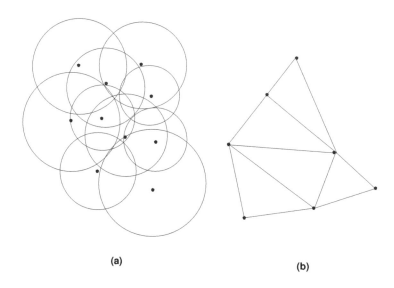

(a)   (b)

Figure 9.17: Comparison of (a) mesh-free approach using nodes and regions of influence with (b) the nodes and elements used in linear interpolation.

The approaches considered to date have used linear interpolation or linear curve fitting. When presented with a set of data, it can be rather limiting to restrict consideration to an assumption of a linear variation of strain across the system or between the particle centroids. O'Sullivan et al. (2003) proposed a local, non-linear approach that uses elements of the algorithms for mesh free methods (e.g. Li and Liu (2000)). The specific method considered by O'Sullivan et al. was the reproducing kernel particle method (RKPM) (Liu et al., 1995). Similarly to the finite element method, the meshless methods were developed to model materials as continua.

The linear triangulation approach is contrasted with the mesh-free approach in Figure 9.17. Where the triangular elements are used to calculate the strain, each particle forms a vertex of at least three triangles and the strains within each of these triangles are calculated independently. In the meshless approach, each particle is assigned its own "region of influence" over which it contributes to the calculated strains. Typically the region of interest is either a circular (2D) or spherical (3D) region (Figure 9.17(b)). Conceptually, it is easy to imagine that within its region of influence the particle's contribution to the strain field should decrease as the distance from the particle centroid increases. A number of wavelet functions can capture this type of response, while also meeting the requirement that the interpolation function describing the displacement field be continuous and differentiable over the region of influence so that the strains can be calculated. The advantage of this method is that it provides a smooth interpolation basis capable of capturing the high deformation gradient field that can exist within the shear bands (and hence the strain field), while also eliminating the high inter-element variation in strain values associated with the triangulation (or other cell-based) homogenization approaches.

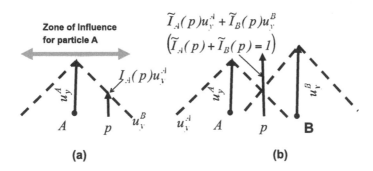

Figure 9.18: Schematic diagram illustrating principles of local, non-linear interpolation

The concept of the local, non-linear interpolation approach is given in Figure 9.18. As illustrated in Figure 9.18(a) if a point P is within the zone of influence of particle A, then its displacement will be the product of the function I_A evaluated at point P and the displacement of particle A (in this case u_y^A), i.e. $u_y^P = I_A(P)u_y^A$. The function I_A describes how the influence of particle A on the calculated displacements varies within its zone of influence. While a linear function is chosen to represent the distribution of the influence in Figure 9.18, non-linear wavelet functions (e.g. cubic or Gaussian functions) can also be used. If, as illustrated in Figure 9.18(b), the point P is also within the zone of influence of particle B then the displacement at P will be given by $u_y^P = \tilde{I}_A(P)u_y^A + \tilde{I}_B(P)u_y^B$, where the influence functions I_A and I_B are scaled so that $\tilde{I}_A + \tilde{I}_B = 1$.

In summation form, the displacement vector, $\mathbf{u}^x$, at an arbitrary point x with position vector $\mathbf{x}$, can be expressed in terms of the nodal or particle displacements, $\mathbf{u}^p$, at nodal positions $\mathbf{x}^p$ as follows

$$\mathbf{u}^x \simeq \sum_{i=1}^{N_p} K_\varrho(\mathbf{x} - \mathbf{x}^p)\mathbf{u}^p \Delta V^p \tag{9.68}$$

where N_p is the number of nodes, i.e. particles, whose zones of influence include the point, ΔV^p is the nodal weight, and the term $K_\varrho(\mathbf{x} - \mathbf{x}^p)$ is given by

$$K_\varrho(\mathbf{x} - \mathbf{x}^p) = C_\varrho(\mathbf{x} - \mathbf{x}^p)\Phi_\varrho(\mathbf{x} - \mathbf{x}^p) \tag{9.69}$$

where $C_\varrho(\mathbf{x} - \mathbf{x}^p)$ is a correction function to reduce the interpolation error, $\Phi_\varrho(\mathbf{x} - \mathbf{x}^p)$ is the compact kernel function, and ϱ is the dilation parameter that defines the size of the window function. The compact kernel function is generally defined as:

$$\Phi_\varrho(\mathbf{x} - \mathbf{x}^p) = \frac{1}{\varrho}\Phi\left(\frac{\mathbf{x}-\mathbf{x}_p}{\varrho}\right)\begin{cases} > 0; \frac{\mathbf{x}-\mathbf{x}^p}{\varrho} \leq 1 \\ = 0; \frac{\mathbf{x}-\mathbf{x}^p}{\varrho} > 1 \end{cases} \tag{9.70}$$

The correction function, $C_\varrho(\mathbf{x} - \mathbf{x}^p)$, is given by

$$C_\varrho(\mathbf{x} - \mathbf{x}^p) = \mathbf{P}\left(\frac{\mathbf{x} - \mathbf{x}^p}{\varrho}\right)\mathbf{b}\left(\frac{\mathbf{x}}{\varrho}\right) \tag{9.71}$$

where the vector $\mathbf{P}(\mathbf{x})$ is a given function and the vector $\mathbf{b}(\mathbf{x})$ is an unknown function that is sought to suit the local particle distribution.

For the implementation presented here the window function, $\phi(x)$, is a wavelet function given by Daubechies (1992). In one-dimensional form the function is given by:

$$
\begin{aligned}
\phi(x) &= \tfrac{1}{6}(x+2)^3 & -2 \leq x \leq -1 \\
\phi(x) &= \tfrac{2}{3} - x^2(1+x/2) & -1 \leq x \leq 0 \\
\phi(x) &= \tfrac{2}{3} - x^2(1-x/2) & -1 \leq x \leq 0 \\
\phi(x) &= \tfrac{1}{6}(x-2)^3 & -2 \leq x \leq -1 \\
\phi(x) &= 0 & otherwise
\end{aligned}
\tag{9.72}
$$

In higher dimensions the shape functions are simply calculated as follows:

$$
\begin{aligned}
\phi(x,y) &= \phi(x)\phi(y) \\
\phi(x,y,z) &= \phi(x)\phi(y)\phi(z)
\end{aligned}
\tag{9.73}
$$

The two-dimensional shape function described in Equation 9.72 is plotted in Figure 9.19, the first derivatives of this function (used in the calculation of the displacement gradients) are illustrated in Figure 9.20.

Figure 9.19: Diagram of wavelet function used in non-linear interpolation (2D)

The incremental volume, ΔV^p, (Equation 9.75), associated with each particle p is calculated by triangulating the system and calculating the incremental volume as

$$
\Delta V^p = \frac{1}{N_v} \sum_{k=1}^{N_T} \Delta \Omega_k
\tag{9.74}
$$

351

Partial Derivative With Respect to x

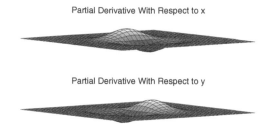

Partial Derivative With Respect to y

Figure 9.20: Diagram of first partial derivatives of wavelet function used in non-linear interpolation (2D)

where $\Delta\Omega_k$ represents the area of a triangle/tetrahedron k with a vertex at point p, N_T is the total number of triangles or tetrahedra with vertices at point p, and N_v is the number of vertices per triangle or tetrahedra ($N_v = 3$ in 2D and $N_v = 4$ in 3D).

The displacement gradients are then calculated by considering the gradient of the shape function $K_\varrho(\mathbf{x} - \mathbf{x}^p)$, i.e.

$$\frac{\partial \mathbf{u}(\mathbf{x})}{\partial x} \simeq \sum_{i=1}^{N_p} \frac{\partial K_\varrho(\mathbf{x} - \mathbf{x}^p)}{\partial x} \mathbf{u}^p \Delta V^p \tag{9.75}$$

A schematic diagram of the approach used to implement the local, non-linear interpolation is illustrated in Figure 9.21. A rectangular grid is generated to serve as a referential continuum discretization over the volume of particles under consideration, i.e. the strains are calculated at the grid points. The area of influence of each particle is a multiple of the particle radius.

For each particle the distance d to each grid point is calculated. If $d < w_r$ for that particle, the contributions to the interpolated grid displacements are evaluated using Equation 9.72, in combination with Equation 9.75. The variable x is given by d/w_r. If $w_r < d < 2w_r$ the contribution to the interpolated grid displacement is calculated using Equation 9.72, in combination with Equation 9.75. The displacement values at grid points located a distance of more than $2w_r$ from the particle under consideration are not influenced by that particle.

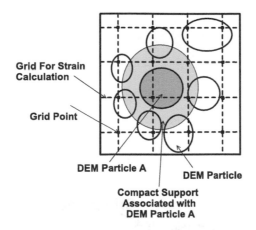

Figure 9.21: Illustration of grid used in local, non-linear interpolation

Wang et al. (2007) also proposed a grid-based approach to calculate incremental displacements and strains. Their approach differs from the non-linear approach proposed by O'Sullivan et al. (2003) as each grid point is assigned only to one particle. Referring to Figure 9.22, the displacement $\mathbf{u}^g$ of each grid point depends on both the particle displacements and rotations (as noted above) and is given by

$$u_x^g = u_x^p + d\left[\cos(\theta^0 + \omega) - \cos(\theta^0)\right] \\ u_y^g = u_y^p + d\left[\sin(\theta^0 + \omega) - \sin(\theta^0)\right] \tag{9.76}$$

This approach is therefore significantly easier to implement than the approach proposed by O'Sullivan (2002). However, in contrast to other approaches, the measured displacement field is not continuous as the translational displacement of the grid points equals the displacement of their associated particle.

To illustrate the applicability of the non-linear interpolation approach reference is made to a simulation carried out by Cheung (2010). The sample tested in the simulation had a diameter of 40mm and a height of 80mm. It comprised 12,622 spheres with sphere radii being uniformly distributed between 0.88mm and 1.32mm. The simulation was performed using the commercial code

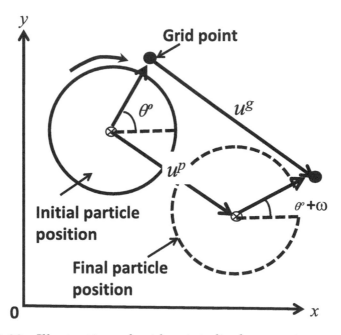

Figure 9.22: Illustration of grid point displacements assumed by Wang et al. (2007)

PFC 3D . The spheres were firstly randomly generated, then two horizontal platens and a cylindrical wall were created to bound the specimen. By moving these walls in a controlled manner using a servo-controlled technique, the specimen was brought to an initial isotropic stress state of 10MPa. Once the initial isotropic stress state was achieved, parallel bonds were installed at all the inter-particle contacts that were found at this stage. The parallel bond model is described by Potyondy and Cundall (2004) and is also considered in Chapter 3. These parallel bonds were therefore uniformly distributed throughout the specimen. The sample was then subject to a triaxial compression test where the cylindrical wall forming the lateral boundary to the specimen was continuously adjusted to maintain a constant horizontal stress while the top boundary moved downwards.

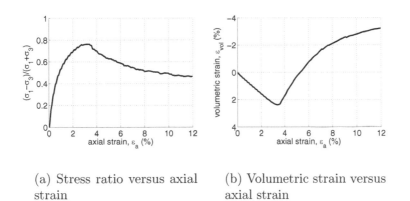

(a) Stress ratio versus axial strain

(b) Volumetric strain versus axial strain

Figure 9.23: Macro-scale stress-strain response for cemented sand specimen (Cheung, 2010)

The overall specimen response is illustrated in Figure 9.23. The specimen reached a peak stress ratio of 0.76 at 3.5% axial strain. Strain softening was observed after peak and the rate of softening decreased with increasing axial strain. The volumetric strain results showed that the specimen contracted prior to peak and dilated during strain softening.

The local volumetric and shear strains at axial strain values of 5% and 10% are presented in Figure 9.24, considering a representative "slice" through the centre of the sample. Referring to Figure 9.25 there is a clear correlation between the points where the strain is a maximum and the zone of most intense disturbance within the sample (where the bonds break and there are notably high particle rotations). It is clear that by calculating the strains, the heterogeneity of the specimen response along the localization can be appreciated, giving insight into the post-peak response of brittle granular materials. The extent of the variation in the response along the localization is not evident in the other plots. There is scope to analyse the constitutive response along the shear band in more detail by coupling the local strain calculations with local measurements of stress.

Validation and comparison of strain calculation algorithms

As described by Bagi (2006) amongst others, evaluations of strain calculation methods have compared the strains calculated with the homogenization method and the global, overall strains. An example of such a validation is given in Figures 9.26 and 9.27 where the validations of the 3D implementation of the triangulation-based linear interpolation approach and the local non-linear approach proposed by O'Sullivan and Bray (2003b) are considered. A servo-controlled triaxial compression test on a sample of 9,000 spheres enclosed within 6 rigid boundary walls was considered. The overall specimen response is illustrated in Figure 9.26(b). The average displacement gradients ($\overline{a}_{11}$, $\overline{a}_{22}$, and $\overline{a}_{33}$) obtained using the rotational discretization described above closely approximated overall, global values as illustrated in Figure 9.27.

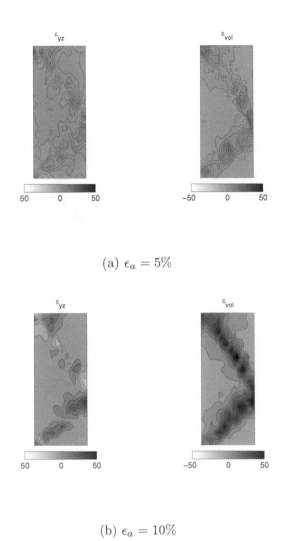

(a) $\epsilon_a = 5\%$

(b) $\epsilon_a = 10\%$

Figure 9.24: Local volumetric and shear strains for cemented sand model at $\epsilon_a = 5\%$ and $\epsilon_a = 10\%$ (Cheung, 2010)

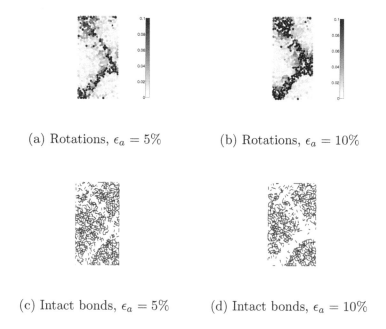

(a) Rotations, $\epsilon_a = 5\%$ (b) Rotations, $\epsilon_a = 10\%$

(c) Intact bonds, $\epsilon_a = 5\%$ (d) Intact bonds, $\epsilon_a = 10\%$

Figure 9.25: Particle-scale response for cemented sand specimen: particle rotations and intact bonds (Cheung, 2010)

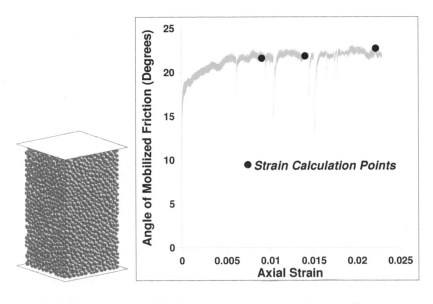

(a) Initial particle configuration

(b) Specimen response for a 3D servo-controlled triaxial test.

Figure 9.26: Validation of 3D kinematic averaging approaches

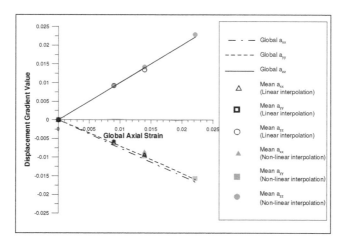

Figure 9.27: Results of validation of three-dimensional kinematic averaging approaches: Comparison of global displacement gradients and average displacement gradients for linear and non-linear interpolation with "rotational" discretization.

Chapter 10

Analysis of Particle System Fabric

Experimental observation of the way sand grains are packed around each other and how this packing evolves during loading is difficult. In a DEM simulation detailed information on the internal structure of a granular material can be accessed at any point. As already established in Chapter 8, the topology of the system of particles and contacts is highly complex. A means to unravel or make sense of these data is needed to establish the link between the particle scale mechanics and the overall macro-scale response. This Chapter describes procedures to quantify the internal structure of a granular material, considering the packing density of the of particles and the preferential orientations of the particles and contacts. Many techniques can be applied to analyse the data available for output from a DEM simulation; here the more commonly used approaches are presented and the information provided should hopefully point DEM analysts in the right direction as they try to make sense of their simulation results.

10.1 Conventional Scalar Measures of Packing Density

In traditional soil mechanics the parameters void ratio, specific volume and porosity are used to quantify the density of packing of the granular material. While these parameters are defined in any undergraduate soil mechanics textbook (e.g. Craig (2007) or Atkinson (2007)), they are repeated here for clarity and in particular for the benefit of DEM users from other disciplines. As outlined in Chapter 12, the response of a granular material is largely determined by its "state." The state is quantified by considering the void ratio (or specific volume) and the mean effective stress. Just as in experimental geomechanics, knowledge of the void ratio is important to be able to interpret the overall response of the material in DEM simulations.

The measures of packing density that are commonly used in geomechanics are based upon the relationship between the volume of solid particles in the material and the overall volume occupied by the granular materials. In experimental work, the dry mass of the material is measured. If the density of the solid material is known, the volume of solid soil particles can be calculated. (If unknown it can be measured using pycnometer testing.) The void ratio, normally denoted e is then given by

$$e = \frac{V_v}{V_s} \tag{10.1}$$

where V_v is the volume of voids and V_s is the volume of solid soil particles. The specific volume, denoted v, is the total volume occupied by the material per unit solid volume and is given by

$$v = \frac{V}{V_s} \tag{10.2}$$

The parameters v and e are thus related as

$$v = 1 + e \tag{10.3}$$

The final basic characterization of packing density is the porosity n, which is defined as the ratio of the volume of voids to the total volume occupied by the soil, i.e.

$$n = \frac{V_v}{V} \qquad (10.4)$$

The porosity and void ratio values are related as follows

$$n = \frac{e}{1+e}$$
$$e = \frac{n}{1-n} \qquad (10.5)$$

Outside of geomechanics the term "solid volume fraction" is often used and is defined as (e.g. Bedford and Drumheller (1983))

$$\nu = \frac{V_s}{V} \qquad (10.6)$$

All of these approaches give (scalar) measurements of the material packing density. They can all be easily calculated in DEM simulations as the volumes and areas occupied by the particles are known. The packing density will vary with the particle size distribution. In a real soil with a broad range of particle sizes, the smaller particles will occupy the voids between larger particles, giving lower void ratios than in the case of the almost uniformly sized particles often considered in DEM analyses. Furthermore, the particle geometry will influence the range of attainable void ratios. As noted in Chapter 12, many DEM simulations are two-dimensional. Therefore measures of the porosity, void ratio and specific volume will be in terms of area, rather than volume. The range of attainable void ratios differs for 2D and 3D materials, with void ratio values usually being smaller for 2D simulations. For example in two dimensions the most dense packing configuration for uniform disks (hexagonal packing) has a void ratio of 0.103, while in the three-dimensional case for uniform spheres (both hexagonal close packing or face-centred-cubic packing) the minimum void ratio is 0.4 (i.e. a packing density of 0.7405).

The packing within the assembly will be inhomogeneous. Marketos and Bolton (2010) highlighted the fact that the packing density will be influenced by the presence of flat boundaries and a local decrease in porosity will occur close to the boundary. Kuo and Frost (1996) present data indicating measurable particle-scale

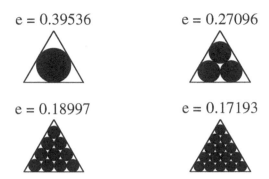

Figure 10.1: Illustration of void ratio variation with size of measurement volume

heterogeneity within sand samples by measuring the void ratios (in 2D) within resin-impregnated specimens of real sand. If a shear band forms, additional heterogeneity will be introduced. Referring to the discussion on heterogeneity in Chapter 9, the concept of a representative volume element (RVE) applies to calculation of void ratio and quantification of the material microstructure. This is simply illustrated in Figure 10.1 which indicates how the void ratio of specimens of uniform disks placed within an equilateral triangle with the same arrangement varies as a function of the number of spheres. Munjiza (2004) considered this problem in a series of DEM simulations and showed that for specimens of uniform spheres deposited under gravity, the measured packing density tended to increase as the ratio of the sphere diameter to the container side length decreased. Therefore to get statistically representative measures of packing density care must then be taken to ensure that the RVE used is sufficiently large. Referring to Figure 10.2, just as was the case when calculating stress, when measuring the void ratio in subregions within the assembly, the intersection of particles with the region boundary should be considered (Bardet and Proubet, 1991).

DEM simulations allow for alternative definitions of void ratio. Referring to Section 8.5, within the granular material the

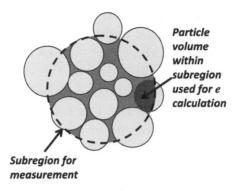

Figure 10.2: To accurately calculate void ratio within the sample the intersection of the disks with the boundary edge should be considered

force transmission is highly heterogeneous and there are particles that do not participate in stress transmission. Recognizing this Kuhn (1999) proposed an effective void ratio including only those particles that participate in stress transmission when calculating the volume of solids.

10.2 Coordination number

The void ratio quantifies the particle packing density without explicit consideration of the particulate structure; only the total mass of particles, the solid particle material density and the overall material volume are required. The coordination number quantifies the number of contacts per particle in the material and it gives a measure of the packing density or packing intensity at the scale of the particles. The simplest definition of the coordination number Z is

$$Z = 2\frac{N_c}{N_p} \tag{10.7}$$

where N_c is the total number of contacts and N_p is the number of particles. The number of contacts is multiplied by 2 as each contact is shared between two particles. While it is standard practice to refer to the void ratio as e and the porosity as n, the notation used to describe the coordination number varies. For example Rothenburg and Kruyt (2004) denote the coordination number using the symbol Γ, while here the symbol Z is used following the notation of Thornton (2000). The coordination number is the most basic particle-scale measure of the material structure and it can easily be determined from DEM simulation data. Referring to Chapter 3, the total number of contacts N_c considered in calculation of the coordination number should include only engaged contacts, and not the potential contacts between particles that are close, but not actually transmitting an inter-particle force.

Just as the evolution of void ratio is typically quantified and analysed when considering data from physical laboratory experiments, it is usual to record the evolution of Z during DEM simulations of element tests. As an example, the variations of both vertical strain and coordination number with displacement in a simulation of the direct shear test (Cui and O'Sullivan, 2006) are illustrated in Figure 10.3. Note that at a strain level of 20%, while the vertical strain data indicate that the sample continues to dilate, the coordination number appears to have reached a more constant value.

Modified, or refined, definitions of the coordination number exist. For example Thornton (2000) defined a mechanical coordination number Z_m to be

$$Z_m = 2\frac{N_c - N_p^1}{N_p - (N_p^1 + N_p^0)} \qquad (10.8)$$

where N_p^1 is the number of particles with one contact, and N_p^0 is the number of particles with no contacts. These particles cannot participate in transmitting stress through the material and are often termed "rattlers" or "floaters." Kuhn (1999) applied a higher level of discrimination in selecting particles for calculation of his coordination number Z_p, termed the *effective coordination number*. Referring to Figure 10.4, Kuhn identifies the pendant, island, peninsula and isolated particles as particles that do not participate in the load-bearing framework. He excludes these particles

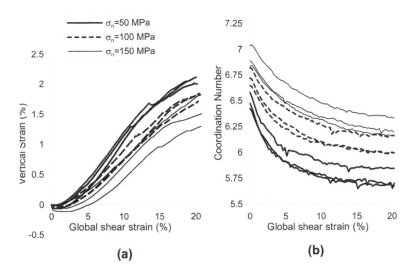

Figure 10.3: Macro-scale deformation and coordination number variation in direct shear test simulations Cui and O'Sullivan (2006)

and their contacts when calculating Z_p.

Other means to quantify the intensity of contacts in the material have been proposed in addition to the coordination number Z. For example Rothenburg and Bathurst (1989) considered the contact density, denoted m_v which is defined to be

$$m_v = \frac{2N_c}{V} \qquad (10.9)$$

where V is the volume of material under consideration.

Degree distribution

A concept that is complementary to the coordination number is the degree distribution. As noted in Chapter 8, if a granular material is considered from the perspective of network analysis, each particle is a node and each contact is a connection. The connectivity or degree of a particle is the particle's own coordination number, i.e. the number of contacts it participates in. Kuhn (2003a) uses the term *valance* for this parameter. The degree distribution, $P_c(k)$, is a function that gives the probability that a given particle

367

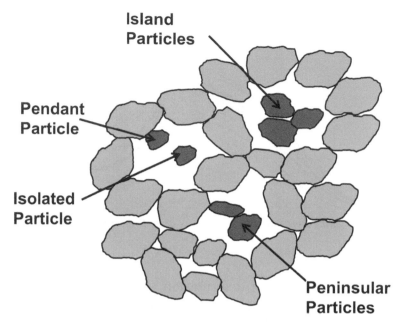

Figure 10.4: Definition of participating particles by Kuhn (1999)

will have k contacts. The average degree equals the coordination number, i.e.

$$Z = \sum_{k=1}^{\infty} kP_c(k) \tag{10.10}$$

As demonstrated by Summersgill (2009), for example, in a polydisperse material (i.e. a granular material with a range of sizes), the individual particle coordination number values tend to decrease with particle size. Consequently the smaller particles are significantly less likely to be participating in the strong force chains in comparison with the larger particles. Oda et al. (1980) suggested looking at the standard deviation of the particle coordination numbers to assess the degree of heterogeneity in the system.

Another interesting measure is described by Wouterse et al. (2009) who introduced the term "caged" to describe the case of particles whose movement is curtailed by neighbouring particles. The caging number is defined as the minimum average number of contacts required to immobilize a particle, assuming these contacts are randomly located.

Experimental determination of Z

An understanding of the relationship between the coordination number and the overall material response predates the development of DEM. For two-dimensional photoelastic systems the coordination number can be relatively easily determined by visual inspection. Experimental determination of Z for three-dimensional materials is more challenging. Oda (1977) described a study using glass ballotini in drained triaxial compression tests. At different points in the tests a black paint was permeated into the voids between the particles; this sample was then drained, leaving a small amount of paint at each contact. Then, following drying, samples of particles were extracted and the number of contact points observed on each particle was noted. Using this methodology Oda demonstrated that there is a correlation between the average coordination number combined with the standard deviation of the individual particle coordination numbers and the peak angle of shearing resistance (angle of internal friction) for the material. This experimental study was clearly time-consuming and tedious and, while it includes data useful for verifying DEM results, it is also a good case study to illustrate why DEM is such an attractive tool for geomechanics studies. More recent experimental studies on coordination number have used micro-computed tomography (μCT) scans of specimens of sand impregnated with resin (Hasan and Alshibli, 2010).

Relation between e and Z

Given the importance of void ratio (e) in geomechanics, it is logical to seek a relationship between Z and e. Obviously the coordination number will increase as the void ratio decreases and vice versa. Oda (1977,1999b), Oda et al. (1980) and Chang et al. (1989) all cite experimental studies where the coordination number of a real (physical) granular materials was determined and compared with

369

the void ratio. Based upon the data obtained in these studies relationships were been proposed between these e and Z; for example Field (1963) related the coordination number and the void ratio as follows:

$$Z = \frac{12}{1+e} \qquad (10.11)$$

Mitchell (1993) gave the following empirical equation to relate coordination number Z and porosity n for the case of uniform rigid spheres as follows:

$$Z = 26.386 - \frac{10.726}{n} \qquad (10.12)$$

Thinking about a real sand, and also about ideal DEM particles, it is difficult to accept that a single analytical expression can describe the relationship between Z and either e, v or n for all granular materials. The relationship will clearly depend on the particle morphology, both shape and surface roughness, the variations in particle morphology within the material, as well as the distribution of particle sizes. Some data to support the idea that e and Z are not simply related is given by Rothenburg and Kruyt (2004), who in a series of biaxial compression test simulations, demonstrated that the relationship between void ratio and coordination number is sensitive to the anisotropy of the contact orientations, and that the use of a single relationship may not be appropriate. Using micro-computed tomography scans on real sand specimens, Hasan and Alshibli (2010) found that the relationship between e and Z for their sand differed from the empirical expressions developed using ideal materials. The data of Rothenburg and Kruyt (2004) and Thornton (2000) suggest that just as at large strain samples tend towards a critical void ratio, where the density remains constant as the material is sheared, a "critical coordination number" also exists.

Outside of geomechanics, the relationship between void ratio and coordination number has been considered in studies of jamming of flowing particles. The "jamming transition" is considered to be the point where the material transitions from a fluid-like behaviour to a solid-like behaviour (e.g. O'Hern et al. (2003)).

Coordination number for real sands

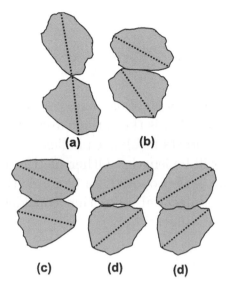

Figure 10.5: Schematic diagram of different types of contacts, after Barton (1993)

As noted in Chapter 4, where smooth convex particles are used, contact between particles is restricted to a single point. The definitions of conforming and non-conforming contacts were introduced in Chapter 3. The idea of a non-conforming, single point contact between particles does not hold for real sands. In the case of densely packed angular particles the contact areas may represent a substantial proportion of the particle surface areas. Then describing the density of inter particle contacts simply using a coordination number is probably inadequate. Figure 10.5 illustrates different types of contacts that can be observed in real sands (drawing on ideas presented by Barton (1993)). Figure 10.5(a) illustrates a tangential contact, (b) a straight contact, (c) a concavo-convex contact, (d) a non-continuous contact and (e) a sutured contact where the particles are connected into each other like jigsaw pieces.

Analysis of micro-computed tomography data on real sand (e.g. Fonseca et al. (2010)) provides 3D evidence of the contact configurations described by Barton (1993). Fonseca et al. (2010) propose that rather than using Z, it may be more appropriate to use a contact index parameter CI to quantify the contact intensity. This contact index is defined as

$$CI = \frac{1}{N_p} \sum_{i=1}^{N_p} \frac{1}{Sp_i} \sum_{j=1}^{N_{c,i}} Sc_j \qquad (10.13)$$

where N_p is the total number of particles considered, Sp_i is the surface area of particle i, Sc_j is the surface area of contact j and $N_{c,i}$ is the number of contacts involving particle i. In two dimensions the surface areas will be lengths. Other metrics for application to real materials have been proposed; for example Barton (1993) proposed the use of a tangential index that considers the percentage of contacts that are tangential.

Redundancy of packing

Civil engineers will be familiar with the concept of redundancy; in a redundant structure there are additional load-carrying elements beyond those needed for basic stability. This idea of redundancy has also been applied in granular mechanics. For example, Maeda (2009) describes 2D packings where the coordination number exceeds 3 as being "hyper-static." Rothenburg and Kruyt (2004) related the stability of the material structure to the coordination number, arguing that in the case of a frictionless system static equilibrium can only be achieved if at least $2N$ contacts exist within a system of N disks (i.e. Z should be ≥ 4). In the case of frictional contacts where shear contact forces impart a moment to the system, static equilibrium or stability requires a minimum coordination number of $Z = 3$.

More recently, Kruyt and Rothenburg (2009), quantified the redundancy of the system by comparing the total number of equilibrium equations with the degrees of freedom at the contacts. In their 2D study they restricted consideration to the particles with 1 or more contacts (i.e. zero-contact rattlers were excluded). The number of elastic contacts (i.e. contacts where the shear force does not exceed the shear resistance given by Coulomb friction) is given by N_c^{el}, while the number of frictional (sliding) contacts is given by N_c^{fr}, and $N_c^{\mathrm{el}} + N_c^{\mathrm{fr}} = N_c$, where N_c is the total number of contacts. The redundancy factor, R, is then given by the ratio of the number of degrees of freedom at the contact points divided by the number of governing equations, as follows

$$R = \frac{2N_c^{\mathrm{el}} + N_c^{\mathrm{fr}}}{3(N_p - N_p^0)} \qquad (10.14)$$

where N_p^0 is the number of rattlers and $R \geq 1$. In a biaxial compression test Kruyt and Rothenburg found that the initial R value of about 1.4, rapidly dropped to 1 and remained at this value as shearing progressed, indicating that the system tended towards a state of static equilibrium at larger strain values. Note that other definitions of redundancy exist, (e.g. Satake (1999)).

10.3 Contact Force Distribution

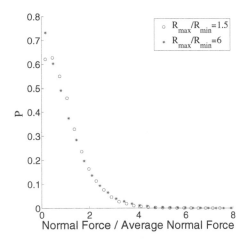

Figure 10.6: Example probability distributions for normal contact force data

The coordination number indicates only the contact intensity and gives no information on the contact force magnitudes. As already outlined in Chapter 8, there is a significant variation in the magnitude of the contact forces. A number of researchers have analysed the contact force data by considering the probability distribution of the magnitude of the contact forces. An example plot is given in Figure 10.6. The probability of a contact transmitting a given force is plotted against the vertical (logarithmic) axis, while along the horizontal axis the contact force values are given (usually normalized by the average contact force). Two simulations of isotropic compression of polydisperse spheres with different size ranges are considered. Clearly there are a large number of contacts transmitting very small forces.

Probability distribution functions can be fitted to the data. Radjai et al. (1996) analysed the distribution of contact forces in an assembly of polydisperse disks deposited under gravity and then subject to shearing. They found that for forces that exceeded the average force, an exponential function gave a good fit to the simulation data. Thornton (1997b) examined the distribution of the contact forces within his specimens of rigid spheres and compared the data with various analytical expressions for the probability distribution. Thornton observed a change in shape in the distribution, with a exponential distribution being valid if only contacts transmitting more than twice the average force are considered. The work of Voivret et al. (2009) is particularly interesting from a geomechanics perspective as they consider the sensitivity of the force distribution to the particle size distribution for highly polydisperse 2D disk specimens. They found that the range of contact force values increases as the particle size distribution increases.

In relation to the tangential component of the contact forces, a number of authors have given consideration as to whether the contacts are sliding or not. For example Cundall et al. (1982) observed that particle sliding tends to occur outside of the strong force chains. Thornton and Antony (2000) and Thornton (2000) considered the proportion of sliding contacts in their system. They found that following an initial increase, the proportion of sliding contacts remains essentially constant after a small amount (1%) of deviator strain and that the proportion seemed to be independent of the initial packing density.

374

10.4   Quantifying Fabric at the Particle Scale

Motivation

It is generally accepted that the density of a sand at a particular stress level is the most important factor determining its behaviour. (This is considered again in Chapter 12, where the critical state soil mechanics framework is introduced). The void ratio and (at a particle scale) the coordination number are scalar measures used to quantify the material packing density. However, as already illustrated in Chapter 8, under certain conditions the contacts in granular material display clear biases in their orientation. Furthermore, when deposited under gravity particles tend to fall so that their long-axis orientation is horizontal. Some means of quantifying or describing the extent and influence of general biases in the topology or geometry of the particulate assembly is needed. As clarified by Mitchell (1993), the term *fabric* "refers to the arrangement of particles, particle groups and pore spaces in the soil." The terms "fabric" and "structure" are often used in an interchangeable manner; however, Mitchell suggests that *structure* be used to account for the effects of fabric, composition and inter-particle forces (including bonding) on soil response.

Experimental evidence has clearly indicated that soil strength and stiffness are anisotropic (i.e. they vary depending on the direction in which the soil is deformed). Early evidence of anisotropic soil response was given in the cubical cell experiments of Arthur and Menzies (1972). In their experiments they varied the direction of loading relative to the orientation of the sample during particle deposition, and found differences of well over 200% in the axial strains taken to reach a given (prepeak) stress ratio. The extent of a soil's anisotropy can be examined in significant detail in a hollow cylinder apparatus (HCA) where the principal stress axes orientation can be controlled. For example, using HCA tests, Zdravkovic and Jardine (1997) showed the non-linear stiffness characteristics of a quartzitic silt to be dependent on the stress path direction and the orientation of the major principal stress axis to the vertical. The effects of fabric are evident even in "simple" granular materials. Shibuya and Hight (1987) found (again using a HCA) that isotropically consolidated, spherical glass ballotini show strongly anisotropic yielding and failure characteristics. Kuwano and Jardine (2002) observed differences in the horizontal and vertical elastic stiffnesses of ballotini specimens using bender elements mounted on triaxial specimens. In 2D, Oda et al. (1985) considered specimens of elliptical particles with differing initial orientations in their experimental (photoelastic) study. Li and Li (2009) and Dean (2005) cite additional examples of a experimental studies of anisotropic soil response from a DEM perspective.

A significant amount of DEM data exists that show the influence of anisotropy on the material response. In their two-dimensional DEM simulations Mahmood and Iwashita (2010) considered the sensitivity of the overall response and the evolution of the internal material fabric to the initial particle orientations in a series of DEM simulations with elliptical particles; in this work the inherent fabric was considered. Yimsiri and Soga (2010) induced a preferential orientation in their contact force network by preshearing (i.e. they formed a stress-induced fabric) and found this had a substantial influence on the material response. In fact, understanding the observed sensitivity of the material response to either the preferred particle or contact orientations, or the evolution of these preferential orientations during deformation, has been the focus of almost all DEM analyses that have calculated a fabric parameter for the material.

It should be recognized that while the preferential orientations are important, other fabric effects can influence soil response. Rather than focussing on the direction of loading relative to the particle orientations, some experimental studies have considered the effect of sample preparation methods on the material response. For example Jefferies and Been (2006) present data from an earlier experimental study where they compared triaxial tests on two reconstituted samples of the same sand. These samples had the same initial void ratio and the same confining pressure, but they had been prepared differently (one using moist tamping, the second using wet pluviaton). It is difficult to directly associate the differences between these methods with different preferred particle orientation; however, notable variations in both the deviator stress response and volumetric strain response were observed. Studies that have demonstrated the influence of sample preparation on the number of cycles to achieve liquefaction in undrained cyclic loading include Nemat-Nasser and Tobita (1982) and Mulilis et al. (1977). In discussing soil liquefaction, Jefferies and Been (2006) suggest the influence of fabric on cyclic soil response, is more marked than the influence of fabric on monotonic or static soil behaviour. Vaid and Sivathayalan (2000) discussed measurable sensitivities to the specimen preparation method in undrained monotonic tests on sand.

Moving outside of the laboratory and thinking about real sands, the situation is even more complex. As noted by Vaughan (1993) additional factors to consider include the stress history (which could be considered in DEM) and bedding (more difficult to model). Vaughan (1993) also argues that almost all natural in-situ soils are subject to a degree of lithification, which has the effect of bonding particles together. While very important, a measure of fabric that gives the preferred particle orientations and the extent of the bias in this direction does not completely describe the material micro-structure.

Anisotropy

As discussed by Barreto (2010), amongst others, in geomechanics anisotropy is classified to be either inherent, induced or initial. The inherent anisotropy is the result of the depositional process; while the inherent anisotropy is influenced by the geometry of grains even spherical grains may develop anisotropy during deposition (Oda, 1972). Casagrande and Carrillo (1944) defined induced anisotropy to be the strain-induced particle reorientation associated with changes in stresses. The initial anisotropy of a sand in situ represents the anisotropy that has developed both during deposition and over the geological stress history of the deposit.

It is conceptually easy to relate the response anisotropy to the preferential orientation of the particles in the material. From a micromechanics perspective, when analysts talk about quantifying the material fabric, they are most likely thinking about quantification of a type of "orientation fabric", using the term adopted by Oda, (e.g. Oda (1977)). Oda applied the term "orientation fabric" to consider only preferential orientations of the particles and the intensity of the bias in the preferential direction. This concept of an "orientation fabric" is extended here to include the orientations of the contact forces and the branch vectors.

To quantify anisotropy, a set of reference axes are chosen and then either the orientations of the particles (e.g. long-axis orientation), Figure 10.7(a), the orientations of the vectors linking the centroids of contacting particles (the branch vectors), Figure 10.7(b), or the orientation of the contact normals, Figure 10.7(c), is considered. For spherical or circular particles the branch vector orientations and the contact normal orientations will be the same.

Note that there are some slight differences in the meanings attributed to the term "branch vector" in the literature. Here the branch vector is taken to be the vector, l^c joining the centroids of the two particles contacting at contact c is adopted following Bagi (1999b). However, Luding et al. (2001) and Latzel et al. (2000) use the term "branch vector" to describe the vector, that they denoted l^{pc}, joining the centroid of particle p to its contact c (i.e. c is restricted to be of the contacts involving particle p).

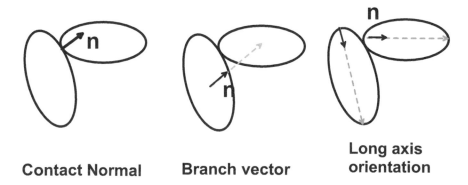

Contact Normal **Branch vector** **Long axis orientation**

Figure 10.7: Different orientation vectors that can be used to quantify fabric

The challenge associated with quantifying fabric is to interpret the orientation data to provide a meaningful measure of any preferred orientation of the vectors. Broadly speaking two approaches are used to quantify fabric: the fitting of curves to rose diagrams or the fabric tensor approach. Both approaches are outlined here. Whichever method is used, the anisotropy quantified will be a measure of the frequency of particles being oriented in the most preferential orientation relative to the frequency of particles having the least preferential orientation. In DEM the fabric has most often been quantified by considering the contact normal orientations, and the discussion on the fabric measures will focus on contact normal orientations. However the mathematical techniques can be applied to any data set of unit vectors. These methods can be applied to the results of DEM analyses and also two-dimensional images or 3D micro-computed tomography data sets. Many of these approaches were initially used in soil mechanics applications to analyse images of thin sections of soil samples, or to analyse 2D photoelastic experiments.

10.5 Statistical Analysis of Fabric: Histograms of Contact Orientations and Curve Fitting Approaches

To create a polar histogram or rose diagram to visualize the distribution of orientations of a set of vectors, an angular interval is selected to define the "bin" size. Then each of these bins is plotted as the segment of a circle whose radius is proportional to the number of contacts oriented within the angles defining the bin limits. The contact force network for a specimen of 2D disks in an isotropic stress state is illustrated in Figure 10.8(a) and the corresponding histogram is illustrated in Figure 10.8(b). There is no apparent bias in the number of contacts oriented in any particular direction. Some analysts prefer to use a conventional linear histogram instead of a rose diagram to illustrate the distribution of the contact orientations. From an interpretative perspective this approach is suited to analysis of two-dimensional or axisymmetric systems, however as illustrated in Figure 10.9, three-dimensional rose diagrams can be created.

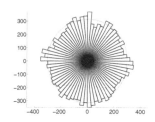

(a) Contact force network   (b) Histogram of contact normals

Figure 10.8: Illustration of a histogram of contact forces for an isotropic assembly of 9,509 2D disks

Plotting the histograms based solely on information on the contact orientations cannot provide information on the relative magnitudes of the contact forces oriented in each direction. O'Sullivan et al. (2008) showed that a simple shading can be applied to each bin in the histogram so that the magnitude of the average force in a given direction, as well as the number of contacts oriented in that direction, can be ascertained from a single plot, this idea is illustrated in Figure 10.10.

The histograms or rose diagram plots of the orientation data can be used to analyse the data quantitatively. This analysis is achieved by fitting an analytical function to the histogram data and the parameters of the function quantify the intensity of the anisotropy and the preferred orientation. The basic idea is that the

380

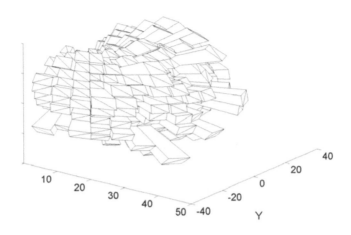

Figure 10.9: Illustration of 3D contact orientation histogram

orientations can be described using a probability density function (PDF) $E(\mathbf{n})$. This function tells us the likelihood that a contact (or particle) will have an orientation described by the unit vector $\mathbf{n}$. The integral of this function over the domain must be 1, i.e.

$$\int_{\Omega} E(\mathbf{n})d\Omega = 1 \qquad (10.15)$$

where $d\Omega$ is the differential solid angle in a spherical coordinate system. Oda et al. (1980) described the PFD as the "fabric ellipsoid." The anisotropy can be determined from the shape of the ellipsoid, and the ellipsoid long-axis orientation gives the preferential orientation of the contact normals. For an isotropic material the fabric ellipsoid will have spherical symmetry. However for most materials deposited under gravity, the ellipsoid will be axisymmetric about a vertical axis, with the fabric appearing isotropic in horizontal views.

381

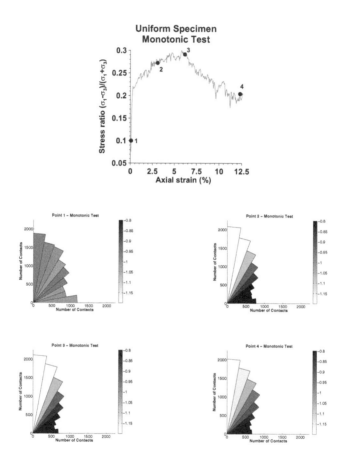

Figure 10.10: Histograms of the normalized contact forces in the vertical plane: Triaxial test simulation on specimens of uniform spheres

382

Fourier analysis of 2D data.

2D fabric analysis is useful when considering either 2D DEM data or axisymmetric systems in 3D. For 3D axisymmetric systems the orientation required for input into a 2D analysis is given by the angle between the vector under consideration and the vertical. Then Equation 10.15 can be expressed as

$$\int_0^{2\pi} E(\theta)d\theta = 1 \tag{10.16}$$

where θ is the inclination to the reference axis. Then, $\int_{\theta_1}^{\theta_2} E(\theta)d\theta$ gives an estimate of the number of vectors with orientations between the angles θ_1 and θ_2. This function can thus be related to the rose diagram or histogram for the data set of vectors. Assuming the contact normals are oriented so that $0 \leq \theta \leq 2\pi$, the function $E(\theta)$ must be a periodic function with a period of 2π. Consequently it can be expanded using the following Fourier series

$$E(\theta) = \frac{1}{2\pi}\left(a_0 + \sum_{k=1}^{\infty}(a_k\cos(k\theta) + b_k\sin(k\theta))\right) \tag{10.17}$$

Oda (1999a) states that a_k and b_k are zero for odd values of k.

Rothenburg and Bathurst (1989) used a Fourier series expansion containing only two terms. The Rothenburg and Bathurst equation is given by

$$E(\theta) = \frac{1}{2\pi}[1 + a\cos2(\theta - \theta_a)] \tag{10.18}$$

where a is a parameter defining the magnitude of fabric anisotropy and θ_a defines the direction of the fabric anisotropy or the principal fabric. The analysis can be extended to include additional terms in the Fourier series expansion; however, interpretation becomes less elegant.

Barreto et al. (2008) outline the steps to determine the parameters a and θ_a. Taking the data set of contact normal orientations each vector is binned into an angular interval $\Delta\theta$ centred around an angle θ_i. The number of contacts in bin i is then given by $\Delta N_c(\theta_i)$ and is normalized by the product $N_c \Delta \theta$, where N_c is the total number of contacts. For each angular interval the value of $\Delta N_c(\theta_i)$ is related to the Fourier parameters as follows:

$$\frac{\Delta N_c(\theta_i)}{N_c \Delta \theta} = \frac{1}{2\pi}\left[1 + a\cos 2\left(\theta_i - \theta_a\right)\right] \qquad (10.19)$$

Equation 10.19 can be rewritten as

$$\frac{\Delta N_c(\theta_i)}{N_c \Delta \theta} = \frac{1}{2\pi} + \frac{a\cos 2\theta_a}{2\pi}\cos 2\theta_i + \frac{a\sin 2\theta_a}{2\pi}\sin 2\theta_i \qquad (10.20)$$

A least squares approximation can be used to determine the coefficients $\frac{a\cos 2\theta_a}{2\pi}$ and $\frac{a\sin 2\theta_a}{2\pi}$, from which the preferred fabric orientation, θ_a, and the anisotropy, a, can be calculated.

Using this approach, the value of a determines whether θ_a gives the major or minor principal fabric direction. For $a > 0$, the θ_a value gives the major principal fabric direction, and for $a < 0$, θ_a gives the minor principal fabric orientation. The magnitude of a can vary between 0 and 1 and relates to the degree of anisotropy. Figure 10.11 illustrates the pattern of the contact normal distributions for various values of a and θ_a. Figure 10.12 plots both the distribution of contact normals and the best fit curve for data obtained from two 3D DEM simulations.

A curve-fitting approach can also be applied to the orientations of the tangential components of the contact force. As noted by Rothenburg and Bathurst (1989), amongst others, the distribution function takes a slightly different form to the distribution for the contact normal orientations, and is given by:

$$E_t(\theta) = a_t \sin 2(\theta - \theta_t) \qquad (10.21)$$

As for the contact normal orientations, curve fitting exercise can be used to determine the parameters a_t and θ_t.

Fourier analysis of 3D data

The Fourier analysis approach can also be applied to 3D data sets. Referring to Chang et al. (1989), in the case where the fabric is

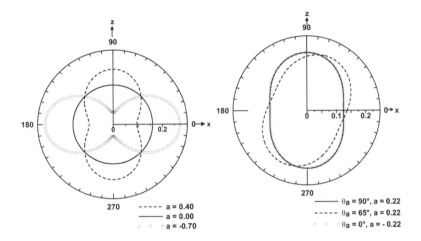

(a) Distribution of contact normals for various anisotropy values

(b) Distribution of contact normals for differing preferred orientations.

Figure 10.11: Illustration of relation between Fourier parameters and contact normal distributions (Barreto et al., 2008)

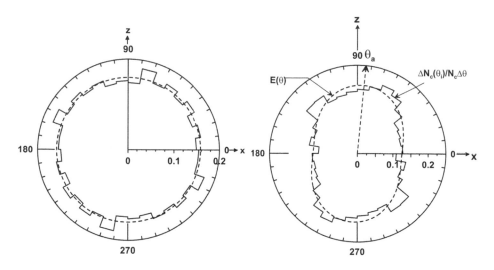

(a) Best fit parameters for isotropic distribution of contact normals

(b) Best fit parameters for anisotropic distribution of contact normals

Figure 10.12: Fourier distribution fitted to contact normal distributions obtained during anisotropic compression of a DEM sample (Barreto et al., 2008)

cross-anisotropic and symmetric about the z-axis as illustrated in Figure 10.13, the function $E(\gamma, \beta)$ is given by

$$E(\gamma, \beta) = \frac{3(1 + a\cos 2\gamma)}{4\pi(3 - a)} \qquad (10.22)$$

where a is the degree of anisotropy $(-1 < a < 1)$.

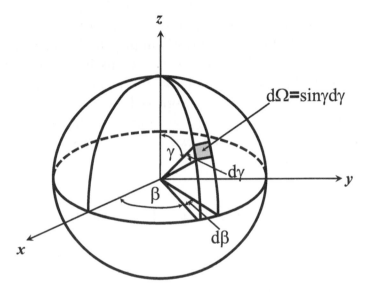

Figure 10.13: Illustration of three-dimensional vector

Chang and Yin (2010) propose an alternative expression for the distribution of the contact orientations in three dimensions that is derived from a truncated form of a continuous three-dimensional spherical harmonic expansion:

$$E(\gamma, \beta) = \frac{1}{4\pi}\left(1 + \frac{a_{20}}{4}\left(3\cos 2\gamma + 1\right) + 3\sin^2\gamma\left(a_{22}\cos 2\beta + b_{22}\sin 2\beta\right)\right)$$
$$(10.23)$$

Note that whether Equation 10.23 or Equation 10.22 is used

$$\int\int E(\gamma, \beta)\sin\gamma d\gamma d\beta = 1 \qquad (10.24)$$

For 3D data, it may be more convenient to analyse the fabric using the fabric tensor approach that is outlined in Section 10.6 below.

Relationship between fabric parameters and overall response

Oda et al. (1980) stated that, after void ratio, the fabric ellipsoid is the second most important index describing the structure of a granular material. The work of Rothenburg and Bathurst (1989) is important as it established a quantitative link between the macro-scale stress-strain response and the fabric parameters obtained by applying the Fourier series approach to the contact normal orientations. As well as using the Fourier parameters a and θ_a introduced above, they fitted two Fourier series to the contact force vectors as follows:

$$\bar{f}_n(\theta) = \bar{f}_{n0}\{1 + a_{nf}\cos2(\theta - \theta_{nf})\} \quad \text{Normal contact forces}$$

$$\bar{f}_t(\theta) = -\bar{f}_{n0}a_{tf}\{\sin2(\theta - \theta_{tf})\} \quad \text{Tangential contact forces}$$

$$(10.25)$$

where θ_{nf} refers to the orientation of the direction of the maximum average force and a_{nf} describes the anisotropy of the contact normal forces, or "magnitude of the directional variation of the average normal forces." In a similar manner a_{tf} describes the anisotropy of the tangential components of the force; however, θ_{tf} defines the direction where the tangential force is zero on average. The parameter $\bar{f}_{n0}$ is the average normal contact force when all groups are given equal weight, i.e.

$$\bar{f}_{n0} = \int_0^{2\pi} \bar{f}(\theta)_n d\theta \qquad (10.26)$$

The parameter $\bar{f}_{n0}$ is introduced in both the equation defining the normal force distribution and the equation describing the tangential force contribution, where it acts as a scaling factor.

As illustrated in Figure 10.14(a), Rothenburg and Bathurst considered the orientation of the three parameters a, a_{nf} and a_{tf} during DEM simulation of biaxial compression of a specimen of 1,000 polydisperse disks subject to biaxial compression. The specimen was initially isotropic and during compression $\sigma_{22} > \sigma_{11}$ and the confining pressure (σ_{11} remained constant. As illustrated in Figure 10.14(b) close agreement was found between the principal stress ratio and the sum $\frac{1}{2}(a + a_n + a_t)$. Note that the formulation presented by Rothenburg and Bathurst (1989) assumes that the directions of contact anisotropy and the principal directions of force are coincident; general conditions where the direction of anisotropy and the principal direction do not coincide were considered by Rothenburg (1980).

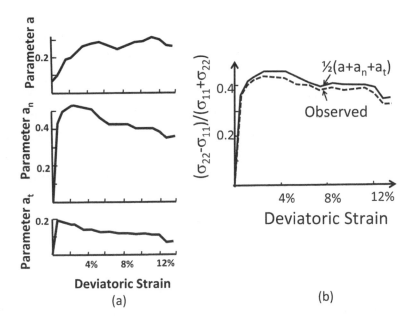

Figure 10.14: Correlation between stress-strain response and fabric parameters in simulations of Rothenburg and Bathurst (1989) (a) Fabric response (b) Stress strain response

10.6 Statistical Analysis of Fabric, the Fabric Tensor

The fabric tensor can be determined directly from the data set of orientation vectors and used to determine the preferred orientation and the magnitude of the anisotropy. As shown below, the values of anisotropy and the preferred orientations obtained using the fabric tensor are equivalent to those obtained using the Fourier series approach. The most commonly used definition of the fabric tensor is the second-order fabric tensor. In tensorial notation the second-order fabric tensor (for contact orientations) is given as

$$\Phi_{ij} = \frac{1}{N_c} \sum_{k=1}^{N_c} n_i^k n_j^k \tag{10.27}$$

where n_i^k is the unit vector describing the contact normal orientation and the summation takes place over the N_c contacts in the system. In granular mechanics the definition of the fabric tensor is generally attributed to Satake(1978, 1982). Ng (2004a) refers to the earlier work of Scheidegger (1965). Fabric tensors for the particle orientations, the branch vector orientations or the void orientations could easily be obtained by substituting the relevant unit vectors into Equation 10.27.

Higher-order fabric tensors can be determined, for example the fourth-order fabric tensor is given by

$$\Phi_{ijkl} = \frac{1}{N_c} \sum_{k=1}^{N_c} n_i^k n_j^k n_k^k n_l^k \tag{10.28}$$

These higher-order fabric tensors result in multidimensional arrays that can be difficult to interpret. Kanatani (1984) is a useful theoretical reference for those attempting to develop an advanced understanding of the fabric tensor, including higher order fabric tensors. Generally in geomechanics related DEM simulations, consideration is restricted to the second-order fabric tensor defined in Equation 10.27, and when the term fabric tensor is used, it normally refers to this second-order tensor.

In two dimensions, expansion of Equation 10.27 gives a two-dimensional matrix:

$$\begin{pmatrix} \Phi_{xx} & \Phi_{xy} \\ \Phi_{yx} & \Phi_{yy} \end{pmatrix} = \frac{1}{N_c} \begin{pmatrix} \sum_{k=1}^{N_c} n_x^k n_x^k & \sum_{k=1}^{N_c} n_x^k n_y^k \\ \sum_{k=1}^{N_c} n_y^k n_x^k & \sum_{k=1}^{N_c} n_y^k n_y^k \end{pmatrix} \qquad (10.29)$$

If the orientation of the vector to the horizontal (x) axis is given by θ, then the normal vector for contact k is given by

$$n^k = \begin{pmatrix} n_x^k \\ n_y^k \end{pmatrix} = \begin{pmatrix} \cos\theta^k \\ \sin\theta^k \end{pmatrix} \qquad (10.30)$$

Then if contact k makes an angle θ^k to the x axis, the fabric tensor is given by

$$\begin{pmatrix} \Phi_{xx} & \Phi_{xy} \\ \Phi_{yx} & \Phi_{yy} \end{pmatrix} = \frac{1}{N_c} \begin{pmatrix} \sum_{k=1}^{N_c} \cos^2\theta^k & \sum_{k=1}^{N_c} \cos\theta^k \sin\theta^k \\ \sum_{k=1}^{N_c} \sin\theta^k \cos\theta^k & \sum_{k=1}^{N_c} \sin^2\theta^k \end{pmatrix} \qquad (10.31)$$

As $\cos^2\theta^k + \sin^2\theta^k = 1$ and the tensor is normalized by the total number of contacts, then the trace of the contact tensor is 1 (i.e. $\Phi_{xx} + \Phi_{yy} = 1$).

The fabric tensor is directly related to the distribution of vector orientations described above. It can be determined from the a_2 and b_2 coefficients from the Fourier fit to the vector orientations (Equation 10.17) as follows:

$$\begin{pmatrix} \Phi_{xx} & \Phi_{xy} \\ \Phi_{yx} & \Phi_{yy} \end{pmatrix} = \frac{1}{4} \begin{pmatrix} a_2 + 2 & b_2 \\ b_2 & -a_2 + 2 \end{pmatrix} \qquad (10.32)$$

Similarly Oda (1999a) states that a relationship exists between the coefficients a_2, a_4, b_2, and b_4 and the fourth-order fabric tensor. Similar relationships exist for the higher-order fabric tensors and the higher-order terms in the Fourier Series expansion.

In three dimensions the second-order contact fabric tensor is given by

$$
\begin{pmatrix}
\Phi_{xx} & \Phi_{xy} & \Phi_{xz} \\
\Phi_{yx} & \Phi_{yy} & \Phi_{yz} \\
\Phi_{zx} & \Phi_{zy} & \Phi_{zz}
\end{pmatrix}
=
$$

$$
\frac{1}{N_c}
\begin{pmatrix}
\sum_{k=1}^{N_c} n_x^k n_x^k & \sum_{k=1}^{N_c} n_x^k n_y^k & \sum_{k=1}^{N_c} n_x^k n_z^k \\
\sum_{k=1}^{N_c} n_y^k n_x^k & \sum_{k=1}^{N_c} n_y^k n_y^k & \sum_{k=1}^{N_c} n_y^k n_z^k \\
\sum_{k=1}^{N_c} n_z^k n_x^k & \sum_{k=1}^{N_c} n_z^k n_y^k & \sum_{k=1}^{N_c} n_z^k n_z^k
\end{pmatrix}
$$

(10.33)

where N_c is the number of contacts and (n_x, n_y, n_z) is the unit vector describing the contact normal orientation. Referring Figure 10.13, in 3D the normal vector for contact k can be related to the angles β and γ as

$$
\mathbf{n}^k =
\begin{pmatrix}
n_x^k \\
n_y^k \\
n_z^k
\end{pmatrix}
=
\begin{pmatrix}
\cos\beta^k \sin\gamma^k \\
\sin\beta^k \sin\gamma^k \\
\cos\gamma^k
\end{pmatrix}
$$

(10.34)

The fabric elements of the fabric tensor are then given by

$$
\begin{aligned}
\Phi_{xx} &= \sum_{k=1}^{N_c} \cos^2\beta^k \sin^2\gamma^k \\
\Phi_{xy} &= \sum_{k=1}^{N_c} \cos\beta^k \sin\beta^k \sin^2\gamma^k \\
\Phi_{xz} &= \sum_{k=1}^{N_c} \cos\beta^k \sin\gamma^k \cos\gamma^k \\
\Phi_{yx} &= \sum_{k=1}^{N_c} \cos\beta^k \sin\beta^k \sin^2\gamma^k \\
\Phi_{yy} &= \sum_{k=1}^{N_c} \sin^2\beta^k \sin^2\gamma^k \\
\Phi_{yz} &= \sum_{k=1}^{N_c} \sin\beta^k \sin\gamma^k \cos\gamma^k \\
\Phi_{zx} &= \sum_{k=1}^{N_c} \cos\beta^k \sin\gamma^k \cos\gamma^k \\
\Phi_{zy} &= \sum_{k=1}^{N_c} \sin\beta^k \sin\gamma^k \cos\gamma^k \\
\Phi_{zz} &= \sum_{k=1}^{N_c} \cos^2\gamma^k
\end{aligned}
$$

(10.35)

As in the two-dimensional case the trace of the fabric tensor is 1 (i.e. $\Phi_{xx} + \Phi_{yy} + \Phi_{zz} = 1$). Both the two-dimensional and three-dimensional fabric tensors are symmetric (i.e. $\Phi_{ij} = \Phi_{ji}$).

As in the two-dimensional case, the curve fitted parameters and the 3D fabric tensor can be related. Referring to Figure 10.13, Equation 10.15 can be expressed as

$$
\Phi_{ij} = \int_0^{2\pi} \int_0^{\pi} n_i n_j E(\gamma, \beta) \sin\gamma \, d\gamma \, d\beta
$$

(10.36)

In three dimensions, knowing the contact normal distribution function the fabric tensor can be calculated as (Oda (1982), Yimsiri and Soga (2010)):

$$\Phi_{ij} = \int_{\Omega} n_i n_j E(n) d\Omega \tag{10.37}$$

where n_i is the contact normal in direction i, $E(n)$ is the contact normal distribution function (spatial probability density function of n) Ω is the unit sphere and $d\Omega$ is the elementary solid angle.

As outlined by Yimsiri and Soga (2010), if the distribution of contact orientations for an axisymmetric system takes the form of Equation 10.22 then the 3D fabric tensor is given by

$$
\begin{pmatrix}
\Phi_{xx} & \Phi_{xy} & \Phi_{xz} \\
\Phi_{yx} & \Phi_{yy} & \Phi_{yz} \\
\Phi_{zx} & \Phi_{zy} & \Phi_{zz}
\end{pmatrix}
=
\begin{pmatrix}
\frac{3a-5}{5(a-3)} & 0 & 0 \\
0 & \frac{3a-5}{5(a-3)} & 0 \\
0 & 0 & \frac{-(5+a)}{5(a-3)}
\end{pmatrix}
\tag{10.38}
$$

where the anisotropy is given by a. This expression assumes that the preferred or principal orientations are aligned with the coordinate axes. Referring back to Equation 10.23, Chang and Yin (2010) include an expression for a 3D second-order fabric tensor whose elements are related to the coefficients a_{20}, a_{22}, and b_{22}.

The orientation data required for calculation of the fabric tensor are available for output from DEM simulations; for example $\mathbf{n} = \frac{\mathbf{f_n}}{|\mathbf{f_n}|}$ where $\mathbf{f_n}$ is the normal contact force vector. Furthermore the contact force will act with equal magnitude in opposite directions on both contacting bodies, and so we can restrict consideration of the contact force orientations to $0 \leq \theta \geq 2\pi$ and $0 \leq \phi \geq 2\pi$. Similarly consideration of the particle orientations and branch vectors can also be restricted to $0 \leq \theta \geq \pi$ and $0 \leq \phi \geq \pi$.

Interpretation of fabric tensor using eigenvalue analysis

The fabric tensor is an abstract concept and it is difficult to visualize. Perhaps the best place to start is to recognize the similarity between the fabric and stress tensors. Both these tensors

393

are second-order and symmetric. Just as the principal stresses and their orientations can be determined from the stress tensor, once the fabric tensor is known the preferred orientations and the magnitude of the anisotropy can be calculated. The magnitude of the major fabric is given by Φ_1, the minor fabric is given by Φ_3, and in three dimensions, the intermediate fabric is given by Φ_2, i.e. $\Phi_1 > \Phi_2 > \Phi_3$. In the case of a two-dimensional planar DEM analysis (e.g. using disks) the Φ_2 parameter does not exist, while in an axisymmetric system either $\Phi_2 = \Phi_3$ or $\Phi_1 = \Phi_2$.

Ng (2004a) provides a useful verbal description of these principal fabric parameters. He states that they represent the degree of clustering of the orientation data, along each of the three preferred fabric directions, i.e. they tell us the extent to which the particles or contacts under consideration are oriented along the principal fabric directions. In the same way, the principal stresses act along specific principal stress orientations. In the case of fabric, knowing these characteristics of the fabric tensor, we can develop an idea of the shape of the distribution of the particle or contact orientations under consideration.

Just as the eigenvalues of the stress tensor give the principal stresses and their orientations, information on the principal fabric parameters can be determined by eigenvalue decomposition of the fabric tensor. The eigenvectors will be orthogonal unit vectors, i.e. the principal fabric directions are oriented at right angles to each other. The extent of the bias in the most preferential direction of fabric orientation is given by the largest eigenvalue and the corresponding eigenvector gives the direction of the principal fabric component.

Alternative analytical approaches can be used to obtain the principal fabric values. The expressions used in continuum mechanics (e.g. Mase and Mase (1999)) to calculate principal stresses and their orientations can be applied to calculate the principal fabric components and their orientations. For a 2D or axisymmetric system, the principal fabric components for a two-dimensional system are then given by

$$\begin{pmatrix} \Phi_1 \\ \Phi_2 \end{pmatrix} = \tfrac{1}{2}\left(\Phi_{xx} + \Phi_{yy}\right) \pm \tfrac{1}{2}\sqrt{\left(\Phi_{xx} - \Phi_{yy}\right)^2 + \Phi_{xy}^2} \quad (10.39)$$

Similarly the expressions used to determine the principal stress orientations are given in most undergraduate solid mechanics texts (e.g. Gere and Timoshenko (1991)) and these can be applied to determine the principal fabric orientations.

Once calculated, the principal fabric components, Φ_1, Φ_2, and Φ_3, can be used to describe the magnitude or intensity of the anisotropy. While a consensus on how the magnitude of anisotropy should be quantified is not yet fully developed, all approaches tend to use these principal values. In three dimensions, a fully anisotropic fabric will be manifested by three distinct eigenvalues, while a transversely anisotropic or cross-anisotropic fabric will yield only two distinct eigenvalues.

Analysis of fabric eigenvalues in two dimensions

For two-dimensional systems or three-dimensional transversely anisotropic (cross-anisotropic) systems, consideration can be restricted to the major and minor fabric components, Φ_1 and Φ_3. (Here the minor fabric component is denoted Φ_3 for both 2D and 3D data). Thornton (2000) quantified the anisotropy by considering the difference $\Phi_1 - \Phi_3$. This approach was also adopted by Cui and O'Sullivan (2006). Similarly, Oda (1999a) cited Curray (1956) who proposed the use of the parameter $\frac{\Phi_1 - \Phi_3}{2}$. Maeda (2009) described the use of a slightly different form of the deviator fabric, which he calls the "deviator fabric intensity", this is given as the product of the coordination number and the deviator fabric, i.e. for his 2D simulations it is $Z(\Phi_1 - \Phi_3)$.

Instead of looking at the difference between the principal eigenvalues of the fabric tensor, some authors have considered the ratio of these two components. For his 2D simulations, Bardet (1994) quantified anisotropy as the ratio $\frac{\Phi_{yy}}{\Phi_{xx}}$; this expression gives a valid estimation of anisotropy when the principal fabric is oriented in the vertical (y) direction (as is the case in biaxial compression tests). Ibraim et al. (2006) directly considered the ratio of the principal fabric values, i.e. $\frac{\Phi_1}{\Phi_2}$.

Comparison of second-order fabric tensor and Fourier series approach to calculating fabric

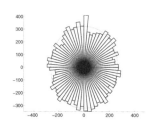

(a) Contact force network   (b) Histogram of contact normals

Figure 10.15: Illustration of a histogram of contact forces for an anisotropic assembly of 9,509 2D disks, $\sigma_1/\sigma_2 = 1.94$

To demonstrate the equivalence of the anisotropy obtained using the fabric tensor and the anisotropy obtained by fitting a Fourier series to the distribution data, two sample 2D analyses are considered. Figure 10.15 gives the distribution of contact forces and the histogram for the specimen considered in Figure 10.8. The specimen has been subject to biaxial compression and these data are the contact normal orientations close to the point where the peak stress is mobilized. The fabric tensor is given by

$$\Phi = \begin{pmatrix} 0.4748 & 0.0014 \\ 0.0014 & 0.5252 \end{pmatrix} \tag{10.40}$$

Eigenvalue analysis gives the values of $\Phi_1 = 0.5252$ and $\Phi_3 = 0.4748$ for the major and minor fabric parameters, with the major fabric having an orientation of 88.4° to the horizontal. The anisotropy given by $\Phi_1 - \Phi_3$ equals 0.0504. Using the curve fitting approach described above to obtain the Fourier parameters, yields values of $a = -0.1003$ and $\theta_a = -1.6°$. As $a < 0$ this gives a preferred major fabric orientation of $90° - 1.6° = 88.4°$ to the horizontal, and $|a| \approx 2\,(\Phi_1 - \Phi_2)$.

The findings are similar when a more anisotropic data set was considered. Figure 10.16 illustrates the distribution of orientations of a set of vectors obtained when analysing clay fabric from SEM data. In this case the fabric tensor is given by

$$\Phi = \begin{pmatrix} 0.612 & -0.0121 \\ -0.0121 & 0.388 \end{pmatrix} \tag{10.41}$$

Eigenvalue analysis gives $\Phi_1 = 0.613$, $\Phi_2 = 0.387$ and $\Phi_1 - \Phi_2 = 0.226$ and a major fabric orientation of -3.10^o to the horizontal. The Fourier parameters obtained are $a = 0.45$ and $\theta_a = -3.10^o$. This time, as $a > 0$, θ_a gives the major fabric orientation directly. Again $|a| \approx 2\,(\Phi_1 - \Phi_2)$.

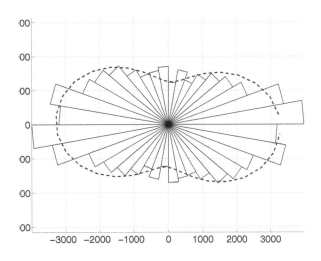

Figure 10.16: Distribution of unit vectors to analyse fabric from SEM images (Wilkinson, 2010)

Analysis of fabric eigenvalues in three dimensions

Interpretation of fabric anisotropy becomes more challenging for the general three-dimensional case. Barreto et al. (2009) proposed quantification of three-dimensional anisotropy or deviator fabric

397

using the following invariant (analogous to the shear stress in the octahedral plane):

$$\Phi_d = \frac{1}{\sqrt{2}}\sqrt{(\Phi_1 - \Phi_2)^2 + (\Phi_2 - \Phi_3)^2 + (\Phi_3 - \Phi_1)^2} \qquad (10.42)$$

Kuo et al. (1998) used a similar expression to assess 3D fabric anisotropy. Ng (2004a) also considered a fully three-dimensional stress state and adopted the approach proposed by Woodcock (1977). In this approach two fabric descriptors β_1 and β_2 are considered:

$$\begin{aligned} \beta_1 &= \ln\left(\frac{\Phi_1}{\Phi_3}\right) \\ \beta_2 &= \ln\left(\frac{\Phi_1}{\Phi_3}\right) / \ln\left(\frac{\Phi_2}{\Phi_3}\right) \end{aligned} \qquad (10.43)$$

Woodcock presented a graphical technique to interpret the fabric eigenvalue data using β_1 and β_2. In Woodcock's notation Φ_1 is denoted S_1, Φ_2 is denoted S_2, and Φ_3 is denoted S_3. As illustrated in Figure 10.17, the value of $\ln\left(\frac{\Phi_2}{\Phi_3}\right)$ is plotted on the horizontal axis, while the value of $\ln\left(\frac{\Phi_1}{\Phi_3}\right)$ is plotted on the vertical axis. If the eigenvalue data plot at the origin the fabric is isotropic and the vectors are uniformly distributed. For an anisotropic fabric the value of β_1 then indicates the extent of the concentration of vectors in the preferred orientation. If the eigenvalue data plot along the vertical axis the fabric is axisymmetric. Otherwise, the β_2 value determines whether the distribution of orientations is a cluster or a girdle.

Alternative definitions of the fabric tensor

Alternative definitions of the fabric tensor exist; for example Luding et al. (2001) propose that in calculating the fabric tensor, the fabric tensor for each particle should initially be calculated (by considering the contact locations relative to the particle centroid), then the average fabric tensor within a defined measurement volume/region should be calculated by considering the sum of the

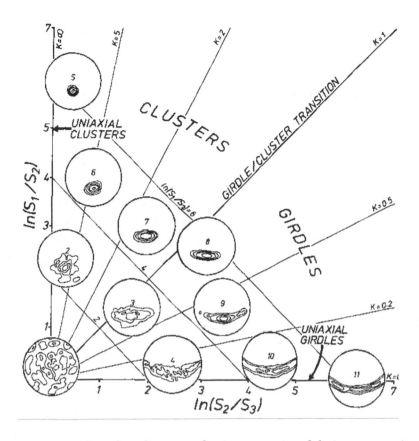

Figure 10.17: Graphical means for interpreting fabric tensor eigenvalues proposed by Woodcock (1977)

particle contributions to the overall fabric, normalized by the particle volumes, and then divided by the overall volume of the measurement region.

Filters can be applied to the data considered for inclusion in the fabric tensor; for example Kuhn (2006) suggests the use of a "strong" fabric tensor, including only those contacts that carry a greater than average force:

$$\Phi_{ij}^s = \frac{1}{N_c^S} \sum_{k \in S} n_i^k n_j^k \tag{10.44}$$

where there are a total of N_c^S contacts within the set S that comprises those contacts transmitting a force exceeding the average force.

Fabric tensor for particle and void orientations

While the particle orientation tensor can simply be calculated by considering the long-axis orientations of the particles (e.g. Ng (2009a)), this definition does not account for the particle geometry. A particle whose intermediate axis length is only slightly bigger than its minor axis length would have the same contribution to the overall tensor as a particle whose intermediate axis length is just smaller than its major axis length. Oda et al. (1985) propose the use of a weighted tensor, $\Phi_{ij}^{\text{particle}}$, for particle orientations that is given by

$$\Phi_{ij}^{\text{particle}} = \frac{1}{\lambda} \sum_{p=1}^{N_p} T_{ki}^p T_{lj}^p S_{kl}^p \qquad (10.45)$$

where N_p is the number of particles and T_{ij} is the orientation tensor for particle p. If x_i^{pb} are the coordinates of the point in a Cartesian coordinate system centred at the particle centroid, $x_j^{\prime pb} = T_{ij} x_i^{pb}$ gives the coordinates of a point on the particle defined in a local coordinate system centred on the particle centroid with axes aligned with the particles principal axes of inertia. The tensor S_{ij}^p is given by

$$S_{ij}^p = \begin{pmatrix} a^p & 0 & 0 \\ 0 & b^p & 0 \\ 0 & 0 & c^p \end{pmatrix} \qquad (10.46)$$

where a^p, b^p and c^p are the major, intermediate and minor half-lengths of the particle respectively. The parameter λ is given by $\lambda = \sum_{p=1}^{N_p} (a^p + b^p + c^p)$.

In their description of analysis of photoelastic experiments, Oda et al. (1985) define a void fabric tensor which can be used to quantify void shape. Referring to Figure 10.18, for the two-dimensional case the analysis involves constructing series of parallel scan lines through an image of the system at inclinations of θ varying between 0° and 180°. At a given orientation, θ, the contribution to the fabric tensor is then given by

$$
\begin{bmatrix}
l_v^\theta \cos^2\theta & l_v^\theta \cos\theta\sin\theta \\
l_v^\theta \cos\theta\sin\theta & l_v^\theta \sin^2\theta
\end{bmatrix}
\tag{10.47}
$$

where l_v^θ is proportional to the total length of the scan line at orientation θ that intercepts the void space, scaled so that

$$
\int_0^{180°} l_v^\theta d\theta = 1
\tag{10.48}
$$

The void orientation fabric tensor is then determined by summing the contributions from each scan line orientation considered. The ratio of the major and minor fabrics $\Phi_1^{\text{void}}/\Phi_3^{\text{void}}$ gives a measure of the void elongations.

Kuo et al. (1998) also used a parallel scan line approach to analyse digitized images of sections through resin-impregnated sand specimens. A finite number of orientations must be considered, and Kuo et al. (1998) suggested that the scan line orientations be varied in 5° increments. This method was developed for analysis of images, and the most straightforward way to apply it to DEM is probably to create binary digital images of the disk system. Where the void fabric is considered for DEM data it is more typical to use the graph-based approaches proposed in the following section.

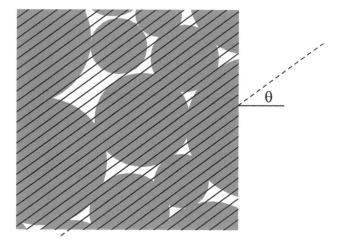

Figure 10.18: System of scanning lines to determine void orientation proposed by Oda et al. (1985)

Use of the fabric tensor to link particle- and macro-scale responses

Having established how to determine the orientation fabric, it remains to discuss how this measure of fabric relates to the macro-scale material behaviour. In their study of experiments on photoelastic oval-shaped rods, Oda et al. (1985) compared the material stress-strain response with the contact normal orientation fabric tensor, the particle orientation fabric tensor and the void orientation fabric tensor. The fabric measure that gave the best correlation with the macro-scale response (quantified using the principal stress ratio σ_1/σ_3) was the ratio of the major to minor fabric for the contact normal orientations. Similarly, in 3D DEM simulations using ellipsoidal particles, Ng (2009a) showed that the contact normal fabric tensor correlates more strongly with the overall material response than the particle orientation tensor. This is hardly surprising as the expression for the fabric tensor ($\Phi_{ij} = \frac{1}{N_c} \sum_{k=1}^{N_c} n_i^k n_j^k$) is very similar to the expression for the stress tensor ($\sigma_{ij} = \frac{1}{V} \sum_{k=1}^{N_c} f_i^k l_j^k$). The unit vector for the contact normal orientation is given by $\mathbf{n}^k = \frac{\mathbf{f}^k}{|\mathbf{f}^k|}$, and for circular and spherical particles the contact normal orientation equals the branch vector orientation, i.e. $\mathbf{n}^k = \frac{\mathbf{l}^k}{|\mathbf{l}^k|}$. Note that Oda and his colleagues were not working in isolation as they developed their understanding of the relationship between the fabric tensor and the stress -strain response. Oda et al. (1980) credit Cowin (1978) and Jenkins (1978) with establishing that a second-order fabric tensor developed from a consideration of the contact normal distribution is a physically meaningful parameter to consider for granular materials.

As discussed further in Chapter 12, numerous DEM simulations have demonstrated a link between the principal values of the contact normal fabric tensor for the contact orientations and the overall stress-strain response. Two examples to illustrate this observation are presented here: Figure 10.19 presents data from a triaxial test simulation, while Figure 10.20 considers a simulation of a direct shear test.

As noted by Oda et al. (1980), the experimental work on photoelastic particles indicated that just as the principal stress axes can rotate during loading, so to do the orientations of the principal axes of the fabric, with the orientation of the major component of the fabric tending towards the principal stress orientation. The evolution of the principal fabric orientation is tied in with the development of strong force chains whose alignment develops during loading so that they can transmit (or resist) the applied stresses. An example of a study that has carefully considered the principal fabric orientation during deformation is Li and Li (2009).

Maeda (2009) observed that the response of the deviator fabric was similar in shape to the stress-strain response and he observed in his simulations the occurrence of a *limit anisotropy* and a *critical anisotropy*. Yunus et al. (2010) also put forward the idea of a critical anisotropy corresponding to the critical state.

Applying the measure of void anisotropy calculated using Equation 10.47 to their photoelastic disk data, Oda et al. (1985) found that upon axial compression the voids tend to orientate themselves with their long axis aligned in the major principal stress direction and the voids themselves become more elongated. Using the graph-based approach to quantify fabric described below, Kuhn (1999) found that during deformation the number of void cells decreased and those that remained became larger and more elongated as particles tended to lose contacts with their neighbours in minor principal stress direction.

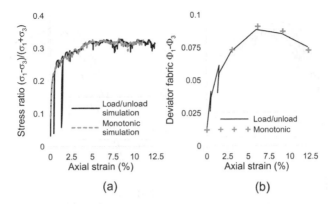

Figure 10.19: Variation in deviator fabric during triaxial test simulations (monotonic test and test involving unload - reload cycles) (a) Macro-scale stress-strain response (b) Deviator fabric variation (O'Sullivan and Cui, 2009b)

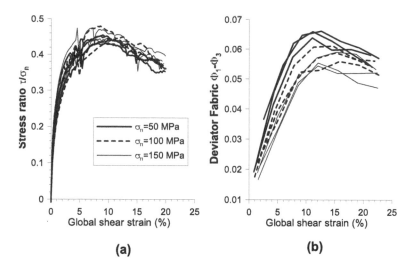

Figure 10.20: Comparison of macro-scale stress ratio and deviator fabric for simulations of a direct shear test (Cui and O'Sullivan, 2006)

10.7 Particle and Void Graphs

The concept of a particle graph was considered in Chapter 9, where Bagi's dual cell system for measuring stress and strain was introduced. There are a few motivations to develop graph representations of granular materials. A graph consists of nodes and cells, and these cells form base units that can be used for the calculation of local stresses and strains or to analyse the material fabric. In addition, it may be possible to use developments in network science to interpret DEM data. Macro-scale data seem to indicate that there might be some type of organized complexity associated with granular materials. If the particle or contact data can be expressed as a network, then the tools used in network analysis and complexity theory can be applied to granular materials.

The idea of using a graph representation for granular materials, including the concept that two, dual graphs exist, i.e. a particle graph and a void graph, is credited to Satake (e.g. Satake(1978,

1992)). Many interesting ideas about using graphs to analyse granular material fabric and deformation have been proposed by Bagi (e.g. Bagi(1996, 2006)) and her particle graph constructs called the material cell and space cell systems were introduced in Chapter 9. Here two additional approaches to construct graph representations of particles are considered, referring primarily to Kuhn (1999) and Li and Li (2009).

Constructing the particle graph

The concepts of a tessellation and the Delaunay triangulation were introduced in Chapter 1. The most straightforward way to construct the particle graph is to create a Delaunay triangulation of the particle centroids and then to delete from the triangulation those edges that connect particles that are not contacting. As illustrated in Figure 10.21, Kuhn (1997) uses a slightly more complex criterion to create his particle graph by restricting consideration to participating particles. The thinner black lines represent the graph formed by Delaunay triangulation on the disk centroids. The thicker lines represent the edges of the particle graph; these edges are identified by removing the isolated, pendular, peninsular and island particles (defined in Figure 10.4) from the Delaunay graph, i.e. only participating particles are included in the particle graph.

Each edge of the graph is then a branch vector, with no specific rule governing the choice of the orientation of the vector (i.e. the vector joining particle a and particle b can point from a to b or b to a arbitrarily). Satake terms this graph the *particle graph*. In the graph the particles correspond to the node points and the branch vectors correspond to the graph edges. The voids then correspond to the cells within the graph. A *void graph* can also be created connecting the centroids of adjacent voids. The particle and void graphs are complementary or dual graphs, with each branch in the void graph being associated with a branch in the particle graph As noted by Li and Li (2009), the definition of a single void in 3D is ambiguous. Therefore when extending this concept to three dimensions, it may be more reasonable to define

407

a minimum "throat width" to define the edges of a given void.

Rather than creating a tessellation based upon the particle centroids, Li and Li (2009) proposed using one based on the contact points. The first step in the graph creation is to carry out a Delaunay triangulation of each particle, with the triangle nodes being the contact points on the surface of the particle. These triangles are merged to form a single solid particle cell for each particle. Then, as illustrated in Figure 10.22(a), the Delaunay triangulation (tessellation) for the entire system of contacts is created. To separate the system into the solid and void elements illustrated in 10.22(b), only those edges in the system tessellation that are also edges of solid particle cells are permitted to remain. Referring to Figure 10.22(c), the cells in the solid cell system are formed by identifying the centres of the void elements, and creating a triangulation of each void element using its centroidal coordinates and the contact points that define the edges of the solid elements enclosing the void element. Any edges that are entirely contained within a solid element are then deleted to form the solid cell system illustrated in Figure 10.22(c). The complementary void cell system (Figure 10.22(d)) is constructed by creating a triangulation based on the centroid of the solid element for each particle and the contact points on the surface of the particle. Then edges that correspond with solid element edges are deleted to form the void cell system.

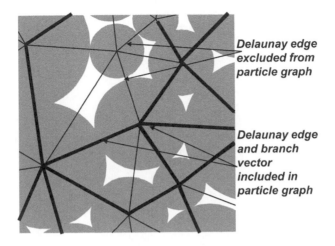

(a) Creation of graph

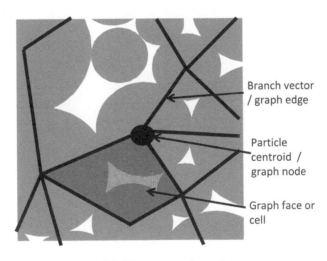

(b) Elements of graph

Figure 10.21: Illustration of the particle graph proposed by Kuhn (1997)

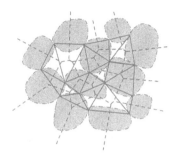

(a) Delaunay triangulation of contacts

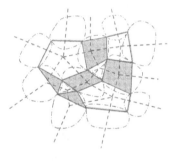

(b) Division of system into solid and void elements

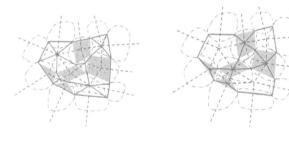

(c) Solid cell system   (d) Void cell system

Figure 10.22: Approach to construct particle and void graphs proposed by Li and Li (2009)

Use of particle and void graphs to analyse material fabric

Once created, the graphs can be analysed to identify features within a particular granular material. The use of graphs to calculate strain has been considered in Chapter 9; for example, Kuhn (1999) made extensive use of his graph structure to identify patterns of deformation within his 2D granular system. Graphs can also be used to analyse fabric.

As noted in Section 10.2, the coordination number can be used to estimate the redundancy of a granular material. Redundancy can be more accurately assessed using the graph construct. For a two-dimensional system, Satake (1999) defined a redundancy number $\hat{r} = s - 3$, where s is the number of sides of a given cell.

Then the mean redundancy number for a system is

$$R = \frac{1}{N_v} \sum \hat{r} \tag{10.49}$$

where N_v is the number of void cells in the system. Satake states that the mean coordination number Z and redundancy number are related as

$$R = \frac{6 - Z}{Z - 2} \tag{10.50}$$

Kuhn (1999) proposes a number of graph-based measures to quantify the topology of the granular material system. He defines the average valence of the system to be

$$m = 2\frac{M_p}{L} \tag{10.51}$$

where M_p is the number of edges in his graph (i.e. the number of branch vectors for the participating particles) and L is the number of cells in his system (these are called *void cells* as a closed loop of edges will surround an effective void). Kuhn presents the Euler formula for his particle graph to be

$$L - M_p + N_p^p = 0 \tag{10.52}$$

where N_p^p is the number of participating particles. As noted above, the effective coordination number is given by

$$Z_p = \frac{2M_p}{N_p^p} \tag{10.53}$$

Combining Equations 10.51, 10.52 and 10.53 the following relationship holds:

$$m = 2 + \frac{4}{Z_p - 2} = \frac{2Z_p}{Z_p - 2} \tag{10.54}$$

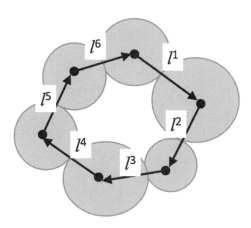

Figure 10.23: Cell branch vectors used by Kuhn (1999) to calculate void anisotropy

As illustrated in Figure 10.23, for each cell the branch vectors are oriented to form a closed loop around the cell. Kuhn (1999) calculates a loop tensor for each cell. This tensor is given as the sum of the dyadic products of the branch vectors for each cell (Tsuchikura and Satake, 1998):

$$K_{ij}^c = \sum_{e=1}^{N_s^c} l_i^e l_j^e \tag{10.55}$$

where l_i^e is the branch vector for edge e and there are a total of N_s^c edges or sides to cell c. The elongation of cell c is given by

$(K_{22}/K_{11})^c$ and the average height-to-width ratio of the cell given by $\left(\sqrt{K_{22}/K_{11}}\right)^c$ is a measure of the average vertical elongation of the cell. The average value of $\left(\sqrt{K_{22}/K_{11}}\right)^c$, for all the cells in the system is a measure of the anisotropy of the system. This measure of void fabric yields results that are in agreement with the findings of Oda et al. (1985); Li and Li (2009) quantified the anisotropy of their void cell system (illustrated in Figure 10.22(d)) by calculating the second-order fabric tensor based on the *void vectors*. They defined these void vectors to be the vectors connecting the centre of each void cell to the contacts along the edge of that void cell.

10.8 Conclusions

The tools presented in this Chapter are a non-exhaustive set of examples that illustrate how the soil fabric, and its evolution, can be quantified from a geometrical point of view. While the approaches are proposed as means to analyse data from DEM simulations, some of these methods can also be applied to analyse 2D images of soil obtained from optical microscopy, scanning electron microscopy or 3D micro-computed tomography data sets. Effective experimental particle-scale image analysis is constantly developing, however information on fabric evolution during deformation in physical tests cannot easily be obtained. Use of image analysis of two-dimensional thin section images or three-dimensional CT data sets can provide information on particle orientations at specific points in the testing sequence. While researchers have been developing techniques to quantify soil fabric based on SEM images over the past few decades (e.g. Tovey (1980), Hattab and Fleureau (2010)), the technology available both to obtain images and to analyse these images is rapidly evolving, and Wilkinson (2010) provides a comprehensive overview of the current state of the art in this area.

This Chapter has, for the most part, focussed on describing method that quantify the fabric of the material in an overall, or average, sense. The existence of particle-scale heterogeneity is obvious from diagrams of the contact force network (e.g. Figure

10.15(a)). Kuhn (2003a) classifies heterogeneity into 5 categories: these include heterogeneity in the material structure, static heterogeneity (stresses and forces) and kinematic heterogeneity (displacements and rotations). The variation in velocities (velocity fluctuations) has been discussed in Chapter 8. Examples of studies that consider kinematic and static heterogeneity include Chang and Misra (1990) and Kuhn (2003a). Of relevance to the description of fabric, Kuhn (2003a) proposes that *topological* heterogeneity can be quantified by considering the coefficient of variation of the valence (i.e. particle-scale coordination number), this can be calculated as the standard deviation of the valence divided by the mean valence. To quantify *geometrical* heterogeneity, Kuhn proposed calculating a fabric tensor for each particle by considering the contacts associated with that particle. The average and the standard deviation of the particle anisotropies calculated from eigenvalues of the particle fabric tensors could then be used to quantify the heterogeneity of the anisotropy.

There are almost as many different methods to analyse the data from DEM simulations as there are research groups in the area. There is, however, a need to adopt a common language to maximize the impact of DEM analyses. Traditional soil mechanics quantifies packing density in terms of porosity or specific volume, i.e. e or v, and describes the evolution of stress using the mean stress p' and the deviatoric stress (q). Numerous studies indicate that if the material anisotropy is quantified by considering the difference between maximum and minimum eigenvalues of the second-order contact fabric tensor, the variation of this anisotropy with strain correlates closely with the stress-strain response. It would seem logical for micro-soil mechanics to adopt the coordination number as a measure of contact density and the deviator fabric $\Phi_1 - \Phi_3$ as a measure of the material anisotropy. This conclusion is, however, largely based on the results of simulations with spherical and convex particles and point contacts. As additional information on more realistic particle geometries becomes available, the contact fabric tensor may need modification to account for contact areas.

Finally, it may be appropriate to question the motivation for quantifying fabric. In the first instance, the methods presented in this Chapter are tools that can be used to propose rational descriptors of the mechanics that underly the often complex macro-scale response of granular materials. These measures may, in the long term, be used in continuum constitutive models. Some continuum constitutive models that include a fabric term have already been proposed, (e.g. Papadimitriou and Bouckovalas (2002)).

Chapter 11

Guidance on Running DEM Simulations

This Chapter includes some observations that may be of use to those running particulate DEM simulations or considering running them in the future. While it is probably an exaggeration to state that there is an "art" to achieving successful simulations, the user certainly needs a bit of time to develop simulation skills prior to embarking on full-scale "production analyses" that will generate the final results. As with any numerical method, the user should also have an understanding of the numerical algorithm and details such as the behaviour of the the contact model used. A complicating factor for DEM simulations, in comparison with continuum methods, is the need to generate the initial packing configuration for the simulations. As noted in Chapter 7, the specimen generation simulation stage where the DEM model is created can be more computationally expensive than the simulation of the boundary value problem of interest. Virtually all DEM analyses simulate a non-linear system; therefore it can be difficult for the analyst to know whether the simulation results are correct or not. (Suggestions on suitable validation problems are given in Section 11.7 below.) These challenges mean that it can be difficult initially for users to develop appropriate techniques used to achieve accurate simulation results. Documented research studies that have used DEM vary in the extent to which the details of the simulation approach are described. The guidance provided by Kuhn (2006) and discussion by Ng (2006) are both useful. As already discussed in Section 5.3, Thornton and Antony (2000) give a particularly clear description of how to use periodic boundaries to achieve a prescribed isotropic stress state and then to shear a sample in a strain-controlled periodic cell simulation. Potyondy and Cundall (2004) discuss the use of DEM to simulate the response of a bonded material and the detailed description of the simulation approach given include comments relevant to DEM analysts simulating the response of unbonded materials. The objective of this Chapter is to provide some preliminary guidance on simulating physical systems using DEM, drawing on personal experience and documented studies. The discussion will, hopefully, be of some use to first time users of DEM in particular.

It is important to be aware that the results obtained in a DEM simulation may be very sensitive to the particular initial geometry of the system to be analysed. The guidance on molecular dynamics simulations by Rapaport (2004) is directly applicable to DEM simulations. He argued that the output of a simulation should be treated using the same statistical methods as used in analysing experiments and that care must be taken to demonstrate reliability of results and adequate sampling. Appropriate validation of the code is also obviously essential.

11.1 DEM Codes

Two DEM codes are associated with the original *Géotechnique* publication outlining the DEM algorithm (Cundall and Strack, 1979a) and the NSF report (Cundall and Strack, 1979b). These codes were called BALL (a 2D code) and Trubal (a 3D code). To date most of the DEM simulations within the geomechanics community have been completed either using the original Trubal code Cundall and Strack (1979a) (including modified versions of Trubal such as ELLIPSE3D: Lin and Ng (1997)) or the "Particle Flow Codes" PFC2D and PFC3D. PFC2D and PFC3D are commercially available and are somewhat linked to Trubal as Peter

Cundall has had significant input in their development. Running simulations using PFC2D or PFC3D requires the analyst to develop skills in the interpreted programming language FISH. While becoming an expert in FISH requires takes time, the flexibility it then offers makes it a very attractive program for both research and industry applications. At the time of writing, use of a second commercial DEM code, EDEM (DEMSolutions, 2009) is becoming more widespread in the process and mining engineering communities in particular. Open source DEM codes include ESyS-Particle Simulation (Weatherley, 2009), YADE (Kozicki and Donz, 2008), "Virtual Geoscience Workbench" (Xiang et al., 2009) and OVAL (Kuhn, 2006). These codes have largely been developed for use in research. The open source LAMMPS Molecular Dynamics software, (Plimpton et al., 2010), includes granular contact models and a granular implementation of LAMMPS, LIGGGHTS has been developed by Kloss and Goniva (2010). Many other DEM codes have been developed within individual research groups. The contact dynamics method (Jean, 2004) is also used within the French geomechanics/particulate modelling community in particular (e.g. Silvani et al. (2009)).

Note that this listing of DEM codes is non-exhaustive and subjective (developed based largely upon familiarity with the codes used in documented geomechanics research studies). Jing and Stephansson (2007) give an overview of the history of the development of DEM and list a number of additional codes. The inclusion of any of the named codes in the above list should not be taken as an endorsement of their accuracy. As with all software, whether developed as a research code or for commercial use, the user take the responsibility to satisfy themselves that the code generates accurate results. Suggestions of DEM validation problems are presented in later in this Chapter.

11.2   Two- or Three-Dimensional Analysis?

An important decision to be made by researchers approaching the use of DEM for the first time is whether to embark on 2D or 3D analyses. Deluzarche and Cambou (2006) include a useful discussion on the implications of using 2D simulations to gain insight into the response of materials with 3D particles. Conceptually particles in a 2D simulation must really be viewed as rods or flat disks (when they are circular). While granular materials of interest to engineers are nearly always 3D, as discussed in Chapter 12, the number of documented 2D DEM studies published each year has tended to outnumber the number of 3D DEM studies. The advantage of 2D simulations is that a 2D simulation will run more rapidly than a 3D simulation with equivalent numbers of particles. This difference in computational effort arises because 2D particles have three degrees of freedom, while 3D particles have six. Furthermore the number of contacts per particle (i.e. the coordination number) is greater in the 3D case as contacts can develop anywhere along the particle surface, rather than being restricted to in-plane contacts, as in the 2D case. The time taken to run a DEM simulation is largely determined by the number of contacts in the system. 3D rigid body dynamic rotational equilibrium is more complex than in the 2D case, as outlined in Chapter 2. Particles in a 2D simulation will not experience out-of-plane forces and only moments acting about axes orthogonal to the analysis plane can be considered. The complexity of 3D systems relative to 2D systems also means that it is easier to develop a 2D DEM code in comparison with the 3D case. As visualization of particle displacements and contact force networks is much easier, 2D models can be more useful to analysts interested in studying particle-scale mechanics in detail. A final consideration is that commercial 2D DEM codes can sometimes be significantly economically cheaper than 3D codes.

In a general discussion on 2D analysis Brooks (2009) describes two dimensional "flatland" as a space with "just enough room for interesting and useful things to arise." This is certainly the case in DEM analyses in geomechanics. The overall response of a 2D system is qualitatively similar to a 3D system, i.e. phenomena such as the dependence of the material response on the state (stress level and void ratio), hysteresis, dilation upon shearing, etc. are all observed. As a 2D DEM model then captures many of the key complex mechanical response features unique to soil and other granular materials, it can be used with a large degree of confidence to study the internal mechanics of granular materials and to assess how different particle-scale parameters influence the overall material response. The benefit of using two-dimensional analogues of soil has been demonstrated in physical experiments using photoelastic disks (e.g. de Josselin de Jong and Verrujit (1969), Oda et al. (1985), Utter and Behringer (2008)), and experiments using Schneebeli rods (e.g. Ibraim et al. (2010)). DEM studies in 2D, including Rothenburg and Bathurst (1989), Kuhn (1999), Wang and Gutierrez (2010) and Li and Yu (2010), amongst many others, represent contributions that have significantly advanced our understanding of granular material response. 2D DEM simulations provide analysts with a convenient means to develop new ideas about how to model granular materials, for example, Jiang et al. (2005, 2009) demonstrated qualitatively the potential of their contact models in 2D DEM simulations.

DEM analysts and the broader geomechanics community need to carefully consider how to interpret the results of 2D DEM analyses. Undoubtedly meaningful qualitative insight can be gained and valid analyses of the relationship between particle-scale mechanics and the overall macro-scale responses can be carried out. However the geometrical restrictions imposed by reducing a 3D problem to 2D including neglecting the out-of-plane contacts and displacements mean that drawing *quantitative* comparisons with the response of physical 3D materials may not be appropriate. Both physical experiments and DEM simulations have clearly shown that the response of 3D granular materials depends on the three-dimensional stress state, for example the intermediate principal stress influences the overall strength (e.g. from a DEM perspective Ng (2004b), Barreto (2010), Thornton and Zhang (2010)). Cui and O'Sullivan (2006) showed that when globally constrained to plane strain (2D) deformation, there are significant particle movements orthogonal to the direction of movement. These observations have implications in particular for the calibration of 2D DEM models against physical test data on materials with three-dimensional particle geometries.

In their discussion on the use of 2D DEM simulations Deluzarche and Cambou (2006) pointed out that the differences in the contact geometries mean that contact model parameters used in a 2D DEM model cannot be directly related to the material properties of real 3D particles. Furthermore careful consideration needs to be taken when attempting to link 2D and 3D grading curves. As almost anyone who has used 2D DEM will know, the void ratios obtained for the 2D material, calculated by considering the ratio of the area of voids to the area occupied by particles, differ significantly from void ratio values obtained for real 3D materials.

11.3 Selection of Input Parameters

The input parameters for DEM simulations can be classified as being geometrical (particle morphology and particle size distribution) or mechanical (contact force stiffness). As noted by Ng (2006), there are various parametric studies in the literature that have documented the sensitivity of DEM simulation results to particle shape, the contact model parameters and the inter-particle friction coefficient adopted. Firstly considering the choice of particle shape, the constraints on particle geometry used in DEM simulations, along with approaches used to achieve more realistic geometries, are discussed in Chapter 4. Circular and spherical particles are easy to implement in a DEM code and simulations will be faster than for the case where non-circular/non-spherical particles are used. Non-circular and non-spherical particles are more realistic. There is therefore a tradeoff between simulation cost (run time) and physical realism. Many of the studies documenting the development of algorithms to simulate non-circular and non-spherical particles include analysis of the material response to particle geometry. As highlighted in Chapter 4, the significant difference between spherical and non-spherical particles is that for non-spherical particles the normal component of the contact force can impart a moment to the particles and play a role in resisting rotation. Tordesillas and Muthuswamy (2009) discuss the influence of rolling resistance on force-chain stability, and the ability to resist rotation will influence the stability of the strong force chains that dominate the material's mechanical response. Despite this limitation, simulations using assemblies of polydisperse disks and spheres generated results from which meaningful conclusions about real soil response can be drawn (e.g. Rothenburg and Bathurst (1989), Thornton (2000)). In considering the choice of particle geometry, DEM analysts may also find experimental studies that have considered the sensitivity of material response to particle geometry useful (e.g. Cho et al. (2006)).

The second geometrical choice to be made is the particle size distribution (PSD). It is very important to realize that uniform disks and spheres will tend to "crystalize"; 2D uniform disks will tend towards a hexagonal packing, while 3D uniform spheres will tend towards either a face-centred-cubic or rhombic packing. There is a significant difference between the response of uniform disks/-spheres arranged on a regular grid and real soil (O'Sullivan et al. (2002, 2004)), inhibiting the development of general conclusions about the material response from simulations using uniformly sized particles. Geotechnical engineers have a good understanding of the influence of the PSD on the mechanical response of a material and determining the PSD is a basic characterization of any soil. Using DEM Cheung (2010) demonstrated that both the particle-scale and overall response characteristics will be influenced by the PSD. As discussed in Section 7.2, it is possible to replicate the particle size distribution for a physical material in a DEM simulation test; however, the computational cost of the simulations can be reduced by neglecting the finer particles in the distribution. Neglecting the smallest particles is valid if we can assume that these particles do not play a major role in stress transmission.

Friction is a parameter that is both geometrical and mechanical: it depends upon both the surface geometry (surface roughness) and the surface hardness. As outlined in Chapter 3, friction is typically modelled in DEM using a single coefficient of friction (i.e. the difference between static and dynamic friction is not considered) and there is little available data on the coefficient of friction between real soil particles. Researchers including Thornton (2000), Cui and O'Sullivan (2006) and Yimsiri and Soga (2010) have shown that there is a non-linear relationship between the overall shear strength of the assembly of particles and the inter-particle friction coefficient, with the overall strength being more sensitive to changes in this parameter at smaller values of friction than at higher values. For many DEM studies it would be appropriate to establish the extent of the sensitivity for the particular problem of interest.

The selection of the contact model parameters also requires careful consideration. Where a Hertzian contact model is used, the model parameters can be directly inferred from the solid particle material properties. As noted in Chapter 3 the Hertzian/Hertz-Mindlin contact model that is often used in DEM simulations is theoretically derived by considering the interaction of two elastic spheres. In principle, this should be adequate as a simple contact model for soils and other granular materials. This model has been shown to adequately represent the larger strain response (e.g. Cui and O'Sullivan (2006)) however it does not generate data on the small-strain response that matches experimental observations. If the response at the contact between sand particles were Hertizan then the small-strain stiffness of the soil (i.e. either E_{max} or G_{max}) would be proportional to the mean stress, p, raised to the power of $\frac{1}{3}$, i.e. $G_{max} \propto p^{1/3}$. However, experimental evidence suggests that the small-strain stiffness of a soil varies with the square root of the mean stress. Discussion on this discrepancy from a micromechanical perspective is given by McDowell and Bolton (2001), Yimsiri and Soga (2000) and Goddard (1990). While the origin of the discrepancy may be either geometrical (i.e. related to shape) or rheological (i.e. related to the material properties), it seems that further refinement to the DEM models used is needed to allow the stress-dependent nature of granular material stiffness to be accurately modelled.

Where a linear spring is used as a penalty spring to calculate the contact force, the objective is often to select a spring stiffness so that the contact overlap is minimized for the stress level under consideration. The ratio between the normal and tangential contact spring stiffnesses (k_n and k_t respectively) may also influence the overall response. As discussed by Hu et al. (2010), at higher k_n/k_t ratios there is a tendency for the load to be attracted to the stiffer normal contacts; this impacts upon the point at which sliding will initiate at the contacts and hence influences the overall material response.

Many DEM simulations have been carried out to examine general aspects of granular material response without having a direct link to a real sand. These simulations are often 2D and use disks and linear contact springs (e.g. Kuhn (1999), Kruyt and Rothenburg (2009)). The coefficient of friction, the spring stiffnesses, and the particle size distributions are therefore not directly linked to any physical measurement and are selected by the user. In laboratory studies considering sand response, researchers often restrict consideration to "standard" sands such as Leighton Buzzard Sand, Monterey Sand or Hostun Sand. It would seem sensible that future DEM analysts adopt a similar strategy and that, where they are not considering a particular sand, they refer to previously published simulation data to select their input parameters.

It can be particularly difficult to relate the damping parameter applied in a DEM simulation to a physical property. Damping is used to minimize the non-physical vibrations that develop at the contacts because of the elastic nature of the rheological model used to relate the contact forces and displacements. As noted in Chapter 3, where a viscous dashpot is used, this can be related to a coefficient of restitution. However, it is more difficult to relate a mass damping parameter to a physical property of the material. Ng (2006) includes a discussion on the sensitivity of the output results to the damping parameter adopted, illustrating that this parameter can influence both the macro-scale and particle-scale responses. As discussed by O'Sullivan (2002) during deformation there will be significant energy dissipation due to frictional sliding. If mass damping is used in quasi-static simulations, it is best to use a small amount of damping during the specimen generation stage of the analysis, and then to reduce this value to 0 or close to 0 during subsequent deformation of the material. This will minimize the potential for this parameter to unduly influence the overall material response.

As in the case of any numerical model, it is useful to carry out a parametric study to assess the extent of the influence of the input parameters on the overall material response. It is also important to document clearly the input parameters when reporting results of analyses.

11.3.1 Calibration of DEM models against physical test data

Many studies have confirmed the ability of DEM models to capture intrinsic features of granular materials (dilatancy, localization, stress dependence of response, etc.). However, the DEM model simplifies the complexity of the real physical system, most notably in terms of modelling contact, particle geometry, particle deformation and typically the number of particles involved. The philosophy of calibration is therefore to acknowledge these simplifications and rather than developing a material working using measured particle-scale parameters, to "tune" or "calibrate" the DEM model to capture the response observed in physical laboratory tests. Typically in calibration studies the model parameters are then varied to capture the macro-scale response. Outside of geomechanics the goal is often to select an appropriate rolling resistance parameter so that spherical particles can be used to simulate industrial processing of non-spherical particles. Most of the discussion in relation to DEM model calibration in geomechanics to date has considered the application of DEM to simulate rock mass response.

The idea of bonding particles together using appropriate contact models to create a model of rock mass has been introduced in Chapter 3 (Section 3.8) and is also discussed in Chapter 12. The objective in these models is to assign appropriate contact parameters between the disks to achieve a mechanical response that matches the response observed in laboratory tests. Once calibrated appropriately, the model can be applied to simulate more applied boundary value problems. The basic idea of this approach is outlined clearly by Potyondy and Cundall (2004). The DEM models often use 2D disks as their base particles (e.g. Fakhimi et al. (2006), Cho et al. (2007), Yoon (2007), Camusso and Barla (2009)). DEM simulations on unbonded particles indicate the same degree of sensitivity to the intermediate principal stress as exhibited by real soils. The intermediate principal stress will also influence the bonded granular material response. Consequently a 2D model calibrated against triaxial test data may not give an accurate representation of the rock mass response during simulation of a boundary value problem where the stress state is fully three-dimensional. Potyondy and Cundall (2004) examined responses under different stress conditions and demonstrated that a DEM model may capture the strength observed in different physical tests in different manners, giving the example of a DEM model material that captures the material response in uniaxial compression, but underestimates the strength once a confining pressure is applied, and then over estimates the strength in Brazilian tests. It may be that rigorous calibration studies should demonstrate the ability of the model to quantitatively simulate the material response for different types of tests, with different test boundary conditions, as well as the variation in observed response as testing conditions (e.g. stress level) vary.

Wang and Tonon (2010) suggest that most calibrations are essentially "trial and error"; however, most researchers vary each input parameters systematically, and some analysts have developed sophisticated calibration strategies. For example when calibrating their model rock against physical test data, Kulatilake et al. (2001) made an initial guess of the rock contact parameters by matching their contact normal stiffness k_n to the material's Young's modulus E using $E = \frac{k_n}{4R}$ where R is a representative particle radius, and $\sigma_t = \frac{S_n}{4R^2}$ where σ_t is the target tensile stress and S_n is the contact normal strength. Assuming the contact shear strengths and stiffnesses are given by k_t and S_t, they then carried out a parametric study to develop calibration curves to select the optimal contact parameters. The complexity of the system means that there can be interaction between the various parameters and independent variation of parameters may not be sufficient to obtain the best possible match to the physical response. Cheung (2010) used a similar approach to calibrate a DEM model of a reservoir sandstone. Yoon (2007) proposed application of a "design of experiment" (DOE) approach in calibration studies. DOE is a structured, organized method for determining the relationship between factors affecting a process and the output of that process and Yoon includes an overview of different types of DOE approach. DOE was also applied by Favier et al. (2010) to calibrate models using unbonded particles for materials processing applications. Well documented parametric studies are useful as they identify the parameters that should be considered in calibration; for example Schöpfer et al. (2009) illustrated that the rock mass response depends strongly on the particle size distribution used. Other studies that consider calibration of particulate DEM codes to simulate rock mass response include Cho et al. (2007) and Camusso and Barla (2009).

As outlined by Potyondy and Cundall (2004) the parameters that have the most significant influence on the overall material response for the bonded particle model of rock are the particle shape, the particle size distribution, the packing of the particles, and the contact model. Many DEM simulations include a limited range of particle sizes (e.g. (Yoon, 2007): $R_{max}/R_{min} = 1.66$, Fakhimi et al. (2002): $R_{max}/R_{min} = 3$). The parametric studies by Potyondy and Cundall (2004) and Schöpfer et al. (2009) provide useful information. For example, Potyondy and Cundall (2004) simulated Brazilian tests and biaxial/triaxial tests in 2D and 3D with 4 different average diameters in each case. They found that while the particle size did not significantly influence the material stiffness, the 2D and 3D Brazilian strengths and the 3D strength in triaxial compression were sensitive to the particle size. Schöpfer et al. (2009) showed that the overall response is sensitive to the shape of the particle size distribution.

Considering the response of unbonded granular materials, an interesting contribution has been made by Calvetti and his colleagues (Calvetti (2008), Calvetti and Nova (2004), Calvetti et al. (2004), Butlanska et al. (2009)). In their DEM calibration method spherical particles are used to simulate the response of sand and the experimentally determined particle size distribution is matched, neglecting particles that are smaller than the D_5 of the material. Then the particles are completely inhibited from rotating to capture the rotational resistance that exists at the contact between real non-spherical irregular sand particles. Calvetti has found that where rotations are inhibited, the macro-scale friction angle is a linear function of the inter-particle friction. As acknowledged by Calvetti (2008), the inter-particle friction values he uses to simulate sand particles are rather low, with friction coefficients in the range of 0.3 - 0.35. The stiffness (Young's modulus, E) of the assembly of particle was found to be a function of the ratio between the normal contact stiffness and average particle diameter. While the ratio of the tangential to normal contact spring stiffnesses $\frac{K_t}{K_n}$ has only a slight influence on the Young's modulus, this ratio strongly influences the Poisson's ratio, and Calvetti proposed that to have realistic Poissons ratios, the value of $\frac{K_t}{K_n}$ should be close to 0.25. Calvetti (2008) demonstrated that this approach to calibration can capture the response of Ticino and Hostun sand, while Gabrielia et al. (2009) successfully captured the response of Adige sand. One point to note in relation to this approach is that because rotation is inhibited, the particles at rest will be in a state of static equilibrium considering translational motion, but the resultant moment on the particle may not be 0. Consequently the stress tensor for individual particles may not always be symmetric. While Calvetti completely inhibited rotation, other researchers have introduced a rotational resistance (often termed rolling friction) between the DEM particles to compensate for the differences in geometry between a real soil particle and a sphere. The value of rolling friction is then an additional parameter to consider in the calibration study as it is difficult to relate to a particle-scale physical measurement.

The work of Cheng et al. (2003) demonstrated the ability of DEM to quantitatively simulate the mechanical response of individual sand grains. As noted in Chapter 4, crushable sand particles were modelled as assemblies of bonded spheres. The calibration process focussed on simulating the crushing of individual agglomerate between two planar boundaries and comparing the response with equivalent unconfined compression tests on real sand grains. They then confirmed the validity of their model by comparing the response of simulations of isotropic compression tests on an assembly of particles with physical test data. This study is considered in more detail in Chapter 12.

Even as DEM evolves, the restrictions on the number of particles simulated and the ideal nature of the particles and contact models will remain. This has lead Simpson and Tatsuoka (2008) to predict that as DEM develops into a tool that could be applied to field-scale problems, rather than being able to predict response from input of particle-scale input parameters, some form of macro-scale calibration of the DEM model against appropriate laboratory tests will always be necessary. Calibration is not easy and a strategy developed for one sand may not be applicable to a second material. For example, while Cheung (2010) successfully calibrated a model that can simulate the response of one cemented sand (Castlegate) at various stress levels, using the same calibration approach a model of Saltwash sandstone was created and this was less effective at replicating the physical response over a range of stress levels.

11.4   Choice of Output Parameters

The process of planning or designing a DEM simulation includes the selection of the parameters to monitor and output. The data chosen for output should of course allow suitable analysis of the response of the system of grains, but also facilitate an assessment of the performance of the simulation. The DEM analyst is presented with a wide spectrum of parameters for potential output. As noted in Chapter 8, "snapshot" files output at specific time intervals that list the particle positions and contact forces are useful to generate images of the system as it evolves. Measurements of the stresses and fabric of the material (using the methods outlined in Chapters 10 and 9) are typically made at more regular intervals. As writing to disk is computationally expensive, these parameters are typically saved to memory for output at the end of the simulation. To assess the effectiveness of the simulation it is useful to monitor either the average or maximum contact overlap and the energy of the system. (Kuhn (2006) monitors a "dimensionless overlap" that is the ratio between the average overlap at the contact points and the mean particle size, while Ng (2006) considers the maximum overlap between particles.) It is also useful to measure the forces along all boundaries to the system to assess whether the assembly is overall in, or close to, a state of static equilibrium. Where a servo-controlled algorithm is used in conjunction with rigid boundaries (Section 5.2) it is important to output both the target stress and actual stress to assess the success of the servo-controlled algorithm and the control parameters used.

11.5   Number of Particles

The problem of achieving a direct mapping between the numbers of particles in the physical systems under consideration and in the DEM model is daunting. There are very large numbers of particles in even small samples of sand. This can be illustrated by a very simple calculation. Consider a small triaxial test specimen that is 30 mm in diameter and 60 mm high and has a void ratio of 0.65. The volume occupied by soil particles is 16,708 mm^3. If we assume a D_{50} of 200 μm for this material, then a representative particle volume would be $\frac{4}{3}\pi 0.1^3 = 0.004$mm^3. Dividing the total volume of particles by this representative volume, a rough estimate of the number of particles in the specimen is 4 million. Similar calculations have been made by others, for example Dean (2005) estimated that there are about 1 million 1 mm sized particles in a volume of 1 litre. The need to achieve results in realistic run times restricts analysts to consider significantly smaller numbers of particles. Referring to Chapter 12, geomechanics DEM simulations using more than 1 million particles are rare. This is a key constraint on the use of DEM to consider industrial problems in particular. As discussed in Chapter 13, developments in high-performance computing will hopefully contribute greatly to solving this problem. In the meantime, analysts need to carefully consider the implications of the unrealistically small numbers of particles used for their particular application. The importance of considering system size is not unique to particulate DEM, Rapaport (2004) also highlighted the need to understand the implications of the finite size of the system when simulations are used to predict physical responses in molecular dynamics simulations.

In experimental geomechanics, consideration is typically given to the relationship between the particle diameter and the specimen size. For example Jeffries et al. (1990) described triaxial compression tests that examined the response of specimens of dense sand to sample size. It would be rare to see a physical experiment (e.g. a triaxial test) performed on a sample that is less than about 10 - 20 times the size of the largest particle in the system. Marketos and Bolton (2010) consider this from a DEM perspective, citing the guidance of Head (1994), who specified that the minimum ratio of sample size to maximum particle dimension should be 5 when testing compressive strength, 10 for consolidation tests and 12 for permeability tests. This rule is not always strictly adhered to in DEM simulations, with the simulation computational cost obviously limiting the number of particles that can be included in the "virtual" samples. It clearly is important to include enough particles in a DEM simulation so that the response is representative of the material in general and not indicative of that particular disk configuration, i.e. the analyst must confirm whether two simulations with the same particle sizes and shapes, and the same coordination number, but with slightly different initial coordinates, give the same mechanical response. In Figure 11.1(a) and (b) biaxial test simulations were repeated on specimens of 224 and 896 disks respectively (see also O'Sullivan et al. (2002)). For both simulations a small perturbation was introduced by varying the particle size distributions slightly. As illustrated in Figure 11.1, the response of the smaller sample was clearly very sensitive to small variations in the distribution of the radii sizes,

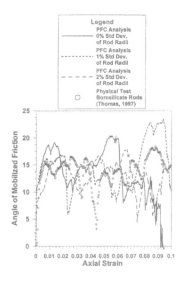

(a) Stress-strain response of samples with 224 disks

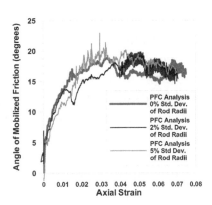

(b) Stress-strain response of samples with 896 disks

Figure 11.1: Illustration of sensitivity of systems with different numbers of particles to small perturbations O'Sullivan et al. (2002)

while the sensitivity of the larger specimen was less marked. The large fluctuations in the response observed in the smaller simulations indicates that the macro-scale response is highly sensitive to a change in only one or two contact positions. In Figure 11.2 the response of two equivalent specimens with 5,728 disks and 12,512 disks can be seen to be very similar. While the results presented in Figure 11.2 indicate that upscaling of DEM results to samples with larger domain sizes is possible, care should be taken in extrapolating these findings to simulations with different boundary conditions, particle geometries or particle size distributions.

Careful consideration of the influence of specimen size on the measured response is given by Kuhn and Bagi (2009). These authors cite the earlier work of Bazant and Planas (1991) who propose the two main mechanisms for introducing size effects in granular materials are a fracture mechanics size effect and a boundary layer effect. Kuhn and Bagi link the fracture mechanics size effect with the fact that the width of a shear band depends on the particle size rather than the specimen size. The boundary layer effect occurs because the particle packing density and fabric close to the boundaries will differ from that in the interior of the specimen and the relative volume of affected material in the boundary zone compared to the total volume will increase as the sample size decreases. In their simulations of tunnel boring machine-soil interaction Melis Maynar and Medina Rodríguez (2005) used a relatively small number of particles for this type simulation (13,100 rigid clusters, each comprising two overlapping spheres), and observed that their simulations were very sensitive to small perturbations (e.g. small changes in the particle packing).

Potyondy and Cundall (2004) make some useful comments regarding the issue of simulation size. They propose a guideline that stresses should be measured over a volume 12 particles in diameter to ensure homogeneity of response. They also provide some interesting data on the relationship between particle size and observed macro-scale response. A parametric study was carried out in which parallel bonded model rock specimens were subject to a Brazilian test and biaxial (2D) and triaxial (3D) tests at 0.1 MPa and 10 MPa. The 2D specimens had dimensions 31.7 mm

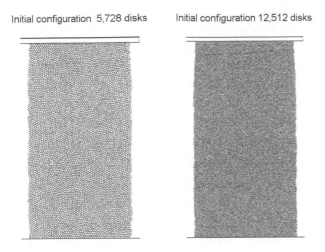

(a) Initial disk configurations for biaxial tests

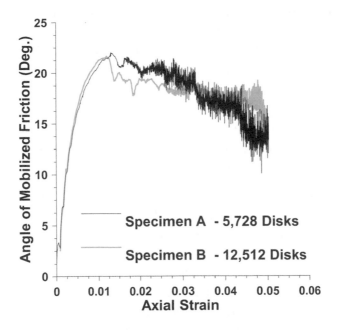

(b) Specimen response for large random biaxial simulations, indicating location of strain measurement points

Figure 11.2: Equivalence of response for two large, dense, two dimensional disk specimens subject to biaxial compression

× 63.4 mm, while the 3D specimens had dimensions 31.7 mm × 31.7 mm × 63.4 mm. The average particle diameter within the specimens was varied and all other input parameters remained constant. Representative findings are summarized in Figure 11.3, where the mean value and the coefficient of variation (based on 10 DEM simulations) are plotted as a function of particle size, considering the overall Young's modulus (Figure 11.3(a)) and friction angle (Figure 11.3(b)). As can be observed the material Young's modulus tends to decrease slightly with increasing particle size in both 2D and 3D simulations, while no clear correlation between the friction angle and the particle size was observed. There is a slight increase in the unconfined compressive strength and cohesion with decreasing particle size, while the tensile stress decreases slightly, and the Poisson's ratio exhibits less sensitivity. While the conclusions of this study may not be generalized to all DEM simulations, the approach taken in this research is thorough and a very good example of the detailed parametric studies that are advisable before drawing conclusions from DEM simulations. A particularly nice aspect of this work is the way the authors repeated the simulations on equivalent specimens (generated using different random number seeds) and then monitored the variation in the observed response. For all of the macro-scale parameters observed there is a clear decrease in the coefficient of variation as the particle size decreases and the number of particles increases.

One way to avoid the problem posed by limited numbers of particles is to use a periodic cell , e.g. Thornton (2000) or Ng (2004b). Where used to simulate element tests this approach also removes the non-uniformities associated with boundary effects, which can affect the macro-scale response measured. Cambou (1999) discusses the fact that where periodic boundary simulations are used the representative volume element (RVE) considered must be sufficiently large. Rapaport (2004) pointed out that where a periodic boundary is used finite size effects arise. Pöschel and Schwager (2005) considered the use of periodic boundaries in hard sphere, event driven particle codes. They stated that if the periodic cell is too small, correlations will develop between opposite sides of the cell. Cundall (1988a) described the repetition of periodic cell

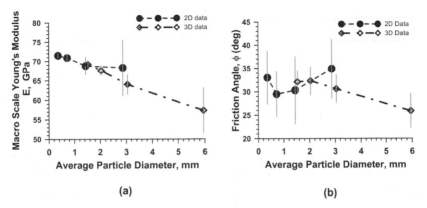

Figure 11.3: Representative results from particle size study of Potyondy and Cundall (2004) (a) Macro-scale Young's modulus versus particle size, (b) Angle of friction versus particle size

simulations with 45, 150 and 1,200 spheres of two sizes subject to triaxial compression to a strain of 60%. He found that the overall sample responses were very similar, with more noise being observed in the 45-sphere specimen. To assess whether a periodic cell size is adequate, the analyst can either check whether the results change significantly if the cell size is increased or alternatively assess whether equivalent results are obtained if the system is perturbed (e.g. a different series of random numbers are used to generate the initial particle positions). However, for his simulations using spheres, Barreto (2010) found 4,000 to be the optimum number of particles for periodic cell simulations; he compared simulations using 2,000, 4,000 and 8,000 particles. The optimum number of particles will certainly depend on the particle size distribution. Thornton and Zhang (2010) used a significantly larger number of particles (27,000) than have been used in other documented studies with periodic cell boundaries.

Simpson and Tatsuoka (2008) acknowledge the suggestion by Cundall (2001) that for simulation of field-scale problems in particular, appropriate scaling laws may be adopted, but that it is difficult to apply this approach to problems with strain localizations. Another alternative is to used a mixed boundary simulation environment. Just as Cui et al. (2007) used circumferential periodic boundaries to simulate axisymmetrical systems, planar periodic boundaries can be used to simulate 3D systems deforming in plane strain conditions. Cook et al. (2004) considered failure of rock mass around a central, circular opening using 2D DEM simulations and used an approach somewhat analogous to mesh refinement in finite element analysis, where small disks were used to simulate the rock mass close to the opening, and the disk size increased moving away from the circular hole. In their model Cook et al. (2004) varied the contact parameters as functions of the disk size to achieve a constant bulk strength for the material.

11.6   Speed of Simulation

A DEM model comprises a large assembly of particles contacting using (most commonly) elastic springs. Care needs to be taken in selecting the speed at which a sample is compressed or sheared. For many geomechanics applications the objective is to simulate a quasi-static response. This means that the system is not flowing and is in, or is close to, a state of static equilibrium. If the speed of deformation is too fast, a dynamic response will be recorded (i.e. stress waves propagating through the system) rather than a static response, as might be intended. Furthermore in a servo-controlled simulation it may appear that the specified stress state is achieved, but in reality the system is responding like a large spring (refer to Figure 2.5) and the equilibrium stress level is less than that instantaneously measured.

To get a feel for the individual contact response in a DEM model it is useful to consider a single degree of freedom system comprising a weightless ball resting on a horizontal boundary where the contact between the ball and the boundary is modelled with a linear spring. If gravity is suddenly turned on, in the absence of energy dissipation, the force in the contact spring will oscillate around the equilibrium position between a value of 0 and twice the ball weight. If damping is applied, the magnitude of the oscillations will decrease and the ball will come to rest. Similarly, when there is a change to the boundary conditions for an assembly of particles, initially there may be an elastic response at many of the contact points, with some of the particles experiencing very high forces. If these forces are very large, the fabric of the assembly can be disturbed, or in an extreme case selected particles may experience such unnaturally high accelerations that they may escape from the boundaries because of excessively large particle displacements occurring within a single time step. As noted above, and emphasized by Kuhn (2006), one way to mitigate this problem is to apply changes in stress state incrementally.

Typically in DEM simulations with rigid walls, it is useful to ensure that the forces measured along the boundaries are in equilibrium, e.g. in a triaxial test simulation the total force on the bottom boundary should be approximately equal in magnitude to the total force along the top boundary throughout the simulation. Where a servo-controlled algorithm is used to control the stress state by moving the system boundaries, ideally stress changes should be applied slowly and incrementally so that the assembly approaches the required stress state in a controlled and monotonic manner. If the target stress is overshot it can be difficult to achieve the desired stress state; the system may simply oscillate around the target value, with the overshooting indicating that the walls are moving too fast. Furthermore there is a risk of inducing undesirable changes to the fabric of the material. Similarly care must be taken when using a servo-controlled algorithm to control the stress state during a simulation and it is essential to check that the desired stress state was achieved and maintained. Carolan (2005) found that maintaining the stress state was more difficult where particle crushing was simulated.

In servo-controlled simulations once the desired specific stress state is attained, it is good practice to halt all boundary motion and run through a number of DEM calculation cycles to ensure that the system is in equilibrium at that required stress level. Kuhn (2006) describes this as allowing a period of "quiescence" For periodic cell simulations Cundall (1988a) described how he set his strain rates to be zero at certain points in his simulations and then monitored the stresses. If large stress changes were noted this was taken to indicate the presence of finite dynamic effects and the test was repeated at a lower strain rate. Kuhn (2006) also acknowledges that when selecting a deformation rate "slow is (usually) better." Accepting that running a simulation at a rate that is too slow will waste time, he found that a trial and error procedure is often needed to select the appropriate deformation rate. He also suggests monitoring the average overlap between the particles as a means to assess whether the speed is appropriate. Allowance for a period of quiescence is also necessary when configurations are read into the program from previous simulations as numerical round-off may result in small variations in the disk positions and in the contact forces. Similarly if there is a change in the boundary conditions, e.g. from rigid boundaries to a stress-controlled membrane, there will be local changes in the forces felt by particles on the exterior of the sample, and a period of DEM cycles without any applied deformation is recommended until a state of static equilibrium is attained.

As acknowledged by many DEM analysts, the time step in a DEM simulation should be significantly small so that a disturbance propagates from a particle only to its nearest neighbours. The strain rate will also influence the amount a particle moves in a given time increment, and so there must be a balance between the time step chosen and the strain rate applied.

Quantitative measures to assess whether the system is in equilibrium exist. Radjai (2009) amongst others proposes a definition for an inertia number that can be used to identify whether a flow is quasi-static. The 3D inertia number I is given by

$$I = \dot{\epsilon}_q \sqrt{\frac{m}{pd}} \qquad (11.1)$$

where $\dot{\epsilon}_q$ is the shear strain rate, m is the particle mass, d is the particle diameter, and p is the confining pressure. In 2D the expression is slightly different:

$$I = \dot{\epsilon}_q \sqrt{\frac{m}{p}} \qquad (11.2)$$

For a simulation to be quasi static the condition $I \ll 1$ should be met, giving an indication that the inertia forces are significantly lower than the contact forces. As outlined by Radjai (2009), this parameter I is derived by considering that the static forces acting on the particle (in 3D) are of the order $f_s = pd^2$ and the impulsive forces due to shear strain, i.e. the inertia forces, are given by $f_i = \frac{md\dot{\epsilon}_q}{\Delta t}$ where m is the average particle mass, Δt is the time scale of the flow and is given by $\Delta t = \dot{\epsilon}_q^{-1}$.

Kuhn (2006) and Ng (2006) consider the ratio of the magnitude of the resultant forces (i.e. the out-of-balance forces) acting on the particles and the magnitude of the average contact force. Kuhn (2006) considers the average moment imbalance acting on the particles as a further indication of a pseudo-static state during simulation. Ng (2006) defines a index to monitor during the simulation as:

$$I_{uf} = \sqrt{\frac{\sum_{p=1}^{N_p}(\mathbf{f}_{res}^p)^2/N_p}{\sum_{c=1}^{N_c}(\mathbf{f}^c)^2/N_c}} \tag{11.3}$$

where $\mathbf{f}_{res}^p$ is the resultant force acting on particle p and $\mathbf{f}^c$ is the contact force for contact c and there are N_p particles and N_c contacts in the system respectively.

11.7 Validation and Verification of DEM Codes

As with any computer program, it is essential that a DEM code be validated. For example in finite element analyses the "patch test" is used to check for convergence and also to check that the programming of the code is correct. In principle, validation of a DEM code can be approached in two ways: analytically or experimentally. Analytical validation using closed form solutions provides information that the model is performing as it should, while experimental validation or verification conforms that the physical material response is captured. Validation exercises are useful to confirm correct implementation of the algorithms, the insensitivity of the program to the hardware platform used in simulations, and that a user is correctly running the code.

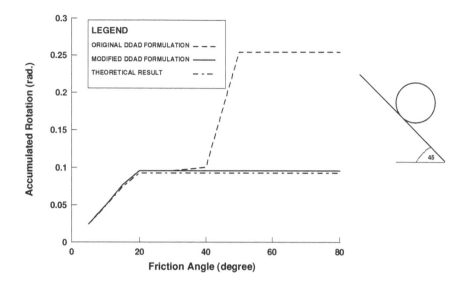

Figure 11.4: Study of ball sliding down an inclined plane (O'Sullivan and Bray, 2003a)

11.7.1 Single-particle simulations

A preliminary check on any DEM code including a rigid wall implementation is to assess whether a single-particle, resting on a horizontal boundary will exhibit simple harmonic motion, with the period being determined by the mass and the contact spring stiffness. Simulating this system also gives an analyst a basic "feel" for the fundamentals of the analysis method and allows them to appreciate its dynamic nature. Simple single-particle systems can be very useful analytical validation tools. For example, O'Sullivan and Bray (2003a) demonstrated that simulating a ball rolling down an inclined plane is a useful test to confirm appropriate implementation of the shear contact model in DEM codes. As illustrated in Figure 11.4 this simple benchmark exercise was instrumental in picking out an error in the shear spring formulation adopted in the implicit DEM code, DDAD, and validating a modified shear spring formulation.

Another example of the use of a simple system in DEM validation is Munjiza et al. (2001) who validated their time integration algorithm for updating the rotations of non-spherical particles in three dimensions by considering the motion of a single non-spherical particle subject to various initial angular velocities. Vu-Quoc et al. (2000) also include some simple analytical validation examples for their ellipsoidal particles.

11.7.2 Multiple particles on lattice packings

The simple single-particle simulations are really only useful to validate implementations of the time integration scheme and particle-boundary interactions. Validation of the ability of the code to simulate the response of an assembly of particles is also necessary. Analytical validation of the ability of DEM codes to simulate the response of multiple interacting particles is challenging as most dense assemblies of particles form statically indeterminate systems. However, expressions for the peak strength of uniform disks and spheres with lattice packings can be found by reference to Rowe (1962) or Thornton (1979). The study by Rowe (1962) included physical experiments on hexagonally packed rods and uniform spheres with lattice packings. Rowe's work is useful for those developing either 2D or 3D DEM codes. He validated analytical expressions he proposed for assemblies of uniform disks and spheres with regular packing configurations against triaxial and biaxial compression tests on assemblies of steel balls and steel rods respectively. In the original documentation of the Trubal code, Cundall and Strack (1979b) used Rowe's triaxial test on face-centred-cubic packed spheres for validation purposes. A detailed description of the simulation approach can be found in the PFC 3D User Manual (Itasca, 2004). In his analytical study Thornton considers assemblies of spheres (3D) and provides expressions for the peak stress ratios under both triaxial and plane strain conditions. Both Rowe and Thornton give expressions for the peak strength as a function of the particle friction, hence a parametric study can be carried out to ensure the appropriate response is observed for a range of friction values. These problems can play a key role in validating new implementations of algorithms in DEM: for example Cui and O'Sullivan (2005) validated their circumferential periodic boundaries by firstly considering the response of a specimen of uniform spheres with a face-centred-cubic packing.

11.7.3 Experimental validation

The analytical expressions proposed by Rowe (1962) and Thornton (1979) are themselves very idealized models of the material response. Consequently comparison of DEM codes against physical tests is also necessary. It is important to draw a distinction between validation tests and calibration tests. When calibrating a model, the real material is tested and the DEM input parameters are varied to match the physical response observed. In contrast, in a validation test a simple granular material is considered and the particle geometry and material properties are accurately modelled in the DEM simulation. A calibrated model is developed to capture features of the overall material response, rather than to study fundamental particle-scale mechanics.

In two dimensions validation studies can be achieved using Schneebeli rods or photoelastic disks. In one of the earliest experimental validation studies Cundall and Strack (1978) used the BALL computer program to simulate two photoelastic simple shear tests on photoelastic disks that had been carried out by Oda and Konishi (1974). O'Sullivan et al. (2002) made some observations that may be useful to others considering two dimensional DEM code validation. If analysts are considering regular packings, it is important to use precision manufactured rods, as the response of these systems are very sensitive to small variations in geometry. If randomly packed polydisperse rods are used then the sample must be large enough to be insensitive to small perturbations to the system as illustrated in Figure 11.1 above. Figure 11.5 illustrates a representative 2D validation study that considered biaxial compression of hexagonally packed disks subject to vacuum confinement, simulations carried out using PFC2D.

In three dimensions, good correlation has been observed between the response of assemblies of steel balls and DEM simula-

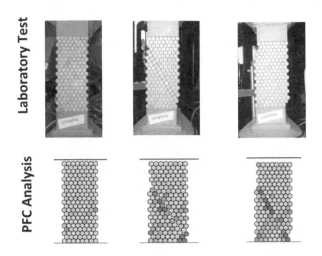

(a) Deformed specimen geometry: comparison of DEM simulation and physical test

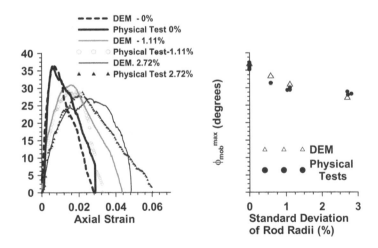

(b) Stress-strain response: comparison of DEM simulation and physical test

Figure 11.5: Biaxial compression of hexagonally packed disks for 2D DEM code validation (O'Sullivan et al., 2002)

tions using spheres (Cui and O'Sullivan (2006), Cui et al. (2007), and O'Sullivan et al. (2008)). Figure 11.6 compares strain-controlled physical cyclic triaxial tests (again with vacuum confinement) with DEM simulations using circumferential periodic boundaries.

Typically validation studies considering the ability of DEM to simulate the response of an assembly of particles focus on the macro-scale, overall response of the assembly. The model's ability to accurately capture the particle-scale mechanics is then inferred from a macro-scale agreement. Studies using photoelastic particles are an obvious method to validate DEM codes at the particle scale. Another potentially promising alternative is the comparison of local strains and deformations obtained by analysis of the DEM data with deformations calculated using analysis of images obtained of 2D Schneebeli rod tests (e.g. using the data developed by Hall et al. (2010)). Developments in micro-computed tomography (e.g. Hasan and Alshibli (2010)) will likely allow future verification of observations on the internal material structure and its evolution that have been made using DEM. There is clearly a need to compare the measurements of particle deformation mechanisms and fabric evolution obtained in DEM with physical experiments. Comparison of images of particle motion during testing with DEM simulations is one possible option, however, this necessitates images with high spatial resolution, and possibly with high temporal resolution. Promising results using this type of approach are presented by Gabrielia et al. (2009) who compared a 3D DEM simulations of a shallow foundation at the top of a slope with a model test on sand. Good agreement was obtained between the vertical load displacement plots for the physical model test and the DEM simulation; considering the internal material response, equivalent distributions of vertical displacements were obtained for the DEM data and the physical test (where the displacement data were obtained using digital image correlation).

446

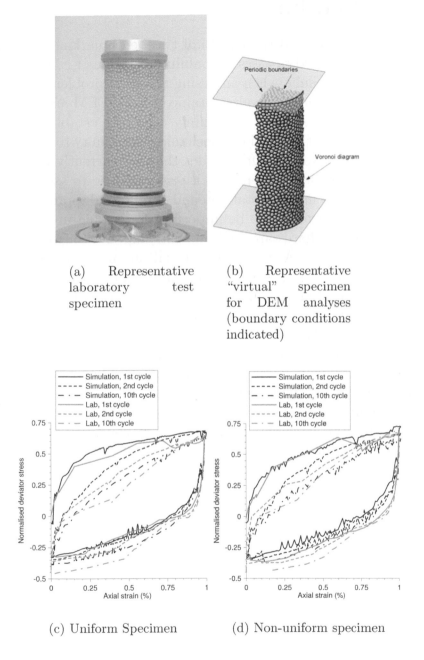

(a)   Representative laboratory   test specimen

(b)   Representative "virtual"   specimen for   DEM   analyses (boundary conditions indicated)

(c) Uniform Specimen

(d) Non-uniform specimen

Figure 11.6: Comparison of macro-scale response observed in the laboratory tests and DEM simulations of strain-controlled cyclic triaxial tests (O'Sullivan et al., 2008)

11.8   Benchmarks

Potts (2003) discussed the results of two benchmarking exercises that were performed to assess the ability of various FEM codes to give equivalent results when analysing the same problem. The results indicated that the results obtained from a FEM analysis can be user dependent. There is certainly a need to establish benchmark tests within the DEM geomechanics community to assess the reliability of the various DEM codes that are currently available. The objective of such a benchmark test would be to assess accuracy of different implementations and simulation approaches. A benchmark test would also be important for testing implementations in different hardware environments and assessing computational performance. As granular materials are inherently complex and highly non-linear it would seem that an initial benchmark test should consider a simple simulation of triaxial/biaxial compression of disks/spheres arranged on lattice packings with known analytical peak strengths, prior to embarking on simulation of a more complex material or boundary value problem. Holst et al. (1999) described a comparative study considering silo filling. When compared, there was a wide scatter in the DEM simulation results. Given the increase in use in DEM since this study was completed, there is certainly scope for a repeated study considering a geomechanics application.

Chapter 12

Use of DEM in Geomechanics

There are two primary motivations to use DEM in geomechanics. Firstly, in applied boundary value problems, discrete element methods can more easily simulate large-deformation problems than continuum-mechanics-based analysis tools. DEM simulations can also capture mechanisms such as arching or erosion that are a consequence of the particulate nature of the material. The second use of DEM is as tool in basic research. A discrete element simulation can probe the material response at a much more detailed scale than can be monitored even in highly sophisticated laboratory tests. In conventional experimental soil mechanics hypotheses about the mechanisms underlying the sometimes highly complex response of soil can be proposed. A DEM model allows us to delve into the inner workings of soil and confirm or reject these hypotheses. To paraphrase Weatherley (2009), in a DEM simulation information which is "hidden" in conventional physical experiments is revealed.

This Chapter firstly gives an estimate of the extent of DEM use in geomechanics, primarily by considering publications in peer-reviewed international journals. Then a review of selected publications documenting DEM simulations of applied boundary value type problems is presented. The final section of this Chapter considers the use of DEM in basic soil mechanics research. As is clear from the data presented in Section 12.1, the number of people using DEM and applying it in a variety of interesting ways is rapidly increasing. The examples of DEM use selected for inclusion here should give readers an idea both of what can be achieved and what has been achieved using particulate DEM. The most recent references used here date from mid 2010, and it is reasonable to anticipate further noteworthy contributions will be published in the near future, building on this growing body of knowledge.

12.1 Extent of DEM use in Geomechanics

DEM is by now established as a research tool across a number of disciplines. Interest in DEM is rising rapidly and assessing the state of the art of DEM use either in geomechanics, or in the broader scientific community, is difficult as the situation is rapidly changing. Zhu et al. (2007) used the Web of Science database to gauge the extent of DEM use across engineering and the physical sciences between 1985 and 2005. To get a more recent assessment of the popularity of DEM the approach was again applied using the ISI Web of Knowledge database (Reuters, 2010). The search for DEM-related publications used the keywords discrete element method/model, distinct element method/model, discrete particle simulation/method/model and granular dynamic simulation. To access the database over the full period, each keyword was input separately as under the "topic" search option; therefore there is a slight risk of double counting of papers. The results of this search are presented in Figure 12.1. The survey indicated that an estimated 2,451 papers relating to discrete element type simulations were published between 1977 and the end of 2009. There has been a continuous increase in the number of papers published per year over the past 20 years, and between 1996 and 2006 there was an almost linear rate of increase with about 18 additional papers being published each year. There has been a more recent sharp increase in the rate of DEM-related publications, with 425 papers being published in 2009. This represents 127 additional papers in comparison with the number of publications from 2008. DEM is clearly a method whose user community is rapidly growing. Direct reference to Zhu et al. (2008) is useful as they cite many publications where DEM has been used both in basic and in applied research across a range of applications, including geomechanics.

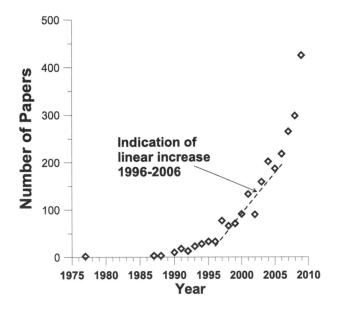

Figure 12.1: Synthesis of data obtained in ISI Web of Knowledge Search, following approach of Zhu et al. (2007)

To determine the level of DEM use in geomechanics research, the contents of nine geotechnical journals were reviewed and a summary database of discrete element related papers published in these journals was created. It was not possible to review every single paper in these journals, rather discrete element publications were identified by reviewing the paper titles and the authors names; therefore it is possible that a few relevant papers may have been omitted. The journals considered were *Géotechnique*, the *International Journal for Numerical and Analytical Methods in Geomechanics*, the *International Journal of Geomechanics* (ASCE), the *Journal of Geotechnical and Geoenvironmental Engineering* (ASCE), *Soils and Foundations*, the *Canadian Geotechnical Journal*, *Computers and Geotechnics*, *Geomechanics and Geoengineer-*

451

ing: An International Journal, and the *International Journal of Rock Mechanics and Mining Sciences.* The time period examined was between the start of 1998 and the end of 2009. A total of 130 papers were identified. This total excludes the use of block DEM codes for rock block stability analyses; for an indication of contributions using this type of analysis refer to Jing and Stephansson (2007) or Bobet et al. (2009).

As illustrated in Figure 12.2(a) the general trend is for the number of publications each year to increase, however this increase has not always been consistent. Similar to the pattern in the broader scientific community (Figure 12.1), there has been a recent surge in DEM related activity in the geomechanics research community. The range of DEM studies published within the geomechanics literature can be broadly classified as documentations of DEM algorithm modifications, validation of DEM models, calibration of DEM models, analyses of the relationship between particle-scale (micro-scale) mechanics and the bulk (macro-scale) material response, development of interpretation techniques, simulations of element tests and simulations of field-scale boundary value problems. About 34 of the papers describe developments to discrete element algorithms, 24 describe the application of discrete element modelling to simulate boundary value problems, while 67 describe discrete element analyses of the micromechanics of soil response including simulations of element tests. In a general review of developments in constitutive and numerical modelling of soil between 1948 and 2008 (focussing primarily on the publication *Géotechnique*) Zdravkovic and Carter (2008) concluded that to date the main use of DEM has been in advancing understanding of material response by simulation of element tests. This reflects a broader trend of numerical simulations becoming an integral part of basic research. However, in contrast to the finite element method, whose use in both applied research and industry is now commonplace, studies that have applied DEM to large-scale boundary value problems are relatively rare.

Figure 12.2(b) is a plot of the number of particles used in published geomechanics DEM simulations versus the year. There has been a significant advancement on the original numbers of parti-

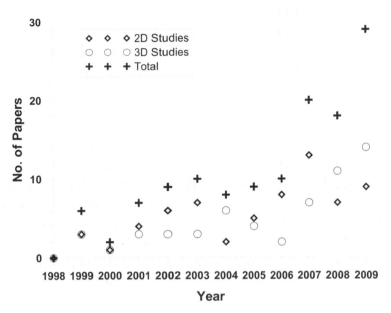

(a) Number of papers published each year

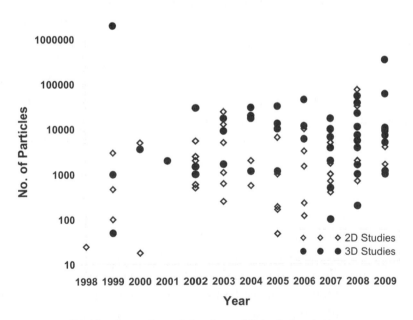

(b) Number of particles in published simulations

Figure 12.2: Summary of data collated in a review of DEM related papers

cles in the simulations of Cundall and Strack (1978) (200 - 1,200 disks), but the numbers of particles remain significantly smaller than those encountered in real applications. The data presented in Figure 12.2(b) indicate that there have been few published studies involving more than 50,000 particles in either 2D or 3D. These are small numbers of particles in comparison with real soil. Consider a very simple estimate: assuming a void ratio of 0.563, there are over 150,000 200 μm spheres in a 10 mm $\times$ 10 mm $\times$ 10 mm cubical specimen. (Refer also to the comments in Section 11.5). While Cundall (2001) predicted that by 2011 a DEM simulation involving 10 million particles would constitute an "easy" problem, referring to Figure 12.2(b) it is clear that within the geomechanics research community the largest simulations are one order of magnitude smaller than this aspiration. It seems that outside of geomechanics larger simulations are more common; for example Cleary (2007) states that his simulations typically consider more than 100,000 particles routinely and refers to simulations including up to 8 million particles. Studies using less computationally intensive two-dimensional DEM simulations outnumbered 3D papers in almost every year until 2007. Special editions of journals focussing on "micro-soil mechanics" and DEM give useful overviews of the types of study that have been completed using DEM. A number of examples of DEM use in geomechanics can be found in the two volume themed issue of *Géotechnique* that was produced in 2010 (Baudet and Bolton, 2010a and 2010b) and in the special edition of *Geomechanics and Geoengineering: An International Journal* entitled "Advances in discrete element methods for geomechanics" (Morris and Cleary, 2009). Other DEM related special editions of journals outside of geotechnics include the *Journal of Engineering Mechanics* (Ooi et al., 2001), *Powder Technology* (Thornton, 2009), *Particuology* (Zhu and Yu, 2008) and *Granular Matter* (Luding and Cleary, 2009). Each of these special editions includes contributions that are likely to be of interest to the geomechanics community and each of these journals regularly publish contributions related to particulate DEM.

12.2 Field-Scale/Applied Boundary Value Simulations

In general, geomechanics studies are application driven. The use of DEM to study soil micromechanics may inform development of more sophisticated and reliable continuum models, and thus have an indirect impact on engineering practice. From an industrial perspective, there is interest in the direct application of DEM to simulate field-scale boundary value problems. An intrinsic feature of discrete element models is their ability to make and break contacts. They are therefore particularly well suited to modelling problems involving large displacements or localizations. The principal challenge in the use of DEM in this way however, is the number of particles that are included in our DEM models. A real boundary value problem will include millions of particles with highly complex and varying geometries. DEM models are therefore simplifications of the physical system that are usually better suited to provide insight into the development of mechanisms, rather than providing quantitative predictions of response.

It is difficult to gauge the current level of DEM use in industry; however published applied research studies are a good indicator of the level of industrial-oriented analyses using DEM. It seems that the use of DEM in geotechnical engineering practice lags its use in process engineering. Zhu et al. (2008) and Cleary (2000, 2007) give a number of examples of applications where DEM has been used. These include flow in hoppers, mixers, drums and mills, pneumatic conveying, pipeline flow, industrial processes such as cyclones, visualization of transient flow mechanisms, examination of breakage rates, analysis of boundary stresses, segregation and mixing rates, and wear rates. Examples of the application of DEM that are relevant to the geomechanics community can be found in geological publications (e.g. Schöpfer et al. (2007) considered the development of faults in rock mass simulated as bonded disks).

In most cases the applied research studies documented in the geomechanics literature have used highly idealized models (small numbers of particles, often 2D) and quantitatively accurate predictions of soil response have been rare. However, the references cited here and in other similar studies are important contributions as they document the first steps towards developing the methodologies for applying DEM to analyse industry-oriented problems. The findings will inform the techniques used in larger scale simulations once the computational resources become more accessible (following the necessary hardware and software developments). Such studies also give an indication of potential future application of DEM in geotechnical engineering.

Simulation of rock mass response

The concept of bonding particles together with tensile-capable contact models to simulate rock mass or cemented sand response was introduced in Chapter 3. As discussed in Chapter 11, the required contact model parameters are determined by calibrating the particulate DEM models against physical tests on rock specimens. Many of the published studies using bonded particle DEM models have simulated element tests rather than applied boundary value problems. Fundamental research using DEM in simulations of element tests is discussed in Section 12.3. Contributions documenting simulations of laboratory tests using bonded particle models are considered separately here, as the objective of these studies has been to demonstrate the ability of DEM to capture rock mass response, rather than to study the fundamental mechanics in detail. These element test simulations form a key part of applied DEM modelling as they are laying the ground for future use of particulate DEM to simulate rock mass response in large-scale field applications. Potyondy and Cundall (2004) give an excellent introduction to the "philosophy" of using this approach to simulate rock mass response; the reviews by Bobet et al. (2009) and Jing and Stephansson (2007) are also useful.

Some of the contact models that have been used in bonded particle simulations were introduced in Chapter 3 (Section 3.8). While a few different approaches have been used to model the cementation at the grain contacts, in all cases the contact model used is capable of transmitting a normal tensile contact force. In the tangential direction a contact shear strength is also specified (i.e. shear resistance is no longer simply governed by Coulomb friction). The contact model is then used to model the cement in a sandstone or a "notional" cement in a granite rock mass. In the parallel bond contact model used by Potyondy and Cundall (2004), moments and forces are transferred at the contact points. Huang and Detournay (2008) used a simpler contact model that does not transmit moment at the contact points; other contact models include the ductile contact bond used by Utili and Nova (2008). In their 2D simulations Li and Holt (2002) and Cho et al. (2007) used non-circular particles composed of disks bonded together, with a distinction being made between inter-granular bonds and intra-granular bonds.

As illustrated in Figure 12.3, Wang and Leung (2008) used small disks to represent the cement phase of the material with a parallel bond contact model used to model the contact between the cementing disks and between the cementing disks and the larger "soil" disks. While this approach is attractive in principle, relating the DEM model parameters to the volume and strength of cement in a real soil is not easy.

The DEM models are often very ideal representations of the real rock mass response whose response they aim to simulate. As noted in Chapter 11, in many cases a 2D DEM model is calibrated against physical tests on a 3D rock. The physical model has elementary particles that can move in 3D and in many situations the in-situ 3D stress state will determine the material response as discussed in Chapter 11. In general, the validity of the bonded-particle approach may depend on the type of rock in question, and engineering judgement should be applied to assess whether a bonded particle model can be applied to the material and problem of interest. It seems reasonable to apply this modelling approach to sandstone, or cemented sands, where bonded particles are a valid approximation of the physical material. Use of this approach to simulate other rock types, such as granite mass may be more difficult to justify.

Potyondy and Cundall (2004) list the features of rock mass response observed in the laboratory, that can be captured using a bonded particle model. Some of the response characteristics are

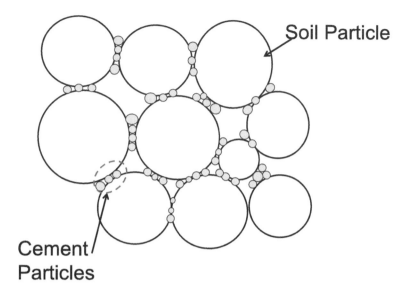

Figure 12.3: Illustration of approach used to model cement by Wang and Leung (2008), with parallel points inserted at contact points between cement and soil particles.

also observed in sand; as discussed in Section 12.3.2 below. The response characteristics most relevant to rock mass include:

1. A continuously non-linear stress-strain response, with either a softening or hardening in the response being observed after yield.

2. A transition from brittle to ductile shear response with increasing confining stress

3. The "spontaneous" appearance of microcracks and localized macro-fractures and emission of acoustic energy.

4. A non-linear strength envelope.

5. A material response that is dependent on the stress regime, with crack patterns being quite different in the tensile and unconfined and confined compressive regimes.

The response of rock is often determined by the joints or faults that divide the rock mass. The block DEM codes UDEC (Itasca, 1998) and DDA (Shi, 1996) are designed to simulate systems of blocks intercepted by planar joints. Simulations of jointed rock mass can also be achieved using particulate DEM. Kulatilake et al. (2001) described triaxial test simulations on an assembly of bonded spheres. Joint planes were created in the triaxial specimens by setting the bond strengths to be very low on planes that intersected the sample; the range of sizes and heterogeneity of the sample meant that the resultant planes were rough. The mechanical response along the rock joints was simulated using direct shear tests. The response observed in the triaxial compression tests was in reasonable agreement (considering both the observed strength and the deformation mode) with the results of physical tests. Wang et al. (2003) used a similar approach to model joints in their 2D model, and again they characterized the joint response using a direct shear test simulation. They then applied the model to simulate the response of a rock slope including a joint. The element test simulation by Park and Song (2009) is interesting as they examine the response at a smaller scale than that considered in other studies. They created a 3D model of a rough joint and simulated a direct shear test along the joint and considered the response of the rock mass along the joint in some detail. Another development in this area is the smooth joint model (Mas-Ivars et al., 2008) which allows smooth joints to be formed in the bonded particle mass, thus removing the influence of the particle geometries on the joint roughness.

The response of rock mass around central, circular openings has been considered by a number of analysts. Petroleum engineers are often concerned with the stability of well bores and perforations stemming from well bores in sandstones. This problem motivated the study of Cook et al. (2004) who created a 2D DEM model of a square sample of bonded disks with a central circular opening. To simulate the impact of fluid flow on the rock mass response, the flow of very small particles towards the central cavity was simulated until failure was observed. The model qualitatively captured fracture patterns observed in physical laboratory experiments on Berea sandstone. Fakhimi et al. (2002) also used bonded disks to model the rock mass material; however, their model includes a softening response, where the tensile strength reduces after yielding. In their model, by controlling the rate of reduction, the brittleness and fracture toughness can be controlled. In their simulations they subjected a rectangular specimen with a central circular opening to biaxial compression and they compared the resultant failure patterns with acoustic emission data from a physical test.

At a larger scale, the potential to use bonded DEM particles to study the response of rock mass to tunnelling or mining has been considered. For example, Potyondy and Cundall (2004) describe a coupled continuum (finite difference) and particulate DEM simulation where, as illustrated in Figure 12.4(a), a section of the rock mass above the tunnel was modelled as a bonded disk assembly. Then, as illustrated in Figure 12.4, they simulated the stress-induced notches (groupings of microcracks) that had been observed in a field case study by applying a strength reduction factor to the parallel bond strengths and then examined the differences in failure patterns (Figure 12.4(b) and (c)). The notch formation process was observed to be sensitive to the particle size. Calvetti et al. (2004) created a 3D model of the rock around 3 abandoned mine tunnels. 9,800 spherical particles were used, and so the average DEM particle was about 1,000 times larger than the real particle size in the cemented sand used to calibrate the DEM model parameters. A weathering process was simulated by reducing the sizes of the bonds in the DEM model. The settlement trough at the ground surface above the tunnels observed in the DEM model was qualitatively in good agreement with the settlement pattern observed in an equivalent finite element model.

Bonded particle DEM models have been used to study the influence of material damage on the overall mechanical response. Schöpfer et al. (2009) included microcracks in their model by removing selected bonds. Potyondy (2007) proposed a modified version of the parallel bond model, called the parallel-bonded stress corrosion (PSC) model, where the radius of the parallel bond varies with stress level and time. This model successfully captured the

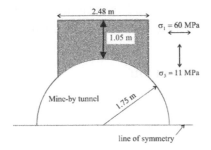

(a) DEM model configuration

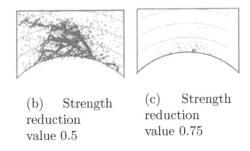

(b) Strength reduction value 0.5

(c) Strength reduction value 0.75

Figure 12.4: DEM simulation of rock mass failure above a tunnel (a) Model (b) and (c) Failure patterns for two strength reduction values (Potyondy and Cundall (2004)

macro-scale response of granite specimens subject to static fatigue tests.

Bonded particle DEM has been applied to look at other types of problems and some more examples are given in the following sections, where applied studies that have used unbonded particles are also considered. Jing and Stephansson (2007) and Bobet et al. (2009) cite references where bonded particle DEM has been applied to look at fracturing and fragmentation due to rock blasting, hydraulic fracturing, rock fracture and faulting.

Machine - soil interaction

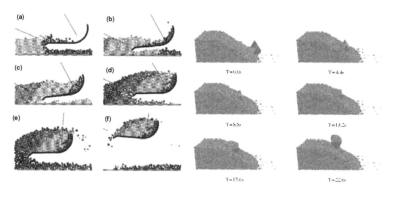

(a) Simulation of drag bucket excavator (Cleary (2000)

(b) Simulation of bucket excavator with angular particles (Nezami et al. (2007)

Figure 12.5: Examples of simulation of machine-soil interaction

The potential for DEM to aid in the prediction of soil-machine interaction and to be used as a tool to improve machine design has been recognized for over a decade, and reflects some of the interest in DEM among process engineers. Horner et al. (1998) simulated soil plowing using particulate DEM, Horner et al. (2001) considered soil-vehicle interaction, while Cleary (2000) considered the influence of particle geometry on filling a dragline excavator (Figure

12.5(a)). Nezami et al. (2007) also considered the issue of particle geometry in their simulations of interaction of a front-end loader bucket and gravel, using 3D angular shapes to model the gravel particles (Figure 12.5(b)). Melis Maynar and Medina Rodríguez (2005) simulated tunnelling using an earth pressure balance machine with a 3D particulate DEM model. The soil was simulated using cluster particles made up of two overlapping spheres, and bonds were created between the particles. Their results showed agreement with field observations; however, the simulation results were very sensitive to small perturbations. As already mentioned in Chapter 11, this was most likely a consequence of the relatively small number of particles used. At a significantly smaller scale, Huang and Detournay (2008) used 2D bonded disk particles in simulations of indentation and cutting of rock, fracture mechanisms were examined and it was shown that the intrinsic length scale for the material can be varied by varying the ratio of the shear to normal bond strength.

Penetration

The potential to use DEM to simulate penetration of rigid bodies into soil has also been recognized for a long time. A notable early study was the simulation of a cone penetration test (CPT) using a system of two-dimensional disks by Huang and Ma (1994). In this simulation only half the problem domain was modelled as symmetry was assumed, and a rigid wall was installed along the axis of symmetry. Jiang et al. (2006) also simulated CPT testing in two dimensions. Butlanska et al. (2009) approached the problem in three dimensions and inhibited rotation of their spherical particles. Despite the relatively small number of particles employed (60,000), when the cone penetration resistance was plotted against depth for different relative densities, the variation in the asymptotic resistance values was similar to experimental results (Figure 12.6(a)). The DEM simulation data then allowed investigation of the internal stress distributions (Figure 12.6(b)). Bruel et al. (2009) also describe simulation of 3D penetration tests. Lobo-Guerrero and Vallejo (2005) simulated pile installation in 2D; their contribution

is notable as they applied their simplified particle failure model
to capture particle crushing around the penetrating pile. Kinlock
and O'Sullivan (2007) demonstrated that plug formation in open-
ended piles can be captured in 2D DEM simulations using disks.
However, their work highlighted the challenge associated with developing a problem domain that is sufficiently large to minimize
the effect of boundaries on the observed response.

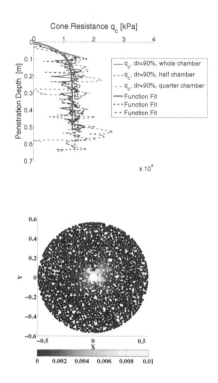

Figure 12.6: (a)Cone resistance versus depth for DEM simulations
of CPT tests (b) Variation in stress around penetrating cone, with
the lighter circles indicating the largest stresses (Butlanska et al.
(2009), Butlanska et al. (2010))

Rockfall, landslides, slope stability

Rockfalls and landslides are inherently large-displacement problems and so seem well suited to DEM analysis. The block discrete element codes UDEC and DDA have been applied to simulate failures involving motion of large blocks of rock (e.g. MacLaughlin et al. (2001)). Particulate DEM (both bonded and unbonded) has also been applied to look at slope stability analyses. In his study Maeda (2009) initially completed a series of biaxial compression tests on unbonded disks to characterize the compression and shearing response of his material by establishing the normal compression line and critical state line for his material. Knowing the critical state parameters, Maeda (2009) then considered the response within different zones in a simulation of dry flow and quantified the internal variation in micro-scale parameters (coordination number and particle velocity), as well as macro/continuum soil mechanics parameters (void ratio and state parameter).

Examples of 2D simulations where the stability of rock slopes has been considered using bonded particle DEM include Utili and Nova (2008) and Wang et al. (2003). Utili and Nova (2008) calibrated their DEM contact properties in rigid blocks and then analysed failure mechanisms for vertical slopes in rock (Figure 12.7). One particularly interesting aspect of their work is the inclusion of a study on the influence of weathering on the failure mechanism, where the rock mass properties were varied to reflect a higher degree of weathering close to the rock surface. The approach used by Wang et al. (2003) differed as they introduced zones of lower strength in the material to simulate joints.

Bertrand et al. (2008) considered the highly discontinuous problem of a rock block impacting a support structure made of a vertically supported wire mesh (for application to road protection). The mesh was modelled as a series of DEM particles connected by tensile-carrying contact bonds. The sensitivity of the force imparted to the supports to the kinetic energy of the impacting rock and the effect of damage to the mesh on the maximum force that can be resisted were both examined.

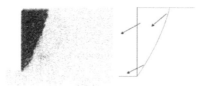

Figure 12.7: Simulation of rock slope stability using DEM, comparison of displacement vectors with upper bound limit analysis solution (Utili and Nova, 2008)

Railway ballast and rockfill

Railway ballast and rockfill are materials of interest in geomechanics that are obviously discrete or discontinuous and have attracted interest from DEM analysts. Lu and McDowell (2006 and 2010) and Hossein et al. (2007) described simulations of ballast response. Lu and McDowell (2010) modelled each ballast particle as a ten-sphere tetrahedon, adding smaller spheres to the edges of the tetrahedra to model the surface asperities using the parallel bond model (Potyondy and Cundall, 2004). They compared their model response with experimental data obtained in large-scale monotonic and cyclic triaxial tests.

Bertrand et al. (2005) used the wire mesh modelling approach to model gabions, as illustrated in Figure 12.8. The rockfill particles themselves were modelled as clusters of overlapping spheres. The DEM model was validated by considering the unconfined compression of a gabion block and comparing with data from an equivalent physical experiment. Deluzarche and Cambou (2006) proposed a DEM model of a rockfill dam, where each rock fill lock is made up of rigid clusters of disks joined together with bonds that have a finite strength. They then used the model to assess damage to the rockfill material during the dam construction, to assess the dam stability and to explore the implications of rockfill damage on the dam deformation.

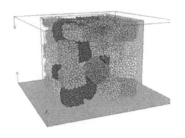

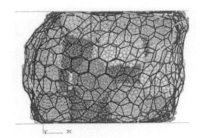

(a) Initial DEM model configuration

(b) Deformed system

Figure 12.8: Use of DEM to simulate gabion structures (a) Initial configuration (b) Deformed mesh at 16% axial strain (Bertrand et al., 2005)

Soil-structure interaction

Apart from the penetration problems described above, there has been limited application of DEM to simulate soil-structure interaction. Calvetti et al. (2004) considered the large-displacement problem of a pipe being pulled through a sandbox to study the damage imparted on pipelines by landslides. The sand particles were modelled as spheres whose rotation was inhibited. The observed displacement mechanisms were qualitatively in agreement with experimental observations. By varying the direction of pipe motion, they could monitor the sensitivity of the force felt by the pipe to the direction of motion and they used their DEM model to develop a failure surface defining the relationships between the horizontal and vertical forces acting on the pile at failure.

In a separate study Calvetti and Nova (2004) used 2D disk particles, again restrained from rotating, to simulate the movement of a retaining wall. The model captured the difference in magnitude between active and passive pressure, with the deformation required to mobilize the full passive resistance exceeding the deformation required to mobilize the full active pressure.

Arching is an important response characteristic associated with granular materials. Jenck et al. (2009) considered the arching

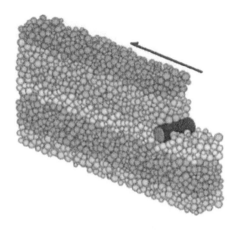

Figure 12.9: DEM model used in simulation of soil pipeline interaction (Calvetti et al., 2004)

mechanism involved when piles support granular embankments overlying soft soil (Figure 12.10). A particularly interesting aspect of this work is the way the authors compared their 2D DEM simulations with physical tests on assemblies of Schneebeli rods and with continuum (finite difference) simulations. While more complex 3D arches develop in the systems that are installed in reality, the good macro-scale agreement of the physical tests and the numerical models, indicates that future 3D DEM simulations will be useful as a tool to inform design of this type of geotechnical structure.

Seepage and soil erosion

As noted by Zhu et al. (2008) the development of coupled fluid-particle DEM algorithms was a key advancement leading to the greater use of DEM to simulate industrial boundary value problems. Chapter 6 has given an introduction to the approaches used to model coupled particle-fluid systems in geomechanics. While, as noted in Chapter 6, water and porewater pressures have a significant influence on soil response, there have been fewer applications

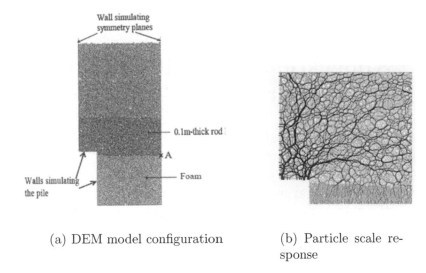

Wall simulating
symmetry planes

— 0.1m-thick rod

×A

Walls simulating
the pile

— Foam

(a) DEM model configuration

(b) Particle scale response

Figure 12.10: DEM simulation of a load transfer platform (Jenck et al., 2009)

of DEM to simulate soil-fluid interaction outside of element test simulations. In some research studies, such as the soil pipe interaction study by Calvetti et al. (2004) noted above, very simple analytical approaches are used to take account of the pore water pressures. An example of a study where a 3D fully coupled DEM model was used is the simulation of seepage beneath a flood wall adjacent to a river by El Shamy and Aydin (2008). As acknowledged by the authors, the numbers of particles used to simulate the subsoil (22,303) was relatively small; however, the distribution of porewater pressures in the soil beneath the flood wall was in good agreement with continuum analysis. The authors present their results as evidence that with refinement of the DEM model, including an increase in the number of particles, realistic seepage analyses will be possible. A second example of an application driving coupled fluid-particle simulation is given by Jeyisanker and Gunaratne (2009), who simulated water seepage through a pavement system that comprised layers of particles with different sizes. Their DEM model used spherical particles and they looked at the variation in flow velocities, pressure and hydraulic gradient within

469

the layered system.

12.3 Application of DEM to Study the Fundamentals of Soil Response

Documented simulations of element tests to study fundamental material response greatly outnumber simulations of field-scale boundary value problems. In physical science and engineering useful models are those that simplify the real physical system while still capturing the main response features, thus allowing detailed analysis to advance knowledge. Particulate DEM meets these criteria. The results of DEM simulations indicate that, even with very simple particle geometries and contact force models, the ensemble overall response of a large assembly of virtual DEM particles captures the key response characteristics of physical granular materials. DEM provides us with a tool to generate the data required to develop rational explanations for complex phenomena observed in the physical laboratory and consequently improve our ability to predict and control soil response. Potyondy and Cundall (2004) draw a distinction between more conventional modelling approaches, where the objective is to reproduce the physical response, and DEM modelling, where the goal is more often to understand the mechanisms driving this response.

Soil and other granular materials behave in a rather unique manner. In fact, it can be argued that granular materials are a fourth state of matter as, depending on the packing density and stress conditions, they can behave like a solid continuum or alternatively they can flow like a liquid. Some specific characteristics of soil or granular material response can be attributed to the particulate nature of the material. These include volume change upon shearing, frictional (or stress-dependent) material strength, stress-dependent stiffness, a wide range of attainable strengths and stiffnesses, the importance of state in determining the material properties, the influence of stress history, rate and ageing effects on the mechanical response, extremely high strain and stress non-linearity, anisotropy, strain-softening associated with shear banding, the importance of effective stress, etc. (see also Simpson and Tatsuoka (2008)).

Soils (and other granular materials) are examples of complex systems, made up of basic units that interact to generate an emergent response that is more complex than the response of the units themselves (e.g. Watts (2004)). This means that a reductional approach to soil mechanics that breaks the material down into its individual particles will not easily provide answers about the overall material response. However, despite the challenges posed by the complexity of the systems, our understanding of the link between the particles, their interactions and the overall "macroscale" soil response has significantly improved since the original development of particulate DEM in the late 1970s.

From a physical perspective, it is slightly naive to state that a granular material is made up of a large system of particles whose individual responses are "simple." Sand grains themselves are very complex, with every particle having its own unique geometry and surface roughness topology. These particles can contain internal flaws and can themselves be damaged, either suffering surface abrasion or complete crushing.

12.3.1 Context and other particle-scale approaches

Before outlining how DEM simulations have advanced understanding soil response, it is important to establish the context for DEM use in geomechanics research. Both prior and subsequent to the development of DEM, a number of research studies have that explicitly considered the particulate nature of soil alternative approaches.

As noted elsewhere in this text (Chapters 8 and 10), key contributions to advancing understanding of the link between particle-scale mechanics and overall granular material response were made using data from physical experiments on disk or rod assemblies. These disks and rods were used to create 2D physical analogue models of soil element tests allowing measurements of particle kinematics that could not be observed in the laboratory. Where

471

opaque, rigid rods are used, they are referred to as Schneebeli rods, and while particle movements can be recorded, measurement of inter-particle forces cannot be made. However, as mentioned in Chapter 8, where photoelastic materials are used, measurements of the particle stresses and inter-particle forces can also be made. Notable early contributions in this area are listed in Chapter 8. Many of the approaches now used to analyse the micromechanics of soil response using DEM were originally developed and applied to interpret photoelastic disk experiments. For example, key contributions demonstrating the correlation of the fabric tensor with the stress-strain response of soil were made by Oda and his colleagues, e.g. Oda et al. (1985). Photoelastic experiments give independent verification of the particle scale mechanics observed in DEM simulations. However, DEM presents advantages over photoelasticity; as highlighted by Gaspar and Koenders (2001), the analysis of a photoelastic assembly to generate data on the contact forces is tedious, and particularly difficult in 3D. On the other hand, referring to Figure 8.20(c), photoelastic experiments provide information on the stress distributions within individual particles; such data cannot be obtained from DEM simulations.

Recent contributions including Ibraim et al. (2010) and Jenck et al. (2009) demonstrate that the use of Schneebeli rods can continue to provide insight into fundamental soil response. Hall et al. (2010) demonstrate that modern image analysis techniques can be used to determine local strains within Schneebeli rod systems; the patterns of deformation revealed by the calculation of local strains resemble the findings of Kuhn (1999).

Granular materials form statically indeterminate systems and general expressions that predict the material response by consideration of the individual particles cannot be analytically derived. However, a number of analytical studies have restricted consideration to uniform particles (disks or spheres) with regular, lattice packing configurations. These types of highly idealized materials were considered by researchers including Rowe (1962), Horne (1965) and Thornton (1979) who used the symmetry of these packings to derive analytical expressions for the peak strengths. As noted in Chapter 11, the expressions proposed by Rowe and

Thornton are useful for DEM code validation. More recently in research to examine particle crushability, Russell et al. (2009) derived expressions for the distribution of stresses within individual particles, again considering regular, lattice packings. Tordesillas and her colleagues (e.g. Tordesillas and Muthuswamy (2009)) have studied force chain stability analytically by considering subsets of particles participating in the strong force chains and hexagonally packed disks.

Conventional constitutive modelling uses theories largely based in continuum mechanics to develop expressions linking the material stresses and strains that can accurately reproduce the macroscale response characteristics observed in the laboratory or the field. Sometimes these models are described as "phenomenological" as the focus is on capturing the overall response, rather than accurately modelling the fundamental, grain-scale mechanisms. An alternative approach to constitutive modelling is to use information on the material fabric and the contact parameters to derive overall constitutive parameters. This approach is called micromechanical continuum modelling or microstructural continuum modelling. Kassner et al. (2005) termed the general process of determining the macro-properties of a many-body system from a knowledge of the individual interactions to be the "forward problem" of statistical mechanics. As explained by Chang and Yin (2010), the idea is that the continuum constitutive response can be determined by using analytical expressions describing the contact orientations, combined with the contact stiffness values (both normal and tangential) and the friction angle at the contacts. The average strain is related to the inter-particle displacements. Key contributions to micromechanical modelling have been by Chang and his colleagues e.g. Chang and Liao (1990), Chang (1993) and Chang and Hicher (2010). This approach to modelling was adopted by Yimsiri and Soga (2000 and 2002) to study small-strain (elastic) response of granular materials. Another continuum mechanics approach that accounts for the particle scale is the Cosserat continuum. As outlined by Kruyt (2003) and Mulhaus et al. (2001), this is a continuum theory that includes point rotations as well as translations (normally in continuum modelling of deformations only translations are considered).

DEM has been applied to research that often appears very abstract in comparison with applied geotechnical engineering. However, the need for theoretical, particle-scale studies on soil response is generally accepted within the geomechanics community. For some time a technical committee (TC) of the International Society of Soil Mechanics and Foundation (ISSMFE) engineering has comprised a group of (mainly research orientated) engineers interested in studying soil response at the particle scale. This technical committee was called "TC35: Geomechanics of Particulate Materials", and was chaired by Professor Malcolm Bolton. Following a restructuring of the ISSMFE technical committees in 2010 the relevant TC name is "TC105 Geo-Mechanics from Micro to Macro." The proceedings of TC35 conferences, such as Hyodo et al. (2006) and Jiang et al. (2010) provide an indication of the way DEM is applied in geomechanics.

Critical state soil mechanics

Prior to outlining how DEM can influence soil mechanics, it is useful to give a general introduction to the basics of critical state soil mechanics, originally described in detail by Schofield and Wroth (1968). As explained by Coop (2009), while the theory of critical state soil mechanics was largely derived to describe the clay response, it can be usefully applied, with some adjustments, to the response of sands. The theory of critical state soil mechanics was formulated using concepts originating in theoretical plasticity, modified to take account of the volume changes during loading exhibited by soils. This framework provides an explanation as to why a soil at a particular density and stress level will behave in a particular way. The key central concept in the critical state mechanics framework is that upon prolonged shearing a soil will tend to an ultimate state where the stresses and volume remain constant. This critical state locus (sometimes called a critical state line) is formed by plotting the void ratio e or specific volume v against the logarithm of the mean effective stress $\ln(p')$. Referring to Been et al. (1991) and Jefferies and Been (2006), for sands the locus of points referred to as the "critical state line" (CSL), is illustrated in Figure 12.11. The CSL is often drawn as a straight line but a curved locus is more appropriate for sands; the diagram shown here follows the geometry given by Jefferies and Been (2006). Muir Wood et al. (2010) used an equation for the curved CSL proposed by Gudehus (1997) and their curved CSL is concave upwards. Ng (2009a) captured the curvature of the critical state line in $e - log(p')$ space for his triaxial DEM simulations on ellipsoidal particles. The critical state line is also associated with a critical shear stress ratio, calculated as $M = \frac{q}{p'}$, where $q = \sigma_1 - \sigma_3$ and p' is the mean effective stress. This stress ratio can also be expressed as a critical state friction angle ϕ'_{cv}.

Referring to Figure 12.11, the response of a sand is shown to be largely governed by the material "state." This state can be quantified by the state parameter ψ, a measure of the difference between the e or v value that the soil will have on the critical state line at the same stress level and the current value of e or v (e.g. Been and Jefferies (1985)). Dense soils ("dry" of critical) have negative values of ψ and, upon shearing, the mobilized stress ratio will increase to a peak stress, exceeding the critical state stress, and subsequently reduce to the critical state value as the material strain softens. Loose soils ("wet" of critical) have positive ψ values and, upon shearing, they will tend to approach the maximum mobilized shear stress monotonically.

Considering the volumetric response, the loose sample will compress upon shearing, and, while the dense sample may initially experience a slight compression, it tends to dilate. The specific volume of both samples will approach the critical state line as shearing progresses. Dilatancy in soils was initially recognized by Reynolds (1885) and Rowe (1962) proposed that a link could be established between the mobilized principal stress ratio and the dilatancy, with dilatancy typically quantified as the ratio of the rates of change of volumetric and deviatoric strain ($D = \dot{\epsilon}_v/\dot{\epsilon}_q$).

A second important concept illustrated in Figure 12.11 is the existence of a normal compression line (NCL) that represents a bound to the possible states that a soil may experience. As outlined by Jefferies and Been (2006) this idea of a single NCL that is parallel to the CSL is not always true for real sands.

The stress-dilatancy relationship, which can be obtained by considering the work equation proposed by Taylor (1948) is very important as couples shear and volumetric strains and takes the following format (in its linear form):

$$\frac{\dot{\epsilon}_v^p}{\dot{\epsilon}_s^p} = M - \frac{q}{p'} \tag{12.1}$$

where M is the critical state friction parameter, p' and q are the mean pressure and deviator stress respectively, and $\dot{\epsilon}_v^p$ and $\dot{\epsilon}_s^p$ are the plastic volumetric and deviator strain rates.

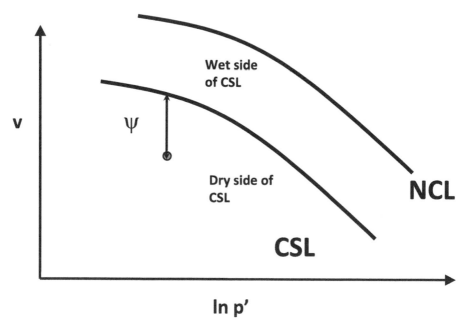

Figure 12.11: Schematic diagram of the critical state line (critical state locus) and normal compression line in $v : \ln(p')$ space (after Jefferies and Been (2006))

Here only a very brief overview of critical state soil mechanics has been provided, more comprehensive descriptions are provided by Wood (1990a) and Jefferies and Been (2006). This discussion has been included for two main reasons. Firstly, the use of a critical state approach has greatly aided the interpretation of geomechanics element tests and the understanding and modelling of soil response. To maximize the potential benefit of DEM simulations to the broader geomechanical community it is therefore worthwhile to considerable the response observed in DEM simulations within this framework. Jefferies and Been (2006) outline the experimental procedures necessary to determine the critical state parameters and these can easily be simulated using DEM. The work of Thornton (2000) and Cheng et al. (2004) (amongst others) has shown that DEM can make significant contributions to our understanding of frameworks for soil response that have largely been developed based upon empirical observations of the overall material response.

A second reason to highlight the critical state framework is that critical state concepts allow the influence of the inherent soil properties, which are invariant with density and stress level, on the mechanical response to be separated from soil state. Referring to Jefferies and Been (2006), the intrinsic properties of a material are its particle geometry, particle size distribution and critical state locus. Acknowledging these two types of contribution to the granular material response is important when interpreting results obtained in DEM simulations. General conclusions about observed response or comparisons between different studies cannot be drawn without knowledge of the material's state. For example if a parametric study is carried out using DEM in which the particle shape or surface properties are varied, the shift in the critical state locus is probably the most reliable index of the change to the material properties.

12.3.2 Demonstration that DEM can capture the micromechanics of response

If DEM is to be used with confidence to study the mechanical behaviour of soil, the ability of DEM to capture the response characteristics typically observed for soil must be established. Some of the mechanical response characteristics specific to granular materials have been listed above. The following lists a number of distinguishing response characteristics of granular materials, including citations of key references indicating that the feature has been successfully captured in DEM simulations (refer also to Cundall (2001)):

1. A "frictional" strength response, where the peak shear stress that can be sustained increases with increasing normal stress (Bolton et al. (2008), Cui and O'Sullivan (2006), Sitharam et al. (2008) and Potyondy and Cundall (2004)).

2. A critical state type response with loose samples contracting upon shearing, dense samples dilating upon shearing and both tending towards unique stress and void ratios at large strain (Thornton (2000), Rothenburg and Kruyt (2004) and Salot et al. (2009)).

3. The sensitivity of peak mobilized stress ratio (peak angle of shearing resistance) to void ratio (Powrie et al. (2005), Rothenburg and Kruyt (2004) and Thornton (2000))

4. A non-linear stress-strain response including yield prior to failure (even where a linear contact model is used) has been observed in almost all particulate DEM element test simulations (Rothenburg and Bathurst (1989), Cheng et al. (2004) and Yimsiri and Soga (2010)).

5. Hysteresis as a consequence of cyclic/repeated loading (e.g. Chen and Hung (1991), Chen and Ishibashi (1990) and Alonso-Marroquin et al. (2008)).

6. Strain softening and development of localizations or shear bands (Bardet (1994), Iwashita and Oda (1998) and Powrie et al. (2005)).

7. Significance of intermediate principal stress and a failure criterion that depends on the magnitude of all three principal stresses (Thornton (2000) and Ng (2004b)).

8. Anisotropy of strength and stiffness (Yimsiri and Soga (2001), Li and Yu (2009) and Yimsiri and Soga (2010)).

9. Stress-dependent elastic stiffness (Holtzman et al., 2008).

10. A phase transformation point that occurs during undrained shearing where the material transitions from having a contractive response to a dilative response (Cheng et al. (2003) and Sitharam et al. (2009)).

11. Non-coaxiality of principal stresses and principal strain increments (Li and Yu, 2009, 2010)).

12.3.3 Overview of key contributions to understanding soil response

Having established the ability of DEM to capture the response trends that one would expect to see in a granular material, DEM can then be used with a degree of confidence to advance understanding of fundamental soil behaviour. Some of the main contributions of discrete element simulations to soil mechanics are as follows:

1. Confirmation of the inhomogeneous nature of stress transmission in granular materials (following early photoelastic studies of de Josselin de Jong and Verrujit (1969)). As discussed in Chapter 8, DEM simulations have repeatedly demonstrated that stress is transmitted across a granular material via a highly heterogeneous network in which the most highly stressed particles form strong force chains aligned in the direction of major principal stress and supported by an orthogonal weaker force network. Graphical illustrations of this concept have been presented by Rothenburg and Bathurst (1989), Masson and Martinez (2001) and Cui et al. (2007), amongst many others. DEM simulations can generate data allowing detailed analysis of the buckling of force chains (e.g. Tordesillas and Muthuswamy (2009)). Thornton (2000) and Rothenburg and Bathurst (1989) observed that the normal contact forces contributed more significantly to the average stress tensor for the assembly than the tangential contact forces. These findings have provided further

evidence for the influence of the force chains on the material response, and add weight to the hypothesis that the buckling of the strong force chains is the key mechanism underlying failure and shear banding in granular materials.

2. A number of parametric studies have considered the influence of inter-particle friction on the overall material response (e.g. Thornton (2000), Powrie et al. (2005), Cui and O'Sullivan(2006), Yimsiri and Soga (2010)). Each of these studies indicates that the peak angle of shearing resistance (often called the peak friction angle) of the material as a whole, is relatively insensitive to the inter-particle friction angle.

3. As noted in Chapter 4, in contrast to disks or spheres, contacting non-spherical or non-circular particles can transmit moment as well as force resulting in a rotational resistance. As demonstrated by contributions from Iwashita and Oda (1998) and Jiang et al. (2005), parametric studies that consider the influence of rotational resistance (sometimes called rolling friction) on identical samples can be carried out using DEM. Findings from studies have indicated that as the resistance to relative rotation of contacting particles increases, there is an increase in the peak and critical state/residual angles of shearing resistance, and this can be tied to the mechanism of strong force deformation (Iwashita and Oda (2000), Tordesillas and Muthuswamy (2009)). It can be difficult to directly compare specimens with differing particle geometries. In these studies typically specimens have had both different initial void ratios as well as different particle geometries. As noted above one possibility is to compare the critical state loci (CSL) for the two materials. Studies in this area include Ng (2009b), who varied the aspect ratio of ellipsoidal particles, and Powrie et al. (2005), who varied the geometry of particles developed by joining two spheres with parallel bonds. The work of Mirghasemi et al. (2002) is a good example of a study that considers the influence of the particle geometry both on the overall response and on the material fabric.

4. Contributions have been made to better understand the response of materials in physical element testing apparatuses. For example DEM simulations have provided insight into the non-uniformities of stress and strain present in element tests; for example Masson and Martinez (2001), Zhang and Thornton (2007), Cui and O'Sullivan (2006) and Wang and Gutierrez (2010) all presented results illustrating the heterogeneity of the contact forces and strains in the direct shear apparatus. Using DEM parametric studies can be carried out to understand the limitations of physical testing apparatuses. For example, the influence of platen friction on the specimen response in triaxial tests was considered by Cui and O'Sullivan (2006) and Powrie et al. (2005). DEM can also be used to improve design of experimental apparatuses; both Zhang and Thornton (2007) and Wang and Gutierrez (2010) examined different configurations for the direct shear apparatus.

5. DEM simulations considering the evolution of localizations and shear bands (e.g. Bardet (1994), Iwashita and Oda (2000), and Powrie et al. (2005)) have shown that particle rotations tend to be significantly large within the shear band, in comparison with the remainder of the sample. Analyses of the local strains (O'Sullivan et al. (2003) and Wang et al. (2007)) have shown that the strains within a shear band are highly heterogeneous with regions of both compressive and dilative strains existing, as observed in plane strain tests on sand by Rechemacher et al. (2010).

6. Numerous DEM simulations have confirmed that the coordination number and the second-order fabric tensor (calculated considering the contact normal orientations) are key descriptors of the material fabric that correlate strongly with the stress-strain response of the material. For example Rothenburg and Bathurst (1989), Rothenburg and Kruyt (2004), Thornton (2000) and Ng (2001), amongst others, have all shown that the evolution of anisotropy calculated by considering maximum and minimum eigenvalues of the second-

order fabric tensor with straining correlates closely with the stress-strain response. Rothenburg and Bathurst (1989), Bardet (1994), Thornton (2000) and Rothenburg and Kruyt (2004) all showed that, upon shearing, the coordination number tends towards a relatively constant value. Thornton (2000) and Rothenburg and Kruyt (2004) showed that for rigid, unbreakable particles, upon shearing, equivalent loose and dense specimens tend towards the same "critical" coordination number.

7. As noted already experimental observation of soil response under fully three-dimensional stress conditions is difficult. Using DEM failure criteria that have been proposed for sand in the literature have been examined in ideal testing conditions by Thornton (2000), Ng (2004b) and Thornton and Zhang (2010), with Thornton and Zhang obtaining a good fit with the 3D failure criterion proposed by Lade and Duncan (2003).

8. Energy considerations are important in the development of continuum constitutive models for soil. For example the Cam Clay yield locus is determined by integration of the flow rule, which is determined from the work done in plastic deformation (e.g. Britto and Gunn (1987)). Some DEM analysts have considered the internal energy of the system in detail. For example, Bardet (1994) found that during strain softening the internal energy dissipation is highest in shear bands. Bolton et al. (2008) calculated the energy dissipated in their DEM simulations of crushable agglomerates and compared the results with the Cam Clay and Modified Cam Clay dissipation functions. Thornton (2000) monitored the kinetic energy to quantify the extent of instabilities induced during shearing.

While the listing of contributions provided here is subjective, its purpose is to give an impression of both the way DEM has advanced understanding of soil response and potential future areas for exploration. As well as facilitating numerical experiments to interrogate the micro-macro link, DEM provides us with a tool to hypothesize on the particle scale mechanics that drive the overall material response. For example, the influence of the particle contact rheology on the overall response of soil can be examined, examples include the studies of Lu et al. (2008) and Gili and Alonso (2002) who considered unsaturated soil response, and Jiang et al. (2005 and 2009) who examined the sensitivity of the overall material response to rolling resistance and particle surface roughness. These numerical experiments can be extended to explore how soil response can be modified and to develop appropriate soil improvement techniques. Prior contributions in this area include the work of Ibraim et al. (2006) who developed a 2D particulate DEM model that can qualitatively capture the response of fibre-reinforced soil. Bonded DEM particles have also been used to explore the use of cementation to mitigate liquefaction hazards by Zeghal and El Shamy (2008).

DEM simulations are convenient tools to answer questions of the type "what would happen if...?" For example Muir Wood et al. (2010) explored the implications of internal erosion (suffusion) on the mechanical behaviour of a soil in a series of 2D simulations where the smallest particles in the assembly were successively removed while maintaining quasi-static conditions and a constant isotropic stress state, and the resultant stress and volumetric responses were observed. These numerical experiments allowed the influence of a change in volumetric state and a change in grading to be monitored and were used to develop a continuum model that can capture the evolution of the overall material response as the particle size distribution changes due to erosion.

One particular advantage posed by DEM simulations is that complex stress states induced in the ground in real physical situations, which are difficult to replicate in a conventional laboratory test, can be simulated. Most notably a DEM simulation can easily achieve a true, three-dimensional stress state with $\sigma_1 \neq \sigma_2 \neq \sigma_3$. To attain these stress conditions in a controlled manner in the laboratory requires the use of complex types of apparatus that are not found in a typical geomechanics laboratory, such as the hollow cylinder apparatus or the true triaxial apparatus. DEM simulations are therefore well placed to have (at the minimum) an indirect influence on engineering practice by playing a role in examining and possibly developing continuum constitutive models.

Research that has used DEM to study fundamental soil response has mainly been restricted to considerations of non-cohesive materials, i.e. particles exceeding about 100 μm. These particles are sufficiently large that the surface attraction forces are negligible in comparison with the particle inertia. DEM simulations of clay are less common and are complicated by both the complexity of the surface interaction forces and the particle geometry. However, researchers including Anandarajah (2003), Lu et al. (2008) and Peron et al. (2009), have explored the use of DEM to simulate clay response.

To illustrate the use of DEM as a tool in basic geomechanics research, three examples of key research studies that have used DEM to significantly advance understanding of granular material response are outlined here as case studies of the benefits of using DEM in fundamental geomechanics research.

12.3.4 Response of assemblies of particles to triaxial and more general stress states

A series of publications by Colin Thornton and his colleagues (Thornton (1997b), Thornton and Antony (1998),Thornton (2000), Thornton and Antony (2000) and Thornton and Zhang (2010)) have all explored the response of assemblies of spherical particles in a periodic cell. The body of research has included triaxial (axisymmetric) and true-triaxial ($\sigma_1 \neq \sigma_2 \neq \sigma_3$) test simulations to examine both failure and pre-failure response. These studies are an excellent example illustrating the ability of DEM to capture the type of response we would expect in a physical system and to allow interrogation of the fundamental mechanics that underly this response.

Thornton and Antony (1998) and Thornton (2000) simulated the triaxial compression at a constant mean stress of dense and loose assemblies of spheres. The simulations included between 3,000 and 8,000 polydisperse spheres with periodic cell bound-

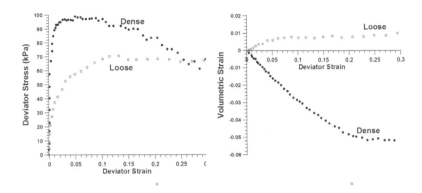

Figure 12.12: (a) Deviator stress response (b) Volumetric stress response for loose and dense specimens of spheres in axisymmetric compression (Thornton, 2000)

aries. As illustrated in Figure 12.12, the differences in the responses observed were similar to the differences observed for equivalent physical tests on loose and dense sands. Upon shearing, the dense sample achieved a peak deviator stress and post-peak strain softening was observed. The loose sample exhibited a strain hardening response. Both samples tended towards a constant-volume condition at large strains, and at large strains their void ratios were similar, i.e. a critical state was almost achieved.

The analysis of the internal material structure revealed that the normal contact forces contributed significantly more to the stress transmission than the tangential contact forces. As acknowledged by Thornton and discussed in Chapter 10, this had been observed in earlier studies including Rothenburg and Bathurst (1989). Thornton (1997b) analysed the distribution of the contact forces within his assemblies. The evolution of deviator fabric during straining is illustrated in Figure 12.13(a). Analysis of the fabric tensor for the greater than average forces and less than average forces revealed that contacts transmitting less than the average forces act to provide lateral stability to the strong force chains (Figure 12.14).

Returning to the comparison of loose and dense material response, as illustrated in Figure 12.13(b), Thornton (2000) found

485

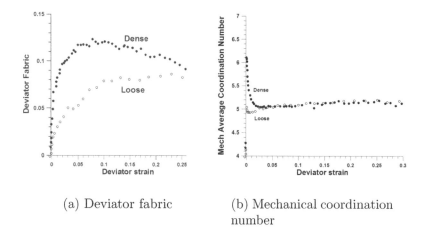

(a) Deviator fabric

(b) Mechanical coordination number

Figure 12.13: Variation in internal structure during axisymmetric compression of assemblies of loose and dense spheres (Thornton, 2000)

that both samples attained the same mechanical coordination number (Equation 10.8) after only a small amount of straining. Rothenburg and Kruyt (2004) observed a similar response in their two-dimensional simulations. In their triaxial stress simulations Thornton and Antony (2000) found that values of the mechanical coordination number for two samples sheared in extension and compression at a constant mean stress were very similar. Two additional simulations (again shearing in triaxial compression and extension) carried out in constant-volume (undrained) conditions also had similar mechanical coordination numbers. Despite starting from the same initial values, the mechanical coordination numbers for the constant-volume simulations were greater those obtained for the constant-mean-stress simulations.

Thornton (2000) described a series of true triaxial type simulations where the principal stress ratio, $b = \frac{\sigma_2 - \sigma_3}{\sigma_1 - \sigma_3}$ was systematically varied. The deviatoric failure states observed were in agreement with the three-dimensional failure criterion proposed by Lade and Duncan (2003). Ng (2004b) also simulated deformation under various b values in a periodic cell with ellipsoidal particles. He considered four different failure criteria and found the failure cri-

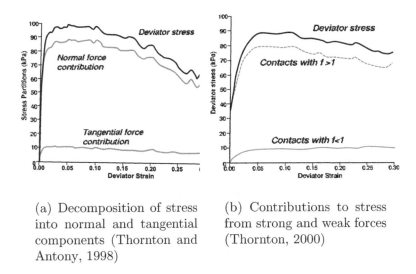

(a) Decomposition of stress into normal and tangential components (Thornton and Antony, 1998)

(b) Contributions to stress from strong and weak forces (Thornton, 2000)

Figure 12.14: Stress partitioning into normal and shear force contributions and strong and weak force network contributions

terion proposed by Ogawa et al. (1974) gave the best match to the simulation data. During a simulation of a constant deviatoric strain test, both the stress tensor and the second-order contact normal fabric tensor was observed to rotate together, while the strain tensor lagged the observed response, indicating non-coaxial behaviour.

Thornton and Zhang (2010) built upon the ideas proposed in the earlier studies, and turned their attention to look at the response of an assembly of 27,000 spheres to deviator strain probes. The deviator strain probes differed in magnitude and direction; however, they all considered the same initial sample. The resulting stress response envelopes had the same geometry as the Lade and Duncan failure surface. Considering the invariants of the fabric tensor, Thornton and Zhang constructed "fabric response envelopes" in a principal fabric coordinate system and observed the response to be that of inverted Lade-type surfaces. A new parameter calculated from the invariants of the fabric tensor was proposed to define these fabric response envelopes.

12.3.5 Particle crushing

Experimental research including Coop and Lee (1993) and Coop et al. (2004) has shown that sands can experience significant particle breakage in both compression and shearing. Coop and Lee (1993) found that a unique relationship exists between the amount of particle breakage that occurs during shearing to a critical state and the value of the mean normal effective stress. These findings, amongst others, have motivated a number of studies into particle crushing using DEM. Approaches to simulate crushable particles using DEM have been described in Chapter 4. Some of the DEM crushing related studies have used very simple DEM models of particle crushing that can provide some insight, while being suited to simulation of boundary value problems (e.g. Lobo-Guerrero and Vallejo (2005)). Here a series of studies that have used agglomerates to simulate the response of silica sand grains are considered. Agglomerates and agglomeration of powders into larger particles are of interest in process engineering and Thornton and Liu (2004) explored the idea of bonding spheres together to simulate agglomerates, while Golchert et al. (2004) used DEM to analyse the influence of the overall agglomerate shape on its breakage behaviour.

As noted in Chapter 4, Robertson (2000), McDowell and Harireche (2002), Cheng et al. (2003, 2004), and Bolton et al. (2008) document the development and validation of the approach to modelling sand particles and subsequent application of the method to analyse the overall material response. Each aggregate of bonded spheres used to simulate a single sand grain is generated following the approach proposed by Robertson (2000). Initially a hexagonal close-packed assembly of bonded spheres is created, reflecting the crystalline nature of the particle material, with selected numbers of spheres being removed to represent the flaws present in real soil particles. At the particle scale there was a two-level confirmation that this approach captured the response of individual sand grains. Cheng et al. (2003) demonstrated that these particles could be calibrated to reproduce the response of single silica sand particles in single particle compression tests. McDowell and Harireche (2002) and Cheng et al. (2003) demonstrated that the size-strength relationship observed in crushing tests on real soil particles can be reproduced, as the Weibull moduli (obtained by analysing the variation in the of failure of an individual distribution of strengths with stress level) were equivalent for both physical tests and numerical experiments.

Cheng et al. (2003) then found that DEM simulations of isotropic compression of assemblies of clusters quantitatively matched the response observed for the silica sand (comparing data where both the void ratios and mean effective stresses were normalized). When triaxial compression test data were considered, the agreement of the model with the physical data was less successful, but the results were broadly consistent. As the validity of the calibrated model was confirmed at both the particle and macro-scales, interpretations of the relationships between particle-scale response and overall response could be made with confidence. Cheng et al. (2003) could then clearly demonstrate the extent to which particle crushing dominates the material response along the normal compression line as well as observing particle crushing in undrained shear. This modelling approach was also used by Cheng et al. (2004) to examine yielding. For DEM samples that were lightly overconsolidated at yield, the yield surfaces were shown to be associated with contours of the percentage of bond breakages or damage events in the material. For samples that were more heavily overconsolidated at yield, the models captured the response trends proposed in published stress dilatancy theories. Most notably the hypothesis proposed by Bolton (1986) that grain breakage can explain the decrease in the peak friction angle with increasing of stress for these samples was confirmed.

Bolton et al. (2008) directly compared the response of specimens of breakable agglomerates and geometrically equivalent unbreakable agglomerates in both isotropic compression and triaxial compression. While the micromechanical analyses by Cheng et al. (2003, 2004) tended to focus on bond breakages only, Bolton et al. (2008) extended the micromechanical analyses to include consideration of coordination number, deviator fabric and the energy stored in each agglomerate. At lower confining pressures the coordination numbers for the breakable and unbreakable agglomerates were similar; as would be expected, they diverged at higher pressures. However in both cases the mean stress was seen to significantly increase the coordination number. The anisotropy of the crushable agglomerates (measured using the second-order fabric tensor) in triaxial compression decreased with increasing confining pressure. When particles break, there is an increase in the number of contacts in the system, resulting in an increase in energy dissipated through friction.

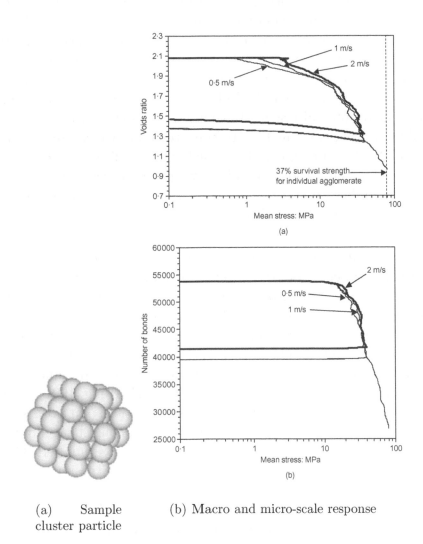

(a)   Sample   (b) Macro and micro-scale response
cluster particle

Figure 12.15: Use of DEM to develop insight into sand particle
crushing (Cheng et al., 2003)

12.3.6   Cyclic loading and hysteresis

To date, most published DEM simulations have considered the response of soil to monotonic loading and deformation. The response of soil to cyclic loading and load reversals is both intriguing and important. Scenarios where soil is subject to cyclic loading include soil beneath road pavements and ballast beneath railway tracks, soil in offshore and wind turbine foundations, soil backfill adjacent to integral bridge abutments, and soil subject to ground shaking during earthquakes. In the previous two examples of the use of DEM in fundamental geomechanics studies, an overview of the development of ideas within two specific research groups was considered; here a more general review of contributions in the area of simulation of cyclic soil response is given.

Considering firstly drained cyclic loading, motivated by earlier experimental studies on ballotini specimens, (Chen et al. (1988) and Ishibashi et al. (1988)), Chen and Ishibashi (1990) and Chen and Hung (1991) described two series of DEM simulations where the response of an assembly of spherical particles with periodic boundaries to both monotonic loading conditions and stress reversals was considered. The initial study (Chen and Ishibashi, 1990) used a relatively small number of particles (155) and a linear contact spring, while the second study (Chen and Hung, 1991) used a Hertzian contact model and about 1,000 particles. In both simulations a hysteretic response was observed and the shear modulus was seen to depend strongly on the coordination number. Chen and Hung (1991) found that upon unloading to the initial stress level the coordination number increases only slightly while upon unloading to an isotropic stress state the fabric anisotropy induced during loading remained essentially intact. Kuhn (1999) subjected his 2D assembly to load reversals and found that particles did not return to their original positions.

As noted in Chapter 11, O'Sullivan et al. (2008) validated the ability of DEM to simulate cyclic response by comparing strain-controlled cyclic triaxial experiments on steel spheres with equivalent DEM simulations. A parametric study considered the variation in overall stresses and in the material fabric during cyclic load-

492

ing. As discussed by both O'Sullivan et al. (2008) and O'Sullivan and Cui (2009a), for the strain amplitudes considered (1%, 0.5%, 0.05%, 0.005%) even after 50 loading cycles, while the macro-scale stress-strain response exhibited little variation from cycle to cycle, a net change in fabric anisotropy and coordination number over each load cycle continued to be observed. Yunus et al. (2010) also considered 3D strain-controlled cyclic loading and, based on their simulation data, they proposed that the fabric anisotropy can be used as an internal variable in phenomenological models (i.e. continuum constitutive models) for soil response.

The findings of Kuhn and of Cui and O'Sullivan indicate that during load reversals there is significant movement of particles to cause changes in contact configuration. In real sands where the particles are not ideal and rigid, when there is a change in contact configurations it is possible for particles to be damaged. In their 2D simulations of railway ballast particles subject to biaxial cyclic loading Hossein et al. (2007) used bonded disk particles to represent the ballast. Their simulations indicated that damage occurred over the initial 2,000 cycles. More realistic ballast particle geometries were achieved by Lu and McDowell (2010) who simulated the particles as tetrahedral of bonded spheres with mini-spheres being bonded to the outside of the tetrahedra to simulate asperities. Monotonic and cyclic triaxial tests were simulated. The inclusion of asperities was found to give a more realistic model response, and the overall response trends gave good agreement with physical test data, considering the sensitivity of volumetric strain to stress level, the magnitude of the permanent deformation and the amount of ballast damage incurred. Lu and McDowell (2006) created a model of ballast material around a sleeper and subjected the ballast to cyclic load reversals by moving the model sleeper. Their results indicated a linear relationship between the cumulative settlement and the logarithm of number of load cycles.

Cyclic triaxial tests do not allow consideration of a range of principal stress orientations. Li and Yu (2010) simulated the response of a 2D material to controlled variation in the principal stress orientation, and applied a number of cycles where the principal stress orientation was systematically varied. They used their results to study the fundamental mechanisms underlying non-coaxial material response. Outside of the geomechanics literature, there have been some documented studies of ratcheting that may also prove interesting (e.g. Garcia-Rojo et al. (2005), Alonso-Marroquin et al. (2008)).

Other examples of the use of DEM to study cyclic response include the simulation of liquefaction using either the constant-volume approach (e.g. Ng and Dobry (1994) or Sitharam et al. (2009)) or a coupled DEM model (e.g. Zeghal and El Shamy (2008) who used the averaged Navier-Stokes approach). These studies have also been considered in Chapter 6.

12.4 Conclusions

Appreciating the significant assumptions and simplifications inherent in DEM, it is both amazing and encouraging that DEM simulations have been shown to be able to capture the major response features of granular materials. The potential benefit that this approach to modelling can bring to the geomechanics community make the application and development of DEM an exciting and satisfying field to work in.

Chapter 13

DEM: Future and Ongoing Developments

DEM is not a new technique; however, a widespread appreciation of the method amongst the geomechanics research community has only recently been achieved and awareness of the technique in geotechnical practice remains limited. The seminal paper on DEM from the perspective of geomechanics was published in 1979 (Cundall and Strack, 1979a). As noted in Chapter 1, molecular dynamics is a technique that shares algorithmic similarities with particulate DEM. Molecular dynamics was originally developed in the 1950s. However, the computational cost of simulations using particulate DEM codes has rendered them relatively inaccessible and unattractive to both researchers and practitioners until relatively recently. Consequently, at the time of writing, and considering its slow uptake in the geotechnical industry in particular to date, it is probably still reasonable to echo the comment of Rapaport (2004) (who considered molecular dynamics) and state that the method is not "fully mature." This final Chapter presents some (subjective) ideas of how DEM use within the geotechnical engineering is likely to evolve in coming years.

In order to see where DEM is going it is useful to see where it has come from. Jing and Stephansson (2007) give a history of DEM development, documenting how it has evolved starting from the original disk and sphere codes BALL and Trubal developed by Cundall and Strack. Drawing on this history as well as other references, the key milestones in the development of DEM can be said to include the hysteretic contact model proposed by Walton and Braun (1986), the extension to non-spherical particles including ellipses and ellipsoids (e.g. Lin and Ng (1997)), the coupling of DEM with fluid (Tsuji et al., 1993), the development of bonded particulate modelling as a means to simulate rock mass response (Potyondy and Cundall, 2004), and the implementation of DEM in a high-performance or parallel computing software environment (e.g. Kloss and Goniva (2010), Weatherley (2009), Kozicki and Donz (2008)). While their objective has been to advance understanding of material response, users of DEM codes have made important developmental contributions by establishing the sensitivity of the response of different systems to the DEM input parameters.

13.1   Computational Power

The computational cost of DEM simulations is the main constraint inhibiting DEM analyses with realistic numbers of particles and realistic particle geometries. As discussed in Section 11.5, this constraint imposes a limitation both on DEM simulations of element tests used in basic research and in simulations of field- or industrial-scale boundary value problems. The amount of computation and hence the processing time will increase at least linearly with the number of particles and the time interval of interest and this obviously also contributes significantly to the associated computational cost. Tackling this problem is a key ongoing challenge for DEM analysts.

The way in which computer hardware is developing, with even processors for desktop PCs having multiple cores, means that DEM codes implemented to run in a parallel or high-performance computing environment will certainly become much more common in geomechanics applications. Up until recently, the argument in favour of implementing numerical algorithms in parallel codes was somewhat muted by the high rate in improvement in the performance of single processors. However, as highlighted by Asanovic et al. (2006), the old rule of a doubling in processor performance every 18 months no longer holds. The future clearly lies in multi-core or many-core architectures, and even possibly with exascale operating systems with 10-100 million processing elements (Kogge, 2008). Both DEM and MD broadly come within the classification of the N-Body Methods identified as one of the "seven dwarfs" of high-performance computing proposed by Colella (2004). These seven dwarfs are seven classes of numerical methods used in physical science that would benefit from parallel implementation. The current rapid developments in the available hardware (including the use of graphical processing units for example) could potentially make simulations involving millions of particles very tractable. Asanovic et al. (2006) have highlighted the fact that the increased complexity of computer hardware architectures poses a significant challenge to code development. Hopefully ongoing software engineering research activities will enable DEM to take advantage of the latest high-performance computing systems as they develop. As highlighted by Asanovic et al. (2006), a key software challenge is to develop a level of abstraction that allows the programmer to implement algorithms without the need for a detailed knowledge of the underlying hardware, while still enabling the user to take full advantage of the power and performance of the multicore system. Research and development considering the use of graphical processing units (GPUs) to accelerate molecular dynamics codes is ongoing, for example Plimpton et al. (2010) includes information on the GPU accelerated version of the molecular dynamics code LAMMPS (Plimpton, 1995). It is likely that particulate DEM codes that also take advantage of GPUs will be developed in the near future.

As discussed by Rapaport (2009) as the computations in MD (and hence DEM) are based on highly localized information (i.e. particles and their adjacent contacts), the algorithms are well suited to implementation in a distributed or high-performance computing environment. Various approaches can be used to implement a DEM algorithm in parallel. Sutmann (2002) gives a relatively broad discussion on options for parallelization of molecular dynamics/DEM codes. One approach considered by both Rapaport (2004) and Pöschel and Schwager (2005) is to divide the system into subdomains or subvolumes. The response of each subvolume is then assigned to a different processor. Information about the particles in adjacent subdomains falling within a margin zone just outside the edge of the subdomain must also be provided. The analyst needs to carefully consider how to choose the size of the subdomain. Examples of parallel implementations of DEM codes can be found amongst the open source research oriented codes that are now available. The Esys DEM code (Weatherley, 2009) uses a domain decomposition approach, with a distributed memory MPI implementation, the DEM code LIGGGHTS (Kloss and Goniva, 2010) which is based upon the MD code LAMMPS (Plimpton, 1995) also uses MPI and a spatial discretization approach. YADE (Kozicki and Donz, 2008) adopts a slightly different approach, using OpenMP and a shared memory strategy. As computer hardware architecture evolves, there will certainly be a number of developments in the approaches adopted to implement DEM codes in high-performance/parallel computing environments.

13.2 Future of DEM

Looking forwards, it is certain that DEM will not replace continuum modelling as a means to predict soil deformations, nor will it replace physical experiments as a means to advance fundamental understanding of soil response. Rather it has established itself firmly as one of the key "tools" available to geotechnical engineers. In their review of likely research developments in geomechanics in the 60-year time period from 2008 to 2068, Simpson and Tatsuoka (2008) surveyed a large number of engineers working in both geomechanics research and in industry. Many respondents to this survey suggested that DEM will play a prominent role in developing future analytical methods in geomechanics. From a geomechanics perspective, when thinking about the future of DEM, key developments will include application of the method to a broader range of problems or research questions, algorithmic/methodological developments and improvements in the realism of DEM models. To maintain a high level of confidence in both the method and its applicability, ongoing careful validation and verification will be needed.

Considering firstly the application of DEM, it is clear that it will continue to be used as a fundamental research tool, looking at the micromechanics of soil response, with the goal in many cases being to use the insight gained to advance constitutive models for continuum modelling. DEM is also well placed to provide information on mechanisms that operate at the particle scale but that cannot easily be observed in the laboratory. In particular there is significant potential for greater exploitation of DEM to simulate soil-water interaction problems. For example, the problem of sand production poses a great economic risk to the petroleum industry and the somewhat similar problem of internal erosion poses a significant risk to dams. There is scope for DEM simulations to inform guidelines to practising engineers considering these problems in the not too distant future. DEM is also very likely to make a significant contribution in the area of soil improvement, the potential in this area has been illustrated in the work by Ibraim et al. (2006) looking at fibre-reinforced soil. As discussed in Chapter 12, at an industrial-scale DEM is most attractive when considering large deformation problems, including penetration, landslide run-out or soil-machine interaction. Roberts (2008) demonstrated that there is scope for DEM to play a role in the development of sensors and transducers embedded in soil. No doubt as awareness of the method grows within the geomechanics community, engineers working both in research and in industry will have countless suggestions on the potential of this method to help them tackle their problems.

Significant future algorithmic developments in DEM are likely to occur outside of geomechanics and the geomechanics DEM community needs to be aware of relevant ongoing developments in powder technology and molecular dynamics. Taking account of suggestions by Simpson and Tatsuoka (2008) and Yu (2004), the specific algorithmic research needed to develop DEM clearly includes the following:

1. Further parametric studies to consider the influence of the model parameters (shape, stiffness, etc.) on the observed response.

2. More realistic contact models and particle morphologies.

3. An increase in the number of particles considered in simulations.

4. The development of more robust models and more efficient computer codes.

5. Improved micro-scale quantification of inter-particle forces and particle fluid interaction forces to inform future model development.

6. Improved theories to relate the macro- and micro-scales.

7. Further developments in relation to coupling particles and fluid, including simulation of multi-phase fluids (e.g. oil and water mixtures).

This list of required studies includes both experimental and numerical research. In considering the future of multi-scale modelling in mechanics in general Kassner et al. (2005) recognized the need for nano- and micro-scale experimentation. As outlined by Kassner et al. there are significant challenges associated with the design and manufacture of the type of small scale apparatus, including accurate control and measurement loads and deformations needed to accurately determine the contact interactions between soil particles. A further challenge is posed by the irregular topology of the surface of real soil particles. A more detailed discussion on the issues involved can be found in Cavarretta (2009).

13.3   Further Reading

The objective of this book has been to provide a general reference for people commencing to use DEM or contemplating the application of DEM to simulate a problem involving granular materials of interest to them. As noted in Chapter 12, there has been a rapid increase in the use of DEM in recent years as evidenced by the number of DEM-related papers that have been published across a range of scientific disciplines as well as publications specifically concerning geomechanics related applications. The breadth of disciplines that use DEM make this a particularly interesting and stimulating area to work in, but for researchers or engineers just starting to use DEM seeking appropriate information can be very daunting.

There are many very interesting DEM related publications in the literature, the most relevant of which will depend on the particular application of interest. However, there are some papers that are likely to be particularly useful to many DEM analysts. The original *Géotechnique* publication by Cundall and Strack (1979a) gives a good overview of the DEM algorithm. While the review papers by Zhu et al. (2007, 2008) approach the use of DEM from a chemical engineering perspective, they contain information that would be of interest to any DEM analyst. Those interested in coupled DEM should refer to Tsuji et al. (1993) or Curtis and van Wachem (2004). While the paper by Potyondy and Cundall (2004) considers application of DEM to simulate the response of rock mass, it includes a particularly clear description of the simulation process, as well as the approaches used to interpret the results and it likely to be useful for analysts considering a range of problem types. Bobet et al. (2009) consider, from a rock mechanics perspective, a number of modelling approaches for discontinuous materials, including particle-based DEM. The two themed issues of *Géotechnique* that considered soil mechanics at the grain-scale include a number of very interesting DEM papers (Baudet and Bolton, 2010a and 2010b). There has also been an issue of *Geomechanics and Geoengineering* devoted to advances in DEM (Morris and Cleary, 2009). Thornton (2000) and Rothenburg and Bathurst (1989) are key references that show how DEM can be used to relate particle-scale mechanisms to overall granular material response. The experimental paper by Oda et al. (1985) is also particularly useful as it gives a useful overview of how fabric can be quantified and related to the granular material's overall mechanical behaviour.

Considering previously published textbooks the publication edited by Oda and Iwashita (1999) provides an excellent introduction to particle-scale mechanics of granular materials, the DEM algorithm and micromechanics interpretation techniques that can be applied to DEM simulation results. The theory manual for the PFC software (Itasca, 2008) includes a very clear description of the DEM algorithm. There have been many conference proceedings including useful DEM related papers. The proceedings of the Powders and Grains series of conferences are particularly interesting as they illustrate the range of applications involving DEM. Reference to the 2009 proceedings is particularly recommended (Nakagawa and Luding, 2009). Readers interested in DEM code development in particular should refer to the textbook by Pöschel and Schwager (2005), who consider code development from the perspective of granular materials, and include the description and code for many DEM related functions in C++. Rapaport (2004) discusses the implementation of a molecular dynamics code and this reference is also likely to be useful to readers interested in development of a particulate DEM code. Munjiza (2004) discusses the development of a combined FEM/DEM algorithm and this book will be of interest both to those interested in DEM code development and in the simulation of particles using non-spherical geometries. Jing and Stephansson (2007) discuss DEM as applied in rock mechanics; they include one chapter on the use of particulate DEM, while also considering the block DEM codes.

For those approaching granular materials from outside of the geotechnical community, there are a number of undergraduate soil mechanics texts that introduce the typical mechanical response characteristics observed in soil, as well the methods used to test and characterize the material. The textbook by Atkinson (2007) is particularly readable and accessible. More in-depth discussions from the perspective of critical state mechanics are given by Wood (1990a) and Jefferies and Been (2006). Mitchell and Soga (2005) approach soil mechanics from a scientific perspective and include many references to particle-scale mechanisms. Geomechanics researchers may find general texts on granular materials such as

Duran (2000) interesting.

13.4 Concluding comments

In his Rankine lecture Potts (2003) presented arguments both in favour and against the use of numerical modelling in geotechnical engineering and queried whether modelling techniques are just advanced toys for academics or genuine tools for routine numerical analyses. Considering this comment, it is useful to summarize the role of particulate DEM in geotechnical engineering. As noted in Chapter 12, granular materials, including soils, are complex systems, where the overall material response emerges due to interactions of simpler base units. This inherent complexity arising from the particulate nature of the material hinders our ability to accurately predict soil response when analysing real problems of industrial, societal or environmental importance. The heterogeneity of the naturally deposited material adds a further level of complexity to the systems.

The underlying goal motivating geomechanics research is to improve our predictive skills, and an improved understanding of soil response will contribute to achieving this goal. At the time of writing it is reasonable to state that the complexity of granular material response is not completely understood. A scientific understanding of a material requires both observation of the overall response via experiments and comprehension of the underlying mechanisms. DEM simulations are uniquely placed to provide detailed data that will facilitate the evolution of our comprehension. Physical and numerical models that allow the exploration of hypotheses and examination of mechanisms will play a key role in achieving this improved understanding. From the perspective of physical science and engineering, a successful model is a simplification of reality that captures the main features of the system of interest while allowing more detailed analysis. Results obtained to date demonstrate that DEM clearly meets these criteria.

There is no doubt that the use of particulate DEM within the research community will continue to grow. This will have an impact on engineering practice either directly or indirectly. As noted in Chapter 12, the application of DEM to large boundary value problems of interest to industry lags its use in basic research significantly. While routine quantitative predictions involving particulate DEM remain in the distant future, in the short to intermediate term the most likely industrial application of particulate DEM in geotechnical engineering will be in qualitative exploration of failure mechanisms involving large deformations.

In 1999, a debate on the future of soil mechanics research was held at Imperial College's "Geotechnics in the New Millennium Symposium." The motion, "that this house believes continuum models are past their sell-by date," was passed. Over a decade later it is clear that DEM is not likely to replace continuum analyses, rather it is a very useful tool to supplement continuum analysis and laboratory experiments to develop our understanding of soil response. While recognizing the limitations and development needs of DEM, when looking to the future of geotechnics Simpson and Tatsuoka (2008) stated that "it is likely that computations that model the non-continuous nature of soils, with discrete load paths, fractures and similar features, will become important and might provide a major step forward."

References

Abbas, A., E. Masad, T. Papagiannakis, and T. Harman (2007). Micromechanical modeling of the viscoelastic behavior of asphalt mixtures using the discrete-element method. *International Journal of Geomechanics 7*(2), 131–139.

Abbire* ady, C. and C. R. I. Clayton (2010). Varying initial void ratios for DEM simulations. *Géotechnique 60*(6), 497–502.

Ahrens, J., B. Geveci, and C. Law (2005). Paraview: An end-user tool for large data visualization. In C. Hansen and C. Johnson (Eds.), *The Visualization Handbook*. Elsevier.

Ai, M. (1985). On mechanics of two dimensional rigid assemblies. *Soils and Foundations 25*(4), 49–62.

Alder, B. J. and T. E. Wainwright (1957). Phase transition for a hard sphere system. *Journal of Chemical Physics 27*, 1208–1209.

Alder, B. J. and T. E. Wainwright (1959). Studies in molecular dynamics. I. General method. *Journal of Chemical Physics 31*, 459.

Alonso-Marroquin, F., H. B. Muhlhaus, and H. J. Herrmann (2008). Micromechanical investigation of granular ratcheting using a discrete model of polygonal particles. *Particuology 6*, 390–403.

Alonso-Marroqun, F. and Y. Wang (2009). An efficient algorithm for granular dynamics simulations with complex-shaped objects. *Granular Matter 11*, 317–329.

Anandarajah, A. (2003). Discrete element modeling of leaching induced apparent overconsolidation in kaolinite. *Soils and Foundations 43*(6), 65–702.

Anderson, T. B. and R. Jackson (1967). Fluid mechanical description of fluidized beds. equations of motion. *I and EC Fundamentals 6*(4), 527–539.

Arthur, J. R. F. and B. K. Menzies (1972). Inherent anisotropy in a sand. *Géotechnique 22*(1), 115–128.

Asanovic, K., R. Bodik, B. C. Catanzaro, J. J. Gebis, P. Husbands, K. Keutzer, D. A. Patterson, W. L. Plishker, J. Shalf, S. W. Williams, and K. A. Yelick (2006). The landscape of parallel computing research: A view from berkeley (technical report no. UCB/EECS-2006-183 http://www.eecs.berkeley.edu/pubs/techrpts/2006/eecs-2006-183.html University of California at Berkeley. Technical report.

Ashmawy, A. K., B. Sukumaran, and V. V. Hoang (2003). Evaluating the influence of particle shape on liquefaction behavior using discrete element modeling. In *Proceedings of the Offshore and Polar Engineering Conference*, pp. 542–550.

Atkinson, J. (2007). *The mechanics of soils and foundations*. Taylor and Francis.

Bagi, K. (1993). A quasi-static numerical model for micro-level analysis of granular assemblies. *Mechanics of Materials 16*(1-2), 101–110.

Bagi, K. (1996). Stress and strain in granular materials. *Mechanics of materials 22*, 165–177.

Bagi, K. (1999a). Some typical examples of tessellation for granular materials. In M. Oda and K. Iwashita (Eds.), *Introduction to mechanics of granular materials*, pp. 2–5. A. A. Balkema.

Bagi, K. (1999b). Stress and strain for granular materials. In M. Oda and K. Iwashita (Eds.), *Introduction to mechanics of granular materials*. A. A. Balkema.

Bagi, K. (2005). An algorithm to generate random dense arrangements for discrete element simulations of granular assemblies. *Granular Matter 7*, 31–43.

Bagi, K. (2006). Analysis of microstructural strain tensors for granular materials. *International Journal of Solids and Structures 43*, 3166–3184.

Bagi, K. and I. Bojtar (2001). Different microstructural strain tensors for granular materials. In N. Bicanic (Ed.), *Proceedings of the Fourth International Conference on Analysis of Discontinuous Deformation*, pp. 111–133. University of Glasgow.

Barber, C. B., D. Dobkin, and H. Huhdanpaa (1996). The quickhull algorithm for convex hulls. *ACM Transactions on Mathematical Software 22*(4), 469–483.

Bardet, J. (1994). Observations on the effects of particle rotations on the failure of idealized granular materials. *Mechanics of Materials 18*, 159–182.

Bardet, J. (1998). Introduction to computational granular mechanics. In B. Cambou (Ed.), *Behaviour of granular materials*, Number 385 in CISM Courses and Lectures. Springer-Verlag.

Bardet, J. and J. Proubet (1991). A numerical investigation of the structure of persistent shear bands in idealized granular material. *Géotechnique 41*, 599–613.

Barreto, D. (2010). *Numerical and experimental investigation into the behaviour of granular materials under generalised stress states*. Ph.D. thesis, Imperial College London.

Barreto, D., C. O'Sullivan, and L. Zdravkovic (2008). Specimen generation approaches for DEM simulations. In S. Burns, P. Mayne, and J. Santamarina (Eds.), *International symposium on deformation characteristics of geomaterials, 4, Atlanta, GA, 22-24 September, 2008.* IOS Press.

Barreto, D., C. O'Sullivan, and L. Zdravkovic (2009). Quantifying the evolution of soil fabric under different stress paths. In M. Nakagawa and S. Luding (Eds.), *Proceedings of the 6th International Conference on Micromechanics of Granular Media Golden, Colorado, 13-17 July 2009*, pp. 181–184. AIP Conference Proceedings.

Barton, M. (1993). Cohesive sands: The natural transition from sands to sandstones. In Anagnostopoulos et al. (Ed.), *Geotechnical Engineering of Hard Soils-Soft Rocks*. A. A. Balkema.

Basarir, H., C. Karpuz, and L. Tutluolu (2008). 3d modeling of ripping process. *International Journal of Geomechanics 8*(1), 11–19.

Baudet, B. and M. Bolton (2010a). Editorial soil mechanics at the grain scale: Issue 1. *Géotechnique 60*, 313–314.

Baudet, B. and M. Bolton (2010b). Editorial soil mechanics at the grain scale: Issue 2. *Géotechnique 60*, 411.

Bazant, Z. P. and J. Planas (1991). *Fracture and size effect in concrete and other quasi-brittle materials*. CRC Press.

Bedford, A. and D. S. Drumheller (1983). On volume fraction theories for discretized materials. *Acta Mechanica 48*, 173–184.

Been, K. and M. Jefferies (1985). A state parameter for sands. *Géotechnique 35*(2), 99–112.

Been, K., M. G. Jefferies, and J. Hachey (1991). The critical state of sands. *Géotechnique 41*(3), 365–381.

Behringer, R., K. E. Daniels, T. S. Majmudar, and M. Sperl (2008). Fluctuations, correlations, and transitions in granular materials: Statistical mechanics for a non-conventional system. *Phil. Trans. R. Soc. A 366*, 493–504.

Belheine, N., J.-P. Plassiard, F.-V. Donze, F. Darve, and A. Seridi (2009). Numerical simulation of drained triaxial test using 3d discrete element modeling. *Computers and Geotechnics 36*, 320–331.

Belytschko, T. (1983). An overview of semidiscretization and time integration procedures. In T. Belytschko and T. Hughes (Eds.), *Computational Methods for Transient Analysis*, Volume 1 of *Computational Methods in Mechanics*. North-Holland.

Belytschko, T., Y. Krongauz, D. Organ, M. Fleming, and P. Krysl (1996). Meshless methods: An overview and recent developments. *Computer Methods in Applied Mechanics and Engineering 139*(1), 3–47.

Belytschko, T., W. Liu, and B. Moran (2000). *Nonlinear Finite Elements for Continua and Structures*. Wiley.

Ben-Nun, O. and I. Einav (2010). The role of self-organization during confined comminution of granular materials. *Philosophical Transactions of the Royal Society of London A 368*, 231–247.

Bertrand, D., F. Nicot, P. Gotteland, and S. Lambert (2005). Modelling a geo-composite cell using discrete analysis. *Computers and Geotechnics 32*, 564–577.

Bertrand, D., F. Nicot, P. Gotteland, and S. Lambert (2008). Discrete element method (DEM) numerical modeling of double-twisted hexagonal mesh. *Canadian Geotechnical Journal 45*(8), 1104–1117.

Bobet, A., A. Fakhimi, S. Johnson, F. Tonon, and M. Yeung (2009). Numerical models in discontinuous media: Review of advances for rock mechanics applications. *Journal of Geotechnical and Geoenvironmental Engineering, ASCE 135*(11), 1547–1561.

Bolton, M. (1986). The strength and dilatancy of sands. *Géotechnique 33*(1), 65–78.

Bolton, M. D., Y. Nakata, and Y. P. Cheng (2008). Micro- and macro-mechanical behaviour of DEM crushable materials. *Géotechnique 58*(6), 471–480.

Bowman, E., K. Soga, and W. Drummond (2001). Particle shape characterization using Fourier descriptor analysis. *Géotechnique 51*(6), 545–554.

Brilliantov, N. V., F. Spahn, J.-M. Hertzsch, and T. Pöschel (2996). Model for collisions in granular gases. *Physical Review E 53*(5), 5382–5392.

Britto, A. and M. Gunn (1987). *Critical State Soil Mechanics via Finite Elements*. Ellis Horwood.

Brooks, M. (2009). To the nth dimension:2D vistas of flatland. *New Scientist 203*(2723), 33.

Bruel, P., M. Benz, R. Gourves, and G. Saussine (2009). Penetration test modelling in a coarse granular medium. In M. Nakagawa and S. Luding (Eds.), *Proceedings of the 6th International Conference on Micromechanics of Granular Media Golden, Colorado, 13-17 July 2009*, pp. 173–176. AIP Conference Proceedings.

Burden, R. L. and J. Faires (1997). *Numerical Analysis*. Brooks Cole Publishing Company.

Butlanska, J., M. Arroyo, and A. Gens (2009). Homogeneity and symmetry in DEM models of cone penetration. In M. Nakagawa and S. Luding (Eds.), *Proceedings of the 6th International Conference on Micromechanics of Granular Media Golden, Colorado, 13-17 July 2009*, pp. 425–428. AIP.

Butlanska, J., C. O'Sullivan, M. Arroyo, and A. Gens (2010). Mapping deformation during cpt in virtual calibration chamber. In M. Jiang (Ed.), *Proceedings of the International Symposium on Geomechanics and Geotechnics: From Micro to Macro (IS-Shanghai 2010)*.

510

Calvetti, F. (2008). Discrete modelling of granular materials and geotechnical problems. *European Journal of Environmental and Civil Engineering 12*, 951–965.

Calvetti, F., C. di Prisco, and R. Nova (2004). Experimental and numerical analysis of soil-pipe interaction. *Journal of Geotechnical and Geoenvironmental Engineering 130*(12), 1292–1299.

Calvetti, F. and R. Nova (2004). Micromechanical approach to slope stability analysis. In *Degradations and Instabilities in Geomaterials*, CISM International Centre for Mechanical Sciences, Number 461, pp. 235–254. Springer.

Calvetti, F., R. Nova, and R. Castellanza (2004). Modelling the subsidence induced by degradation of abandoned mines. In *Continuous and discontinuous modelling of cohesive frictional materials*, Number 137148. Taylor and Francis Group.

Camborde, F., C. Mariotti, and F. Donze (2000). Numerical study of rock and concrete behaviour by discrete element modelling. *Computers and Geotechnics 27*, 225–247.

Cambou, B. (1999). Fundamental concepts in homogenization process. In M. Oda and K. Iwashita (Eds.), *Mechanics of Granular Materials*, pp. 35–39. A. A. Balkema.

Cambou, B., M. Chaze, and F. Dedecker (2000). Change of scale in granular materials. *European Journal of Mechanics. Vol. A (Solids) 19*, 999–1014.

Campbell, C. (2006). Granular material flows an overview. *Powder Technology 162*, 280–229.

Campbell, C. and C. Brennan (1985). Computer simulation of granular shear flow. *Journal of Fluid Mechanics 151*, 167–188.

Camusso, M. and M. Barla (2009). Microparameters calibration for loose and cemented soil when using particle methods. *International Journal of Geomechanics 9*(5), 217–229.

Carolan, A. (2005). Discrete element modelling of direct shear box tests. Master's thesis, Imperial College London.

Casagrande, A. and N. Carrillo (1944). Shear failure of anisotropic materials. *Proceedings of the Boston Society Of Civil Engineering 31*, 74–87.

Cavarretta, I. (2009). *The influence of particle characteristics on the engineering behaviour of sands.* Ph.D. thesis.

Cavarretta, I., M. Coop, and C. O'Sullivan (2010). The influence of particle characteristics on the behaviour of coarse grained soils. *Géotechnique 60*(5), 413–424.

Chang, C. (1993). Micromechanical modeling of deformation and failure for granulates with frictional contacts. *Mechanics of Materials 16*(1-2), 13–24.

Chang, C. and P.-Y. Hicher (2010). An elasto-plastic model for granular materials with microstructural consideration. *International Journal of Solids and Structures 42*(14), 4258–4277.

Chang, C. and C. Liao (1990). Constitutive relations for particulate medium with the effect of particle rotation. *International Journal of Solids and Structures 26*(4), 437–453.

Chang, C. and A. Misra (1990). Application of uniform strain theory to heterogeneous granular solids. *ASCE Journal of Engineering Mechanics 116*(10), 2310–2328.

Chang, C., S. S. Sundaram, and A. Misra (1989). Initial moduli of particulate mass with frictional contacts. *International Journal for Numerical and Analytical Methods in Geomechanics 13*, 626–644.

Chang, C. and Z.-Y. Yin (2010). Micromechanical modeling for inherent anisotropy in granular materials. *ASCE Journal of Engineering Mechanics 136*(7), 830–839.

Chen, Y.-C. and H. Hung (1991). Evolution of shear modulus and fabric during shear deformation. *Soils and Foundations 31*(4), 148–160.

Chen, Y.-C. and I. Ishibashi (1990). Dynamic shear modulus and evolution of fabric of granular materials. *Soils and Foundations 30*(3), 1–10.

Chen, Y.-C., I. Ishibashi, and J. Jenkins (1988). Dynamic shear modulus and fabric: Part I, depositional and induced anisotropy. *Géotechnique 38*(1), 25–32.

Cheng, Y., M. Bolton, and Y. Nakata (2004). Crushing and plastic deformation of soils simulated using DEM. *Géotechnique 54*(2), 131–141.

Cheng, Y., Y. Nakata, and M. Bolton (2003). Discrete element simulation of crushable soil. *Géotechnique 53*(7), 633–641.

Cheung, G. (2010). *Micromechanics of sand production in oil wells.* Ph.D. thesis, Imperial College London.

Cheung, G. and C. O'Sullivan (2008). Effective simulation of flexible lateral boundaries in two- and three-dimensional DEM simulations. *Particuology 6*, 483–500.

Chhabra, R., L. Agarwal, and N. Sinha (1999). Drag on non-spherical particles : An evaluation of available methods. *Powder Technology 101*, 288–295.

Cho, G., J. Dodds, and J. Santamarina (2006). Particle shape effects on packing density, stiffness, and strength: natural and crushed sands. *ASCE Journal of Geotechnical and Geoenvironmental Engineering 132*(5), 591–601.

Cho, N., C. Martin, and D. C. Sego (2007). A clumped particle model for rock. *International Journal for Rock Mechanics and Mining Sciences 44*, 997–1010.

Chopra, A. (1995). *Dynamics of Structures.* New Jersey: Prentice Hall.

Christoffersen, J., M. Mehrabadi, and S. Nemat-Nasser (1981). A micro-mechanical description of granular material behaviour. *ASME Journal of Applied Mechanics 48*, 339–344.

Cleary, P. (2000). DEM simulation of industrial particle flows: case studies of dragline excavators, mixing in tumblers and centrifugal mills. *Powder Technology 109*(1-2), 83–104.

Cleary, P. (2007). Granular flows: fundamentals and applications. In Y. Aste, T. Di Matteo, and A. Tordesillas (Eds.), *Granular and Complex Materials*, pp. 141–168. World Scientific.

Cleary, P. (2008). The effect of particle shape on simple shear flows. *Powder Technology 179*, 144–163.

Colella, P. (2004). Defining software requirements for scientific computing. *DARPAHPCS* Presentation.

Collop, A., G. McDowell, and Y. Lee (2007). On the use of discrete element modelling to simulate the viscoelastic deformation of an idealized asphalt mixture. *Geomechanics and Geoengineering: An International Journal 2*(2), 77–86.

Cook, B. K., M. Y. Lee, A. A. DiGiovanni, D. R. Bronowski, E. D. Perkins, and J. R. Williams (2004). Discrete element modeling applied to laboratory simulation of near-wellbore mechanics. *International Journal of Geomechanics 4*(1), 1927.

Cook, B. K., D. R. Noble, and J. R. Williams (2004). A direct simulation method for particle-fluid systems. *Engineering Computations: International Journal for Computer-Aided Engineering 21*, 151–168.

Coop, M. (2009). Soil properties: Strength. In *Lecture Notes: A Short Course on Foundations*. Imperial College London.

Coop, M. and I. Lee (1993). The behaviour of granular soils at elevated stresses. In *Predictive Soil Mechanics Proceedings of C.P. Wroth Memorial Symposium*, pp. 186–198. Thomas Telford, London.

Coop, M., K. Sorensen, T. Bodas Freitas, and G. Georgoutos (2004). Particle breakage during shearing of a carbonate sand. *Géotechnique 54*(3), 157–163.

Cowin, S. C. (1978). Microstructural continuum models for granular materials. In *US-Japan seminar on continuum mechanical and statistical approaches in the mechanics of granular materials, Sendai, Japan*, pp. 162–170.

Craig, R. (2007). *Craig's Soil Mechanics* (7th ed.). Spon.

Cresswell, A. W. and W. Powrie (2004). Triaxial tests on an unbonded locked sand. *Géotechnique 54*(2), 107–115.

Cui, L. (2006). *Developing a Virtual Test Environment for Granular Materials Using Discrete Element Modelling*. Ph.D. thesis, University College Dublin.

Cui, L. and C. O'Sullivan (2003). Analysis of a triangulation based approach for specimen generation for discrete element simulations. *Granular matter 5*, 135 – 145.

Cui, L. and C. O'Sullivan (2005). Development of a mixed boundary environment for axi-symmetric DEM analyses. In *Powders and Grains 2005: Proceedings of the 5th International Conference on Micromechanics of Granular Media*, pp. 301–305. A.A. Balkema.

Cui, L. and C. O'Sullivan (2006). Exploring the macro- and micro-scale response of an idealised granular material in the direct shear apparatus. *Géotechnique 56*, 455 – 468.

Cui, L., C. O'Sullivan, and S. O'Neil (2007). An analysis of the triaxial apparatus using a mixed boundary three-dimensional discrete element model. *Gotechnique 57*(10), 831–844.

Cundall, P. (1987). Distinct element models of rock and soil structure. In E. Brown (Ed.), *Analytical and Computational Methods in Engineering Rock Mechanics*. Allen and Unwin.

Cundall, P. (1988a). Computer simulations of dense sphere assemblies. In M. Satake and J. Jenkins (Eds.), *Micromechanics of Granular Materials*, pp. 113–123. Elsevier.

Cundall, P. (1988b). Formulation of a three-dimensional distinct element model part I. a scheme to detect and represent contacts in a system composed of many polyhedral blocks. *International Journal of Rock Mechanics and Mining Sciences and Geomechanics Abstracts 25*(3), 107–116.

Cundall, P. (2001). A discontinuous future for numerical modelling in geomechanics? *Geotechnical Engineering Proceedings of the Institution of Civil Engineers 149*(1), 41–47.

Cundall, P., A. Drescher, and O. Strack (1982). IUTAM conference on deformation and failure of granular materials. In *Numerical experiments on granular assemblies: measurements and observations*, pp. 355–370.

Cundall, P. and O. Strack (1978). *The distinct element methods as a tool for research in granular media, Part I, Report to NSF.*

Cundall, P. and O. Strack (1979a). A discrete numerical model for granular assemblies. *Géotechnique 29*(1), 47–65.

Cundall, P. and O. Strack (1979b). *The distinct element methods as a tool for research in granular media, Part II, Report to NSF.*

Cundall, P. A. and R. D. Hart (1993). Numerical modeling of discontinua,. In A. Hudson (Ed.), *Comprehensive Rock Engineering*, pp. 35–39. Pergamon Press.

Curray, J. (1956). Analysis of two-dimensional orientation data. *Journal of Geology 64*, 117–131.

Curtis, J. and B. van Wachem (2004). Modeling particle-laden flows: A research outlook. *American Institute of Chemical Engineers Journal 50*, 2638 – 2645.

Dantu, P. (1957). Contribution a l'etude mechanique et geometrique des milieux pulverulents. In *Proceedings of Fourth International Conference of Soil Mechanics and Foundation Engineering, London*, pp. 144–148.

Dantu, P. (1968). Etude statistique des forces intergranulaires dans un milieu pulverulent. *Géotechnique 18*(1), 50 –55.

Das, N., P. Giordano, D. Barrot, S. Mandayam, A. K. Ashmawy, and B. Sukumaran (2008). Discrete element modeling and shape characterization of realistic granular shapes. In *Proceedings of the Eighteenth (2008) International Offshore and Polar Engineering Conference*, pp. 525–533.

Daubechies, I. (1992). *Ten Lectures on Wavelets*. Society for Industrial and Applied Mathematics.

de Josselin de Jong, G. and A. Verrujit (1969). Etude photoelastique d'un empilement de disques. *Cah. Grpe fr. Etud. Rheol. 2*, 73–86.

Dean, E. T. R. (2005). Patterns, fabric, anisotropy and soil elastoplasticity. *International Journal of Plasticity 21*, 513–571.

Dedecker, F., M. Chaze, P. Dubujet, and B. Cambou (2000). Specific features of strain in granular materials. *Mechanics of Cohesive-Frictional Materials 5*, 173–193.

Delaney, G., S. Inagaki, and T. Aste (2007). Fine tuning DEM simulations to perform virtual experiments with three dimensional granular packings. In Y. Aste, T. Di Matteo, and A. Tordesillas (Eds.), *Granular and Complex Materials*, pp. 141–168. World Scientific.

Deluzarche, R. and B. Cambou (2006). Discrete numerical modelling of rockfill dams. *International Journal for Numerical and Analytical Methods in Geomechanics 30*(11), 1075–1096.

DEMSolutions (2009). Edem http://www.dem-solutions.com/ accessed july 2009.

Desrues, J., G. Viggiani, and P. Bsuelle (2006). *Advances in X-ray Tomography for Geomaterials*. John Wiley.

Di Benedetto, H., F. Tatsuoka, D. Lo Presti, C. Sauzeat, and H. Geoffroy (2005). Time effects on the behaviour of geomaterials. In H. Di Benedetto, T. Doanh, H. Geoffroy, and C. Sauzeat (Eds.), *Deformation Characteristics of Geomaterials Recent Investigations and Prospects*. A. A. Balkema.

Di Felice, R. (1994). The voidage function for fluid-particle interaction systems. *International Journal of Multiphase Flow 20*(1), 153–159.

Di Renzo, A. and F. P. Di Maio (2004). Comparison of contact-force models for the simulation of collisions in DEM-based granular flow codes. *Chemical Engineering Science 59*, 525–541.

Dobson, C., G. Sisias, C. Langton, R. Phillips, and M. Fagan (2006). Three dimensional stereolithography models of cancellous bone structures from μCT data: testing and validation of finite element results. *Journal of Engineering in Medicine 220*(3), 481–484.

Doolin, D. (2002). *Mathematical structure and numerical accuracy of discontinuous deformation analysis*. Ph.d. thesis, University of California, Berkeley.

Drescher, A. and G. de Josselin de Jong (1972). Photoelastic verification of a mechanical model for the flow of a granular material. *Journal of the Mechanics and Physics of Solids 20*(5), 337–340.

Duran, J. (2000). *Sands, powders, and grains : An introduction to the physics of granular materials*. New York: Springer.

Duran, O., N. P. Kruyt, and S. Luding (2010). Analysis of three-dimensional micro-mechanical strain formulations for granular materials: evaluation of accuracy. *International Journal of Solids and Structures 47*, 251–260.

Duttine, A. and F. Tatsuoka (2009). Viscous properties of granular materials having different particle shapes in direct shear. *Soils and Foundations 49*(5), 777–796.

El Shamy, U. and F. Aydin (2008). Multiscale modeling of flood-induced piping in river levees. *ASCE Journal of Geotechnical and Geoenvironmental Engineering 134*(9), 1385–1398.

El Shamy, U. and T. Gröger (2008). Micromechanical aspects of the shear strength of wet granular soils. *International Journal for Numerical and Analytical Methods in Geomechanics 32*(14), 1763–1790.

El Shamy, U. and M. Zeghal (2005). Coupled continuum-discrete model for saturated granular soils. *Journal of Engineering Mechanics (ASCE) April 2005*, 413–426.

Ergun, S. (1952). Fluid flow through packed columns. *Chemical Engineering Progress 48*(2), 89–94.

Fakhimi, A., F. Carvalho, T. Ishida, and J. Labuze (2002). Simulation of failure around a circular opening in rock. *International Journal of Rock Mechanics and Mining Sciences 39*, 507–515.

Fakhimi, A., J. Riedel, and J. F. Labuz (2006). Shear banding in sandstone: Physical and numerical studies. *International Journal of Geomechanics 6*(3), 185–194.

Favier, J., M. Abbaspour-Fard, M. Kremmer, and A. Raji (1999). Shape representation of axi-symmetrical, non-spherical particles in discrete element simulation using multi-element model particles. *Engineering Computations 16*(4), 467–480.

Favier, J., D. Curry, and R. LaRoche (2010). Calibration of DEM material models to approximate bulk particle characteristics. In G. Meesters, C. Hauser-Vollrath, and T. Pfeiffer (Eds.), *Proceedings of the 6th World Congress on Particle Technology (WCPT6) (CD ROM)*, Nuremberg, Germany.

Favier, J. F., M. H. Abbaspour-Fard, and M. Kremmer (2001). Modeling nonspherical particles using multisphere discrete elements. *ASCE Journal of Engineering Mechanics 127*(10), 969–1074.

Feng, Y., K. Han, and D. Owen (2003). Filling domains with disks: An advancing front approach. *International journal for numerical methods in engineering 56*(5), 699–731.

Feng, Y. T., K. Han, and D. Owen (2007). Coupled lattice boltzman method and discrete element modelling of particle transport in turbulent fluid flows: Computational issues. *International Journal for Numerical Methods in Engineering 72*, 1111–1134.

Ferrez, J.-A. (2001). *Dynamic Triangulations for Efficient 3-D simulation of Granular Materials*. Ph. D. thesis, Ecole Polytechnique Federal de Lausanne.

Field, W. (1963). Towards the statical definition of a granular mass. In *Proceedings of 4th Australia and New Zealand Conference on Soil Mechanics*, pp. 143–148.

Fonseca, J., C. O'Sullivan, and M. Coop (2009). Image segmentation techniques for granular materials. In M. Nakagawa and S. Luding (Eds.), *Proceedings of the 6th International Conference on Micromechanics of Granular Media Golden, Colorado, 13-17 July 2009*, pp. 223–226. AIP Conference Proceedings.

Fonseca, J., C. O'Sullivan, and M. Coop (2010). Quantitative description of grain contacts in a locked sand. In K. Alshibli and A. H. Reed (Eds.), *Applications of X-Ray Microtomography to Geomaterials: GeoX 2010*, pp. 17–25. Wiley.

Frost, J. and D.-J. Jang (2000). Evolution of sand microstructure during shear. *ASCE Journal of Geotechnical and Geoenvironmental Engineering 116*, 116–130.

Fung, Y. (1977). *A First Course in Continuum Mechanics*. New Jersey: Prentice-Hall.

520

Gabrielia, F., S. Cola, and F. Calvetti (2009). Use of an up-scaled DEM model for analysing the behaviour of a shallow foundation on a model slope. *Geomechanics and Geoengineering 4*(2), 109–122.

Garcia, X., J.-P. Latham, J. Xiang, and J. Harrison (2009). A clustered overlapping sphere algorithm to represent real particles in discrete element modelling. *Géotechnique 59*(9), 779–784.

Garcia-Rojo, R., F. Alonso-Marroquin, and H. J. Herrmann (2005). Characterisation of the material response in granular ratcheting. *Physical Review E 72*(4), 041302.

Gaspar, N. and M. Koenders (2001). Micromechanic formulation of macroscopic structures in a granular medium. *ASCE Journal of Engineering Mechanics 127*(10), 987–993.

Gere, J. and S. P. Timoshenko (1991). *Mechanics of materials* (3rd ed.). Chapman and Hall.

Gidaspow, D. (1994). *Multiphase flow and fluidization.* Academic Press, San Francisco.

Gili, J. A. and E. E. Alonso (2002). Microstructural deformation mechanisms of unsaturated granular soils. *International Journal for Numerical and Analytical Methods in Geomechanics 26*(5), 433–468.

Goddard, J. (1990). Nonlinear elasticity and pressure-dependent wave speeds in granular media. *Proceedings of the Royal Society of London A 430*, 105–131.

Goddard, J. (2001). Delaunay triangulation of granular media. In N. Bicanic (Ed.), *Proceedings of ICADD-4, Fourth International Conference on Discontinuous Deformation.* University of Glasgow.

Golchert, D., R. Moreno, M. Ghadiri, and J. Litster (2004). Effect of granule morphology on breakage behaviour during compression. *Powder Technology 143-144*, 84–96.

Goldberg, D. (1991). What every computer scientist should know about floating-point arithmetic. *Computing Surveys, Association for Computing Machinery 23*(1), 5–48.

Golub, G. and C. F. Van Loan (1983). *Matrix-computations*. Oxford: North Oxford Academic.

Greenwood, J. A., H. Minshall, and D. Tabor (1961). Hysteresis losses in rolling and sliding friction. *Proceedings of Royal Society of London A 259*, 480–507.

Grimmett (1999). *Percolation* (Second ed.). Springer-Verlag.

Gudehus, G. (1997). Attractors, percolation thresholds and phase limits of granular soils. In R. Behringer and J. Jenkins (Eds.), *Powders and Grains 1997*, pp. 169–183.

Gutierrez, G. A. (2007). *Influence of late cementation on the behaviour of reservoir sands*. Ph.D. thesis, Imperial College London.

Häggströ, O. and R. Meester (1996). Nearest neighbour and hard sphere models in continuum percolation. *Random Structures and Algorithms 9*, 295–315.

Hall, S. A., D. MuirWood, E. Ibraim, and G. Viggiani (2010). Localised deformation patterning in 2D granular materials revealed by digital image correlation. *Granular Matter 12*(1), 1–14.

Han, K., Y. Feng, and D. Owen (2005). Sphere packing with a geometric based compression algorithm. *Powder Technology 155*, 33–41.

Harkness, J. (2009). Potential particles for the modelling of interlocking media in three dimensions. *International Journal for Numerical Methods in Engineering 80*, 1573–1594.

Hasan, A. and K. Alshibli (2010). Experimental assessment of 3D particle to particle interaction within sheared sand using synchontron microtomography. *Géotechnique 60*(5), 369–380.

Hattab, M. and J.-M. Fleureau (2010). Experimental study of kaolin particle orientation mechanism. *Géotechnique 60*(5), 323–332.

Head, K. (1994). *Manual of Soil Laboratory Testing: Vol. 2: Compressibility, Shear Strength and Permeability* (2nd ed.), Volume 2. London: Pentech Press.

Hentz, S., D. L., and F. V. Donze (2004). Identification and validation of a discrete element model for concrete. *Journal of Engineering Mechanics 130*(7-8), 709–719.

Hill, R. (1956). The mechanics of quasi-static plastic deformation in metals. In G. Batchelor and R. M. Davies (Eds.), *The G.I. Taylor 70th Anniversary Volume*, pp. 7–31. Cambridge University Press.

Hogue, C. (1998). Shape representation and contact detection for discrete element simulations of arbitrary geometries. *Engineering Computations 15*(10), 374–389.

Holst, J. M. F., J. M. Rotter, J. Y. Ooi, and G. H. Rong (1999). Numerical modelling of silo filling, II: Discrete element analysis. *Journal of Engineering Mechanics 125*(1), 104–110.

Holtzman, R., D. B. Silin, and T. W. Patzek (2008). Mechanical properties of granular materials: A variational approach to grain-scale simulations. *International Journal for Numerical and Analytical Methods in Geomechanics 33*(3), 391–404.

Hoomans, B., J. Kuipers, W. Briels, and W. Van Swaaij (1996). Discrete particle simulation of bubble and slug formation in a two-dimensional gas-fluidized bed: A hard-sphere approach. *Chem. Engng. Sci. 51*, 99–108.

Hori, M. (1999). Two theories of micromechanics. In M. Oda and K. Iwashita (Eds.), *Mechanics of Granular Materials*, pp. 39–45. A. A. Balkema.

Horne, M. (1965). The behaviour of an assembly of rotund, rigid, cohesionless particles. II. *Proceeedings of the Royal Society of London. Series A, Mathematical and Physical Sciences 286*(1404), 79–97.

Horner, D. A., A. R. Carrillo, J. Peters, and J. West (1998). High resolution soil vehicle interaction modelling. *Mechanics Based Design of Structures and Machines 26*(3), 305–318.

Horner, D. A., J. Peters, and A. R. Carrillo (2001). Large scale discrete element modeling of vehicle-soil interaction. *ASCE Journal of Engineering Mechanics 127*(10), 1027–1032.

Hossein, Z., B. Indraratna, F. Darve, and P. Thakur (2007). DEM analysis of angular ballast breakage under cyclic loading. *Geomechanics and Geoengineering: An International Journal 2*(3), 175–181.

Houlsby, G. (2009). Potential particles: a method for modelling non-circular particles in DEM. *Computers and Geotechnics 36*(6), 953–959.

Hu, M., C. O'Sullivan, R. R. Jardine, and M. Jiang (2010). Stress-induced anisotropy in sand under cyclic loading potential particles: a method for modelling non-circular particles in DEM. In M. Jiang (Ed.), *Proceedings of the International Symposium on Geomechanics and Geotechnics: From Micro to Macro (IS-Shanghai 2010)*.

Huang, A.-B. and M. Y. Ma (1994). An analytical study of cone penetration tests in granular material. *Canadian Geotechnical Journal 31*(10), 91–103.

Huang, H. and E. Detournay (2008). Intrinsic length scales in tool-rock interaction. *International Journal of Geomechanics 8*, 39–44.

Hyodo, M., H. Murata, and Y. Nakata (2006). *Geomechanics and Geotechnics of Particulate Media Proceedings of the International Symposium on Geomechanics and Geotechnics of Partic-*

ulate Media, Ube, Japan, 12-14 September 2006. Taylor and Francis.

Ibraim, E., J. Lanier, D. Muir Wood, and G. Viggiani (2010). Strain path controlled shear tests on an analogue granular material. *Géotechnique 60*(7), 545–559.

Ibraim, E., D. Muir Wood, K. Maeda, and H. Hirabayashi (2006). Fibre-reinforced granular soils behaviour: numerical approach. In M. Hyodo, H. Murata, and Y. Nakata (Eds.), *Proceedings of the International Symposium on Geomechanics and Geotechnics of Particulate Media*, Ube, Yamaguchi, Japan, pp. 443–448. Taylor and Francis.

Ishibashi, I., Y.-C. Chen, and J. T. Jenkins (1988). Dynamic shear modulus and fabric: Part II, stress reversal. *Géotechnique 38*(1), 33–37.

Itasca (1998). *UDEC (Universal Distinct Element Code) 3.00.* Minneapolis Minnesota.

Itasca (2004). *PFC2D 3.10 Particle Flow Code in Two Dimensions, Theory and Background volume* (Third ed.). Minneapolis, Minnesota.

Itasca (2008). *PFC3D 4.0 Particle Flow Code in Three Dimensions, Theory and Implementation Volume.* Minneapolis, Minnesota.

Iwashita, K. and M. Oda (1998). Rolling resistance at contacts in simulation of shear band development by DEM. *ASCE Journal of Engineering Mechanics 124*, 285–292.

Iwashita, K. and M. Oda (2000). Micro-deformation mechanism of shear banding process based on modified distinct element method. *Powder Technology 109*, 192–205.

Jean, M. (2004). The non-smooth contact dynamics method. *Computer Methods in Applied Mechanics and Engineering 177*(3-4), 235–257.

Jefferies, M. and K. Been (2006). *Soil Liquefaction: A Critical State Approach.* Taylor and Francis.

Jeffries, M., K. Been, and J. Hachey (1990). The influence of scale on the constitutive behaviour of sand. In *Proceedings of 43rd Canadian Geotechnical Conference, Quebec,* Volume 1, pp. 263–273.

Jenck, O., D. Dias, and R. Kastner (2009). Discrete element modelling of a granular platform supported by piles in soft soil-validation on a small scale model test and comparison to a numerical analysis in a continuum. *Computers and Geotechnics 36,* 917–927.

Jenkins, J. (1978). Gravitational equilibrium of a model of granular media. In *US-Japan seminar on continuum mechanical and statistical approaches in the mechanics of granular materials,Sendai, Japan,* pp. 181–188.

Jensen, R. and D. Preece (2001). Modeling sand production with darcy-flow coupled with discrete elements. In Desai (Ed.), *10th International Conference on Computer methods and Advances in Geomechanics,* Tucson, Arizona, pp. 819–822.

Jerier, J.-F., D. Imbault, F.-V. Donze, and P. Doremus (2009). A geometric algorithm based on tetrahedral meshes to generate a dense polydisperse sphere packing. *Granular Matter 11,* 43–52.

Jeyisanker, K. and M. Gunaratne (2009). Analysis of water seepage in a pavement system using the particulate approach. *Computers and Geotechnics 36,* 641–654.

Jiang, M., J. Konrad, and S. Leroueil (2003). An efficient technique for generating homogeneous specimens for DEM studies. *Computers and Geotechnics 30,* 579–597.

Jiang, M., S. Leroueil, and J. Konrad (2004). Insight into shear strength functions of unsaturated granulates by DEM analyses. *Computers and Geotechnics 31,* 473–489.

Jiang, M., S. Leroueil, H. Zhu, and J.-M. Konrad (2009). Two-dimensional discrete element theory for rough particles. *International Journal of Geomechanics 9*(1), 20–33.

Jiang, M., F. Liu, and M. Bolton (2010). *Geomechanics and Geotechnics: From Micro to Macro, Proceedings of IS-Shanghai 2010: International Symposium on Geomechanics and Geotechnics: From Micro to Macro,*. CRC Press.

Jiang, M., H.-S. Yu, and D. Harris (2005). A novel discrete model for granular material incorporating rolling resistance. *Computers and Geotechnics 32*, 340–357.

Jiang, M. J., H.-S. Yu, and D. Harris (2006). Discrete element modelling of deep penetration in granular soils. *International Journal for Numerical and Analytical Methods in Geomechanics 30*(4), 335–361.

Jing, L. and O. Stephansson (2007). *Fundamentals of Discrete Element Methods for Rock Engineering: Theory and Applications.* Elsevier.

Jodrey, W. and E. Tory (1985). Computer simulation of close random packing of equal spheres. *Physical Review A 32*(4), 2347–2351.

Joe, B. (2003). Geompack++. ZCS Inc., Calgary, AB, Canada,(www.allstream.net/~bjoe/index.htm).

Johnson, K. (1985). *Contact Mechanics.* Cambridge University Press.

Johnson, S. M., J. R. Williams, and B. K. Cook (2008). Quaternion-based rigid body rotation integration algorithms for use in particle methods. *International Journal for Numerical Methods in Engineering 74*, 1303–1313.

Kafui, K. and C. Thornton (2000). Numerical simulations of impact breakage of a spherical crystalline agglomerate. *Powder Technology 109*, 113–132.

Kafui, K., C. Thornton, and M. Adams (2002). Discrete particle-continuum fluid modelling of gas–solid fluidised beds. *Chemical Engineering Science 57*, 2395–2410.

Kanatani, K. (1984). Distribution of directional data and fabric tensors. *International Journal of Engineering Science 22*(2), 149–164.

Kassner, M. E., S. Nemat-Nasser, Z. Suo, G. Bao, J. C. Barbour, L. C. Brinson, H. Espinosa, H. Gao, S. Granick, P. Gumbsch, K.-S. Kim, W. Knauss, L. Kubin, J. Langer, B. C. Larson, L. Mahadevan, A. Majumdar, S. Torquato, and F. van Swol (2005). New directions in mechanics. *Mechanics of Materials 37*(2-3), 231–259.

Kawaguchi, T., T. Tanaka, and Y. Tsuji (1998). Numerical simulation of two-dimensional fluidized beds using the discrete element method (comparison between the two- and three-dimensional models). *Powder Technology 96*(2), 129–138.

Ke, T. and J. Bray (1995). Modeling of particulate media using discontinuous deformation analysis. *Journal of Engineering Mechanics 121*(11), 1234–1243.

Ketcham, R. and W. Carlson (2001). Acquisition, optimization and interpretation of x-ray computed tomographic imagery: applications to the geosciences. *Computers and Geosciences 27*, 381–400.

Kinlock, H. and C. O'Sullivan (2007). A micro-mechanical study of the influence of penetrometer geometry on failure mechanisms in granular soils. In H. Olson (Ed.), *Geo-Denver 2007: New Peaks in Geotechnics, Proceedings ASCE Geo Congress 2007*. ASCE.

Kishino, Y. (1989). Investigation of quasi-static behavior of granular materials with a new simulation method. *Proceedings of JSCE (in Japanese)* (406), 97–196.

Kishino, Y. (1999). *Physical and mathematical backgrounds*, pp. 149–155. A. A. Balkema.

Kitamura, R. (1981a). Analysis of deformation mechanism of particulate material at particle scale. *Soils and Foundations 21* (2), 85–98.

Kitamura, R. (1981b). A mechanical model of particulate material based on stochastic process. *Soils and Foundations 21* (2), 63–72.

Kloss, C. and C. Goniva (2010). Liggghts-a new open source discrete element simulation software. In *Proceedings of the fifth International Conference on Discrete Element Methods, London, UK, 25-26 August.*

Kogge, P. (2008). Exascale computing study: Technology challenges in achieving exascale systems. Technical report, DARPA IPTO.

Kozicki, J. and F. Donz (2008). A new open-source software developed for numerical simulations using discrete modeling methods. *Computer Methods in Applied Mechanics and Engineering 197* (1), 4429–4443.

Kramer, S. L. (1996). *Geotechnical earthquake engineering.* Prentice Hall.

Kremmer, M. and J. Favier (2000). Calculating rotational motion in discrete element modelling of arbitrary shaped model objects. *Engineering Computations 17* (6), 703–714.

Krumbein, W. and L. Sloss (1963). *Stratigraphy and Sedimentation.* San Francisco: W.H. Freeman.

Kruyt, N. (2003). Statics and kinematics of discrete cosserat-type granular materials. *International Journal of Solids and Structures 40* (3), 511–534.

Kruyt, N. and L. Rothenburg (1996). Micromechanical definition of the strain tensor for granular materials. *ASME Journal of Applied Mechanics 118* (11), 706–711.

Kruyt, N. and L. Rothenburg (2009). Plasticity of granular materials: a structural mechanics view. In M. Nakagawa and S. Luding (Eds.), *Proceedings of the 6th International Conference on Micromechanics of Granular Media Golden, Colorado, 13-17 July 2009*, pp. 1073–1076. AIP Conference Proceedings.

Kuhn, M. (2006). Oval and ovalplot: Programs for analyzing dense particle assemblies with the discrete element method: http://faculty.up.edu/kuhn/oval/doc/oval$_0$618.*pdf*.

Kuhn, M. R. (1995). Flexible boundary for three-dimensional DEM particle assemblies. *Engineering Computations 12*(2), 175–183.

Kuhn, M. R. (1997). Deformation measures for granular materials. In *Mechanics of Deformation and Flow of Particulate Materials, Proceedings of ASCE Engineering Mechanics Division*, pp. 91–104.

Kuhn, M. R. (1999). Structured deformation in granular materials. *Mechanics of Materials 31*, 407–429.

Kuhn, M. R. (2003a). Heterogeneity and patterning in the quasi-static behavior of granular materials. *Granular Matter 4*, 155–166.

Kuhn, M. R. (2003b). A smooth convex three-dimensional particle for the discrete element method. *Journal of Engineering Mechanics 129*(5), 539–547.

Kuhn, M. R. and K. Bagi (2009). Specimen size effect in discrete element simulations of granular assemblies. *Journal of Engineering Mechanics 135*(6), 485–492.

Kuhn, M. R. and J. K. Mitchell (1992). The modeling of soil creep with the discrete element method. *Engineering Computations 9*(2), 277–287.

Kulatilake, P., B. Malama, and W. J. (2001). Physical and particle flow modeling of jointed rock block behavior under uniaxial loading. *International Journal of Rock Mechanics and Mining Sciences 38*, 641–657.

Kuo, C. and J. Frost (1996). Uniformity evaluation of cohesionless specimens using digital image analysis. *ASCE Journal of Geotechnical and Geoenvironmental Engineering 122*(5), 390–396.

Kuo, C., J. D. Frost, and J. A. Chameau (1998). Image analysis determination of stereology based fabric tensors. *Géotechnique 48*(4), 515–525.

Kuwano, R. and R. J. Jardine (2002). On the applicability of cross-anisotropic elasticity to granular materials at very small strains. *Géotechnique 52*(10), 727–749.

Kwok, C.-Y. and M. Bolton (2010). DEM simulations of thermally activated creep in soils. *Géotechnique 60*(6), 425–434.

Labra, C. and E. Oñate (2008). High-density sphere packing for discrete element method simulations. *Communications in Numerical Methods in Engineering*, 837–849.

Ladd, R. (1978). Preparing test specimens using undercompaction. *Geotechnical Testing Journal 1*(1), 16–23.

Lade, P. V. and J. M. Duncan (2003). Elastoplastic stressstrain theory for cohesionless soil. *Journal of Geotechnical and Geoenvironmental Engineering ASCE 101*(6), 1037–1053.

Lambe, T. and R. Whitman (1979). *Soil Mechanics*. Wiley.

Lanier, J. and M. Jean (2000). Experiments and numerical simulations with 2D disks assembly. *Powder Technology 109*, 206–221.

Latzel, M., S. Luding, and H. Herrmann (2000). Macroscopic material properties from quasi-static, microscopic simulations of a two-dimensional shear cell. *Granular matter 2*, 123–135.

Li, L. and R. Holt (2002). Particle scale reservoir mechanics. *Oil and Gas Science and Technology 57*(5), 525–538.

Li, S. and W. K. Liu (2000). Numerical simulations of strain localization in inelastic solids using mesh-free methods. *International Jounral for Numerical Methods in Engineering 48*, 1285–1309.

Li, X. and X. Li (2009). Micro-macro quantification of the internal structure of granular materials. *Journal of Engineering Mechanics 135*(7), 641–656.

Li, X. and H. Yu (2009). Influence of loading direction on the behavior of anisotropic granular materials. *International Journal of Engineering Science 47*, 12841296.

Li, X. and H. Yu (2010). Numerical investigation of granular material behaviour under rotational shear. *Géotechnique 60*(5), 381–394.

Li, X., H. Yu, and X. Li (2009). Macromicro relations in granular mechanics. *International Journal of Solids and Structures 46*(25-26), 4331–4341.

Li, Y., Y. Xu, and C. Thornton (2005). A comparison of discrete element simulations and experiments for sandpiles composed of spherical particles. *Powder Technology 160*, 219–228.

Liao, C., T. Chang, D. Young, and C. Chang (1997). Stress-strain relationships for granular materials based on the hypothesis of best fit. *International Journal of Solids and Structures 34*, 4087–4100.

Likos, W. J. (2009). Pore-scale model for water retention in unsaturated sand. In M. Nakagawa and S. Luding (Eds.), *Powders and Grains 2009, Proceedings of the 6th International Conference on Micromechanics of Granular Media*.

Lin, X. and T.-T. Ng (1997). A three-dimensional discrete element model using arrays of ellipsoids. *Géotechnique 47*(2), 319–329.

Liu, W., S. Jun, and Y. Zhang (1995). Reproducing kernel particle methods. *International Journal of Numerical Methods in Fluids 20*, 1081–1106.

Lobo-Guerrero, S. and L. E. Vallejo (2005). DEM analysis of crushing around driven piles in granular materials. *Géotechnique 55*(8), 617–623.

Lobo-Guerrero, S., L. E. Vallejo, and L. F. Vesga (2006). Visualization of crushing evolution in granular materials under compression using DEM. *International Journal of Geomechanics 6*(3), 195–200.

Lu, M. and G. McDowell (2010). Discrete element modelling of railway ballast under cyclic loading. *Géotechnique 60*(6), 459–468.

Lu, M. and G. R. McDowell (2006). Discrete element modelling of ballast abrasion. *Géotechnique 56*(9), 651–655.

Lu, M. and G. R. McDowell (2008). Discrete element modelling of railway ballast under triaxial conditions. *Geomechanics and Geoengineering: An International Journal 3*(4), 257–270.

Lu, N., M. Anderson, W. J. Likos, and G. W. Mustoe (2008). A discrete element model for kaolinite aggregate formation during sedimentation. *International Journal for Numerical and Analytical Methods in Geomechanics 32*(8), 965–980.

Luding, S. and P. Cleary (2009). DEM 2007 editorial. *Granular Matter 11*, 267–268.

Luding, S., M. Latzel, W. Volk, S. Diebels, and H. Herrmann (2001). From discrete element simulations to a continuum model. *Computer methods in applied mechanics and engineering 191*, 21–28.

MacLaughlin, M. (1997). *Discontinuous Deformation Analysis of the Kinematics of Landslides*. Ph.D. thesis, University of California, Berkeley.

MacLaughlin, M., N. Sitar, D. Doolin, and T. Abbot (2001). Investigation of slope-stability kinematics using discontinuous deformation analysis. *International Journal of Rock Mechanics and Mining Sciences 38*, 753–762.

Maeda, K. (2009). Critical state-based geo-micromechanics on granular flow. In M. Nakagawa and S. Luding (Eds.), *Powders and Grains 2009: Proceedings of the 6th International Conference on Micromechanics of Granular Media Golden, Colorado, 13-17 July 2009*, pp. 17–24. AIP Conference Proceedings.

Mahmood, Z. and K. Iwashita (2010). Influence of inherent anisotropy on mechanical behaviour of granular materials based on DEM simulations. *International Journal of Numerical and Analytical Methods in Geomechanics 34*(8), 795–819.

Mark, A. and B. van Wachem (2008). Derivation and validation of a novel implicit second-order accurate immersed boundary method. *Journal of Computational Physics 227*(13), 6660–6680.

Marketos, G. and M. Bolton (2010). Flat boundaries and their effect on sand testing. *International Journal for Numerical and Analytical Methods in Geomechanics 34*, 821–837.

Mas-Ivars, D., D. O. Potyondy, M. Pierce, and P. Cundall (2008). The smooth-joint contact model. In *8th World Congress on Computational Mechanics (WCCM8) 5th European Congress on Computational Methods in Applied Sciences and Engineering (ECCOMAS 2008) Venice, Italy*.

Masad, E. and B. Muhunthan (2000). Three-dimensional characterization and simulation of anisotropic soil fabric. *Journal of Geotechnical and Geoenvironmental Engineering, ASCE 126*(3), 199–207.

Mase, G. and G. Mase (1999). *Continuum mechanics for engineers* (Second ed.). CRC Press.

Masson, S. and J. Martinez (2001). Micromechanical analysis of the shear behavior of a granular material. *Journal of engineering mechanics 127*(10), 1007–1016.

Matsushima, T. and K. Konagai (2001). Grain-shape effect on peak strength of granular materials. In Desai et al. (Ed.), *Computer Methods and Advances in Geomechanics*. A. A. Balkema.

McDowell, G. and M. Bolton (2001). Micro mechanics of elastic soil. *Soils and Foundations 41*(6), 147–152.

McDowell, G. R. and O. Harireche (2002). Discrete element modelling of soil particle fracture. *Géotechnique 52*(2), 131–135.

Melis Maynar, M. J. and L. E. Medina Rodríguez (2005). Discrete numerical model for analysis of earth pressure balance tunnel excavation. *ASCE Journal of Geotechnical and Geoenvironmental Engineering 131*(10), 1234–1242.

Mindlin, R. (1949). Compliance of elastic bodies in contact. *ASME Journal of Applied Mechanics 16*, 259–269.

Mindlin, R. and H. Deresiewicz (1953). Elastic spheres in contact under varying oblique forces. *ASME Journal of Applied Mechanics 20*, 327–344.

Mirghasemi, A., L. Rothenburg, and E. Matyas (1997). Numerical simulations of assemblies of two-dimensional polygon-shaped particles and effects of confining pressure on shear strength. *Soils and Foundations 37*(3), 43–52.

Mirghasemi, A., L. Rothenburg, and E. Matyas (2002). Influence of particle shape on engineering properties of assemblies of two-dimensional polygonal-shaped particles. *Géotechnique 52*(3), 209–217.

Mitchell, J. (1993). *Fundamentals of Soil Behavior* (Second ed.). New York: John Wiley and Sons.

Mitchell, J. and K. Soga (2005). *Fundamentals of Soil Behavior* (Third ed.). New York: John Wiley and Sons.

Morris, J. P. and P. W. Cleary (2009). Advances in discrete element methods for geomechanics. *Geomechanics and Geoengineering, 4*, 1.

Morsi, S. A. and A. J. Alexander (1972). An investigation of particle trajectories in two-phase flow systems. *Journal of Fluid Mechanics 55*(2), 193–208.

Mueth, D. M., H. M. Jaeger, and S. R. Nagel (1997). Force distribution in a granular medium. *Physical Review E 57*(3), 3164–3169.

Muir Wood, D. (2007). The magic of sands-the 20th Bjerrum Lecture presented in Oslo, November 2005. *Canadian Geotechnical Journal 44*, 1329–1350.

Muir Wood, D., K. Maeda, and E. Nukudani (2010). Modelling mechanical consequences of erosion. *Géotechnique 60*(6), 447–458.

Mulhaus, H.-B., H. Sakaguchi, L. Moresi, and M. Graham (2001). Particle in cell and discrete element models for granular materials. In Desi et al. (Ed.), *Computer Methods and Advances in Geomechanics*. A. A. Balkema.

Mulilis, J. P., H. B. Seed, C. K. Chan, J. K. Mitchell, and K. Arulanandan (1977). Effects of sample preparation on sand liquefaction. *Journal of the Geotechnical Engineering Division, ASCE 103*(GT2), 91–108.

Munjiza, A. (2004). *The combined finite-discrete element methods.* John Wiley.

Munjiza, A. and K. R. F. Andrews (1998). Nbs contact detection algorithm for bodies of similar size. *International Journal for Numerical Methods in Engineering 43*(1), 131–149.

Munjiza, A., J. Latham, and N. John (2001). Transient motion of irregular 3D discrete elements. In N. Bicanic (Ed.), *Proceedings of the Fourth International Conference on Analysis of Discontinuous Deformation*, pp. 111–133. University of Glasgow.

Munjiza, A., J. Latham, and N. John (2003). 3D dynamics of discrete element systems comprising irregular discrete elementsintegration solution for finite rotations in 3D. *International Journal for Numerical Methods in Engineering 56*, 35–55.

Murakami, A., H. Sakaguchi, and T. Hasegawa (1997). Dislocation, vortex and couple stress in the formation of shear bands under trap door problems. *Soils and Foundations 37*(1), 123–135.

536

Mustoe, G. and M. Miyata (2001). Material flow analyses of non-circular shaped granular media using discrete element methods. *SCE Journal of Engineering Mechanics 127*(10), 1017–1026.

Nakagawa, M. and S. Luding (2009). *Proceedings of the 6th International Conference on Micromechanics of Granular Media Golden, Colorado, 13-17 July 2009.* AIP Conference Proceedings.

Nemat-Nasser, S. (1999). Averaging theorems in finite deformation plasticity. *Mechanics of Materials 31*, 493–523.

Nemat-Nasser, S. and M. Hori (1999). *Micromechanics: Overall Properties of Heterogeneous Materials.* North Holland.

Nemat-Nasser, S. and Y. Tobita (1982). Influence of fabric on liquefaction and densification potential of cohesionless sand. *Mechanics of Materials 1*(1), 43–62.

Nezami, E., Y. M. A. Hashash, D. Zhao, and J.Ghaboussi (2007). Simulation of front end loader bucket-soil interaction using discrete element method. *International Journal for Numerical and Analytical Methods in Geomechanics 31*(9), 1147–1162.

Ng, T.-T. (2001). Fabric evolution of ellipsoidal arrays with different particle shapes. *Journal of Engineering Mechanics 127*(10), 994–999.

Ng, T.-T. (2004a). Macro- and micro-behaviors of granular materials under different sample preparation methods and stress paths. *International Journal of Solids and Structures 41*, 5871–5884.

Ng, T.-T. (2004b). Shear strength of assemblies of ellipsoidal particles. *Géotechnique 54*(10), 659–670.

Ng, T.-T. (2006). Input parameters of discrete element methods. *ASCE Journal of Engineering Mechanics 132*(7), 723–729.

Ng, T.-T. (2009a). Discrete element simulations of the critical state of a granular material. *International Journal of Geomechanics 9*(5), 209–216.

Ng, T.-T. (2009b). Particle shape effect on macro- and micro-behaviors of monodisperse ellipsoids. *International Journal for Numerical and Analytical Methods in Geomechanics 33*, 511–527.

Ng, T.-T. and R. Dobry (1994). Numerical simulations of monotonic and cyclic loading of granular soil. *Journal of Geotechnical Engineering ASCE 120*(2), 388–403.

Ng, T.-T. and R. Dobry (1995). Contact detection algorithms for three-dimensional ellipsoids in discrete element method. *International Journal for Numerical and Analytical Methods in Geomechanics 19*(9), 653–659.

Oda, M. (1972). Initial fabrics and their relations to the mechanical properties of granular materials. *Soils and Foundations 12*(4), 45–63.

Oda, M. (1977). Co-ordindation number and its relation to shear strength of granular material. *Soils and Foundations 17*(2), 29–42.

Oda, M. (1982). Fabric tensor for discontinuous geological materials. *Soils and Foundations 22*, 96–108.

Oda, M. (1999a). Fabric tensor and its geometrical meaning. In M. Oda and K. Iwashita (Eds.), *Mechanics of Granular Materials*, pp. 19–27. A. A. Balkema.

Oda, M. (1999b). Measurement of fabric elements. In M. Oda and K. Iwashita (Eds.), *Mechanics of Granular Materials*, pp. 226–230. A. A. Balkema.

Oda, M. and K. Iwashita (Eds.) (1999). *Introduction to mechanics of granular materials*. A. A. Balkema.

Oda, M. and H. Kazama (1998). Micro-structure of shear band and its relation to the mechanism of dilatancy and failure of granular soils. *Géotechnique 48*(4), 465–481.

Oda, M. and J. Konishi (1974). Microscopic deformation mechanism of granular material in simple shear. *Soils and Foundations 14*(4), 25–38.

Oda, M., J. Konishi, and S. Nemat-Nasser (1980). Some experimentally based fundamental results on the mechanical behaviour of granular materials. *Géotechnique 30*(4), 479–495.

Oda, M., J. Konishi, and S. Nemat-Nasser (1982). Experimental micromechanical evaluation of strength of granular materials: effect of particle rolling. *Mechanics of Materials 1*, 267–283.

Oda, M., S. Nemat-Nasser, and J. Konishi (1985). Stress-induced anisotropy in granular masses. *Soils and Foundations 25*(3), 85–97.

Ogawa, S., S. Mitsui, and O. Takemure (1974). Influence of the intermediate principal stress on mechanical properties of a sand. In *Proceedings of 29th Annual Meeting of JSCE*, pp. 49–50.

O'Hern, C. S., L. E. Silbert, A. J. Liu, and S. R. Nagel (2003). Jamming at zero temperature and zero applied stress: The epitome of disorder. *Physical Review E 68*(1), 011306.

Okabe, A., B. Boots, K. Sugihara, and S. N. Chiu (2000). *Spatial Tessellations Concepts and Applications of Voronoi Diagrams* (Second ed.). New York: Wiley.

Ooi, J., S. Sture, and M. Hopkins (2001). Editorial, special issue: The statics and flow of dense granular systems, advances in the mechanics of granular materials. *ASCE Journal of Engineering Mechanics 127*(10), 970.

O'Sullivan, C. (2002). *The Application of Discrete Element Modelling to Finite Deformation Problems in Geomechanics*. Ph.D. thesis, University of California, Berkeley.

O'Sullivan, C. and J. Bray (2002). Relating the response of idealized analogue particles and real sands. In *Numerical Modelling in Micromechanics via Particle Methods*, pp. 157–164. A. A. Balkema.

O'Sullivan, C., J. Bray, and M. Riemer (2004). An examination of the response of regularly packed specimens of spherical particles using physical tests and discrete element simulations. *ASCE Journal of Engineering Mechanics 130*(10), 1140–1150.

O'Sullivan, C. and J. D. Bray (2003a). A modified shear spring formulation for discontinuous deformation analysis of particulate media. *ASCE Journal of Engineering Mechanics 129*(7), 830–834.

O'Sullivan, C. and J. D. Bray (2003b). Selecting a suitable time-step for discrete element simulations that use the central difference time integration approach. *Engineering Computations 21*(2/3/4), 278–303.

O'Sullivan, C., J. D. Bray, and S. Li (2003). A new approach for calculating strain for particulate media. *International Journal for Numerical and Analytical Methods in Geomechanics 27*(10), 859–877.

O'Sullivan, C., J. D. Bray, and M. F. Riemer (2002). The influence of particle shape and surface friction variability on macroscopic frictional strength of rod-shaped particulate media. *Journal of Engineering Mechanics 128*(11), 1182–1192.

O'Sullivan, C. and L. Cui (2009a). Fabric evolution in granular materials subject to drained, strain controlled cyclic loading. In M. Nakagawa and S. Luding (Eds.), *Powders and Grains 2009, Proceedings of the 6th International Conference on Micromechanics of Granular Media*, pp. 285–288.

O'Sullivan, C. and L. Cui (2009b). Micromechanics of granular material response during load reversals: Combined DEM and experimental study. *Powder Technology 193*, 289–302.

O'Sullivan, C., L. Cui, and S. O'Neil (2008). Discrete element analysis of the response of granular materials during cyclic loading. *Soils and Foundations 48*, 511 – 530.

540

Painter, B., S. Tennakoon, and R. Behringer (1998). Collisions and fluctuations for granular materials. In H. Herrmann, J.-P. Hovi, and S. Luding (Eds.), *Physics of Dry Granular Media*, Volume 350 of *E: Applied Sciences*. NATO ASI Series: Kluwer Academic.

Papadimitriou, A. and G. Bouckovalas (2002). Plasticity model for sand under small and large cyclic strains: a multiaxial formulation. *Soil Dynamics and Earthquake Engineering 22*, 191–204.

Park, J.-W. and J.-J. Song (2009). Numerical simulation of a direct shear test on a rock joint using a bonded-particle model. *International Journal of Rock Mechanics and Mining Sciences 46*(8), 1315–1328.

Patankar, S. (1980). *Numerical Heat Transfer and Fluid Flow*. Taylor and Francis.

Peron, H., J. Delenne, L. Laloui, and M. El Youssoufi (2009). Discrete element modelling of drying shrinkage and cracking of soils". *Computers and Geotechnics 36*, 61–69.

Peters, J. F., K. R., and R. S. Maier (2009). A hierarchical search algorithm for discrete element method of greatly differing particle sizes. *Engineering Computations 26*(6), 621–634.

Plimpton, S. (1995). Fast parallel algorithms for short-range molecular dynamics. *Journal of Computational Physics 117*, 1–19.

Plimpton, S., P. Crozier, and A. Thompson (2010). LAMMPS molecular dynamics simulator. http://lammps.sandia.gov/index.html, accessed Dec 2010.

Pöschel, T. and T. Schwager (2005). *Computational Granular Dynamics*. Springer-Verlag.

Potapov, A. V., M. L. Hunt, and C. S. Campbell (2001). Liquid-solid flows using smoothed particle hydrodynamics and the discrete element method. *Powder Technology 116*, 204–213.

Potts, D. (2003). Numerical analysis: a virtual dream or practical reality? *Géotechnique 53*(6), 525–572.

Potyondy, D. O. (2007). Simulating stress corrosion with a bonded-particle model for rock. *International Journal of Rock Mechanics and Mining Sciences 44*, 677–691.

Potyondy, D. O. and P. A. Cundall (2004). A bonded-particle model for rock. *International Journal of Rock Mechanics and Mining Sciences 41*(8), 1329–1364.

Pournin, L., M. Weber, M. Tsukahara, J.-A. Ferrez, T. M. Ramaioli, and T. M. Liebling (2005). Three-dimensional distinct element simulation of spherocylinder crystallization. *Granular Matter 7*(2–3), 119–126.

Powrie, W., R. M. Harkness, X. Zhang, and D. I. Bush (2002). Deformation and failure modes of drystone retaining walls. *Géotechnique 52*(6), 435–446.

Powrie, W., Q. Ni, R. M. Harkness, and X. Zhang (2005). Numerical modelling of plane strain tests on sands using a particulate approach. *Géotechnique 55*(4), 297–306.

Radjai, F. (2009). Force and fabric states in granular media. In M. Nakagawa and S. Luding (Eds.), *Proceedings of the 6th International Conference on Micromechanics of Granular Media Golden, Colorado, 13-17 July 2009*, pp. 35–42. AIP Conference Proceedings.

Radjai, F., M. Jean, J.-J. Moreau, and S. Roux (1996). Force distributions in dense two-dimensional granular systems. *Physical review letters 77*(2), 274–277.

Rapaport, D. (2004). *The art of molecular dynamics simulation* (Second ed.). Cambridge University Press.

Rapaport, D. C. (2009). The event-driven approach to n-body simulation. *Progress of Theoretical Physics 178*, 5–14.

542

Rechemacher, A., S. Abedi, and O. Chupin (2010). Evolution of force chains in shear bands in sands. *Géotechnique 60*(5), 343–351.

Remond, S., J. L. Gallias, and A. Mizrahi (2008). Simulation of the packing of granular mixtures of non-convex particles and voids characterization. *Granular Matter 10*, 157–170.

Reuters, T. (2010). Isi web of knowledge database. http://www.isiwebofknowledge.com/, Accessed May 2010.

Reynolds, O. (1885). On the dilatancy of media composed of rigid particles in contact, with experimental illustrations. *Philosophical Magazine 20*, 469–481.

Richefeu, V., M. S. El Youssoufi, R. Peyroux, and R. F. (2008). A model of capillary cohesion for numerical simulations of 3D polydisperse granular media. *International Journal for Numerical and Analytical Methods in Geomechanics 32*(11), 1365–1383.

Roberts, N. (2008). *A distinct element study of how cell action affects soil stress measurements made in sand.* MSc thesis, Imperial College London.

Robertson, D. (2000). *Computer simulations of crushable aggregates.* Ph.D. thesis, Cambridge University.

Rothenburg, L. (1980). *Micromechanics of idealized granular systems.* Ph.D. thesis, Carleton University, Ottawa.

Rothenburg, L. and R. Bathurst (1989). Analytical study of induced anisotropy in idealized granular materials. *Géotechnique 39*(4), 601–614.

Rothenburg, L. and R. Bathurst (1991). Numerical simulation of idealized granular assemblies with plane elliptical particles. *Computers and Geotechnics 11*, 315–329.

Rothenburg, L. and N. Kruyt (2004). Critical state and evolution of coordination number in simulated granular materials. *International Journal of Solids and Structures 41*, 5763–5774.

Rowe, P. (1962). The stress-dilatancy relation for static equilibrium of an assembly of particles in contact. *Proceedings of the Royal Society of London. Series A, Mathematical and Physical Sciences 269*(1339), 500–527.

Russell, A. R., D. Muir Wood, and M. Kikumoto (2009). Crushing of particles in idealised granular assemblies. *Journal of the Mechanics and Physics of Solids 57*, 1293–1313.

Sack, R. (1989). *Matrix Structural Analysis*. Illinois: Waveland Press.

Salot, C., P. Gotteland, and P. Villard (2009). Influence of relative density on granular materials behavior: DEM simulations of triaxial tests. *Granular Matter 11*(4), 221–236.

Satake, M. (1976). Constitution of mechanics of granular materials through graph representation. *Theoretical and Applied Mechanics* (26), 257–266.

Satake, M. (1982). Fabric tensor in granular materials. In P. Vermeer and H. Luger (Eds.), *Proceedings of IUTAM Symposium on Deformation and Failure of Granular Materials*, pp. 63–68. A.A. Balkema.

Satake, M. (1992). A discrete-mechanical approach to granular materials. *International Journal of Engineering Science 30*(10), 1525–1533.

Satake, M. (1999). *Graph representation*, Chapter 1.2.2, pp. 5–8. A.A. Balkema.

Scheidegger, A. (1965). On the statistics of the orientation of bedding planes, grain axes and similar sedimentological data. *US Geological Survey Professional Paper 525*, 164–167.

Schofield, A. and C. Wroth (1968). *Critical State Soil Mechanics*. McGraw Hill.

Scholts, L., P.-Y. Hicher, F. Nicot, B. Chareyre, and D. F. (2009). On the capillary stress tensor in wet granular materials. *International Journal for Numerical and Analytical Methods in Geomechanics 10*(33), 1289–1313.

Schöpfer, M. P., S. Abe, C. Childs, and J. J. Walsh (2009). The impact of porosity and crack density on the elasticity, strength and friction of cohesive granular materials: Insights from DEM modelling. *International Journal of Rock Mechanics and Mining Sciences 46*(2), 250–261.

Schöpfer, M. P. J., C. Childs, and J. J. Walsh (2007). Two-dimensional distinct element modeling of the structure and growth of normal faults in multilayer sequences: 2. impact of confining pressure and strength contrast on fault zone geometry and growth. *Journal of Geophysical Research 112*(B10).

Serrano, A. and J. Rodriguez-Ortiz (1973). A contribution to the mechanics of heterogeneous granular material. In *Proceedings of Symposium of Plasticity and Soil Mechanics, Cambridge*.

Shafipour, R. and A. Soroush (2008). Fluid coupled-DEM modelling of undrained behavior of granular media. *Computers and Geotechnics 35*, 673–685.

Shames, I. and F. Cozzarelli (1997). *Elastic and Inelastic Stress Analysis.* Taylor and Francis.

Sheng, Y., C. Lawrence, B. Briscoe, and C. Thornton (2004). Numerical studies of uniaxial powder compaction process by 3D DEM. *Engineering Computations 21*(2/3/4), 304–317.

Shewchuk, J. (1996). Triangle:engineering a 2D quality mesh generator and delaunay triangulator. In *First Workshop on Applied Computational Geometry*, pp. 124–133. ACM.

Shewchuk, J. (2002). Triangle: A two-dimensional quality mesh generator and delaunay triangulator, (version 1.4). (University of California Berkeley, 2002) (http://www-2.cs.cmu.edu/quake/triangle.html).

Shewchuk, J. R. (1999). Lecture notes on delaunay mesh generation, UC Berkeley Department of Computer Science.

Shi, G. (1988). *Discontinuous deformation analysis, a new numerical model for the statics and dynamics of block systems.* Ph. D. thesis, University of California, Berkeley.

Shi, G.-H. (1996). Manifold method. In M. Salami and D. Banks (Eds.), *Discontinuous Deformation Analysis (DDA) and Simulations of Discontinuous Media.* TSI Press, New Mexico.

Shibuya, S. and D. Hight (1987). A bounding surface for granular materials. *Soils and Foundations 27*(4), 123–156.

Silvani, C., T. Dsoyer, and S. Bonelli (2009). Discrete modelling of time-dependent rockfill behaviour. *International Journal for Numerical and Analytical Methods in Geomechanics 33*(5), 665–685.

Simpson, B. and F. Tatsuoka (2008). Geotechnics: the next 60 years. *Géotechnique 58*(5), 357–368.

Sitar, N., M. MacLaughlin, and D. M. Doolin (2005). Influence of kinematics on landslide mobility and failure mode. *ASCE Journal of Geotechnical and Geoenvironmental Engineering 131*(6), 716–728.

Sitharam, T., S. Dinesh, and N. Shimizu (2002). Micromechanical modelling of monotonic drained and undrained shear behaviour of granular media using three-dimensional DEM. *International Journal for Numerical and Analytical Methods in Geomechanics 26*, 1167–1189.

Sitharam, T., J. Vinod, and B. Ravishankar (2008). Evaluation of undrained response from drained triaxial shear tests: DEM simulations and experiments. *Géotechnique 58*(7), 605–608.

Sitharam, T. G., J. S. Vinod, and B. V. Ravishankar (2009). Postliquefaction undrained monotonic behaviour of sands: experiments and DEM simulations. *Géotechnique 59*(9), 739–749.

Skinner, A. (1969). A note on the influence of interparticle friction on the shearing strength of a random assembly of spherical particles. *Géotechnique 19*(1), 150–157.

Skylaris, C.-K., P. D. Haynes, A. A. Mostofi, and M. C. Payne (2005). Introducing ONETEP: Linear-scaling density functional simulations on parallel computers. *Journal of Chemical Physics 122*(8), 084119.

Sloane, N. J. A. (1998). The sphere packing problem. In *Proceedings of Internat. Congress Math. Berlin, Documenta Mathematica Extra Volume ICM*, Volume 3, pp. 387–396.

Stoyan, D. (1973). Models of random systems of non-intersecting spheres. In *Prague Stochastics 98, JCMF*, pp. 543–547.

Summersgill, F. (2009). The use of particulate discrete element modelling to assess the vulnerability of soils to suffusion. Master's thesis, Imperial College London.

Sutmann, G. (2002). Classical molecular dynamics. In J. Grotendorst, D. Marx, and A. Muramatsu (Eds.), *Quantum Simulations of Complex Many-Body Systems: From Theory to Algorithms, Lecture Notes*, Volume 10, pp. 211–254. John von Neumann Institute for Computing, Jülich, NIC Series.

Suzuki, K., J. P. Bardet, M. Oda, K. Iwashita, Y. Tsuji, T. Tanaka, and T. Kawaguchi (2007). Simulation of upward seepage flow in a single column of spheres using discrete-element method with fluid-particle interaction. *ASCE Journal of Geotechnical and Geoenvironmental Engineering 133*(1), 104–109.

Tamura, T. and Y. Yamada (1996). A rigid-plastic analysis for granular materials. *Soils and Foundations 36*(3), 113–121.

Taylor, D. (1948). *Fundamentals of Soil Mechanics*. John Wiley.

Terzaghi, K. (1936). The shearing resistance of saturated soils. In *Proceedings of the First International Conference on Soil Mechanics*, Volume 1, pp. 54–56.

Thomas, P. (1997). *Discontinuous deformation analysis of particulate media.* Ph.D. thesis, University of California, Berkeley.

Thomas, P. and J. Bray (1999). Capturing nonspherical shape of granular media with disk clusters. *Journal of Geotechnical and Geoenvironmental Engineering 125*(3), 169–178.

Thornton, C. (1979). The conditions for failure of a face-centered cubic array of uniform rigid spheres. *Géotechnique 29*(4), 441–459.

Thornton, C. (1997a). Coefficient of restitution for collinear collisions of elastic-perfectly plastic spheres. *Journal of Applied Mechanics 64*, 383–386.

Thornton, C. (1997b). Force transmission in granular media. *KONA Powder and Particle 15*, 81–90.

Thornton, C. (1999). Interparticle relationships between forces and displacements. In M. Oda and K. Iwashita (Eds.), *Mechanics of Granular Materials*, Chapter 3.4, pp. 207–217. A.A. Balkema.

Thornton, C. (2000). Numerical simulations of deviatoric shear deformation of granular media. *Géotechnique 50*(1), 43–53.

Thornton, C. (2009). Preface to special issue on discrete element methods. *Powder Technology 193*(3), 215.

Thornton, C. and S. Antony (2000). Quasi-static shear deformation of a soft particle system. *Powder Technology 109*, 179–191.

Thornton, C. and S. J. Antony (1998). Quasi-static deformation of particulate media. *Philosophical Transactions of the Royal Society of London A 356*, 2763–2782.

Thornton, C. and L. Liu (2000). DEM simulations of uniaxial compression and decompression. In D. Kolymbas and W. Fellin (Eds.), *Proceedings of International Workshop on compaction of soils, granulates and powders*, pp. 251–261. A.A. Balkema.

Thornton, C. and L. Liu (2004). How do agglomerates break? *Powder Technology 143-144*, 110–116.

Thornton, C. and Z. Ning (1998). A theoretical model for the stick/bounce behaviour of adhesive, elastic-plastic spheres. *Powder Technology 99*, 154–162.

Thornton, C. and K. Yin (1991). Impact of elastic spheres with and without adhesion. *Powder Technology 65*, 153–166.

Thornton, C. and L. Zhang (2010). On the evolution of stress and microstructure during general 3D deviatoric straining of granular media. *Géotechnique 60*, 333–341.

Ting, J. M. (1993). A robust algorithm for ellipse-based discrete element modelling of granular materials. *Computers and Geotechnics 13*, 175–186.

Tordesillas, A. (2007). Stranger than friction: force chain buckling and its implications for constitutive modelling. In Y. Aste, T. Di Matteo, and A. Tordesillas (Eds.), *Granular and Complex Materials*, pp. 95–110. World Scientific.

Tordesillas, A. (2009). Themomicromechanics of dense granular materials. In M. Nakagawa and S. Luding (Eds.), *Proceedings of the 6th International Conference on Micromechanics of Granular Media Golden, Colorado, 13-17 July 2009*, pp. 51–54. AIP Conference Proceedings.

Tordesillas, A. and M. Muthuswamy (2009). On the modeling of confined buckling of force chains. *Journal of the Mechanics and Physics of Solids 57*(4), 706–727.

Tovey, N. K. (1980). A digital computer technique for orientation analysis of micrographs of soil fabric. *Journal of Microscopy 120*, 303–315.

Trussell, R. R. and M. Chang (1999). Review of flow through porous media as applied to head loss in water filters. *Journal of Environmental Engineering 125*(11), 998–1006.

Tsuchikura, T. and M. Satake (1998). Statistical measure tensors and their application to computer simulation analysis of biaxial compression test. In H. Murakami and J. Luco (Eds.), *Engineering Mechanics: A Force for the 21st Century*, pp. 1732–1735. A. A. Balkema.

Tsuji, Y., T. Kawaguchi, and T. Tanaka (1993). Discrete particle simulation of two-dimensional fluidized bed. *Powder Technology 77*, 79–87.

Tsunekawa, H. and K. Iwashita (2001). Numerical simulation of triaxial test using DEM. In Y. Kishino (Ed.), *Powders and Grains 01*, pp. 177–182.

Utili, S. and R. Nova (2008). DEM analysis of bonded granular geomaterials. *International Journal for Numerical and Analytical Methods in Geomechanics 32*(17), 1997–2031.

Utter, B. and R. P. Behringer (2008). Experimental measures of affine and nonaffine deformation in granular shear. *Physical Review Letters 100*(20), 208–302.

Vaid, Y. P. and S. Sivathayalan (2000). Fundamental factors affecting liquefaction susceptibility of sands. *Canadian Geotechnical Journal 37*(3), 592606.

Van Wachem, B. and S. Sasic (2008). Derivation, simulation and validation of a cohesive flow cfd model. *Fluid Mechanics and Transport Phenomena 54*(1), 9–19.

Vaughan, P. (1993). Engineering behaviour of weak rocks: some answers and some questions. In *Proceedings of Int. Symp. Geotechnical Engineering of Hard Soils -Soft Rocks, Athens*.

Viggiani, G. and S. Hall (2008). Full-field measurements, a new tool for laboratory experimental geomechanics. In S. Burns, P. W. Mayne, and J. Santamarina (Eds.), *Deformation Characteristics of Geomaterials*, pp. 3–26. IOS Press.

Voivret, C., F. Radjaï, J.-Y. Delenne, and M. S. El Youssoufi (2009, Apr). Multiscale force networks in highly polydisperse granular media. *Physical Review Letters 102*(17), 178001.

Vu-Quoc, L., X. Zhang, and O. Walton (2000). A 3-D discrete element method for dry granular flows of ellipsoidal particles. *Comput. Methods Appl. Mech. Engrg. 187*, 483–528.

Walton, O. and R. Braun (1986). Viscosity, granular-temperature and stress calculations for shearing assemblies of inelastic, frictional disks. *Journal of Rheology 30*(5), 949–980.

Wang, C., D. Tannant, and P. Lilly (2003). Numerical analysis of the stability of heavily jointed rock slopes using *pfc2d*. *International Journal of Rock Mechanics and Mining Sciences 40*, 415–424.

Wang, C.-Y., C.-C. Chuang, and J. Sheng (1996). Time integration theories for the DDA method with finite element meshes. In M. Salami and D. Banks (Eds.), *Discontinuous Deformation Analysis (DDA) and Simulations of Discontinuous Media*. TSI Press.

Wang, J., J. E. Dove, and M. S. Gutierrez (2007). Discrete-continuum analysis of shear banding in the direct shear test. *Géotechnique 57*(7), 513–526.

Wang, J. and M. Gutierrez (2010). Discrete element simulation of direct shear specimen scale effects. *Géotechnique 60*(5), 395–409.

Wang, Y. and F. Tonon (2010). Calibration of a discrete element model for intact rock up to its peak strength. *International Journal for Numerical and Analytical Methods in Geomechanics 34*(5), 447–469.

Wang, Y. H. and S. C. Leung (2008). A particulate-scale investigation of cemented sand behavior. *Canadian Geotechnical Journal 45*(1), 29–44.

Wang, Y.-H., D. Xu, and T. K. Y. J. (2008). Discrete element modeling of contact creep and aging in sand. *ASCE Journal of Geotechnical and Geoenvironmental Engineering 134*(9), 1407–1411.

Watts, D. J. (2004). *Six Degrees: The Science of a Connected Age.* Vintage Books.

Weatherley, D. (2009). Esys-particle v2.0 users guide. Technical report, Earth Systems Science Computational Centre, University of Queensland.

Weisstein, E. W. (2010). Quaternion. MathWorld–A Wolfram Web Resource. http://mathworld.wolfram.com/Quaternion.html.

Wen, C. and Y. H. Yu (1966). Mechanics of fluidization. *Chemical Engineering Progress Symposium Series 62*, 100–111.

Wilkinson, S. (2010). *The Engineering Behaviour and Microstructure of UK Mudrocks.* Ph.D. thesis, Imperial College London.

Williams, J. and N. Rege (1997). Coherent vortex structures in deforming granular materials. *Mechanics of Cohesive-Frictional Materials 2*, 223–236.

Wood, D. (1990a). *Soil Behaviour and Critical State Soil Mechanics.* A. A. Balkema.

Wood, W. (1990b). *Practical Time Stepping Schemes.* Oxford: Clarendon Press.

Woodcock, N. (1977). Specification of fabric shapes using an eigenvalue method. *The Geological Society of American Bulletin 88*, 12311236.

Wouterse, A., S. Luding, and A. P. Philipse (2009). On contact numbers in random rod packings. *Granular Matter 11*, 169–177.

Xiang, J., J. P. Latham, and A. Munjiza (2009). Virtual geoscience workbench http://sourceforge.net/projects/vgw/develop accessed July 2009.

Xu, B., A. Yu, S. Chew, and P. Zulli (2000). Numerical simulation of the gassolid flow in a bed with lateral gas blasting. *Powder Technology 109*, 13–26.

Xu, B. H. and A. B. Yu (1997). Numerical simulation of the gas-solid flow in a fluidized bed by combining discrete particle method with computational fluid dynamics. *Chemical Engineering Science 52*(16), 2785–2809.

Yimsiri, S. and K. Soga (2000). Micromechanics-based stress-strain behaviour of soils at small strains. *Géotechnique 50*(5), 559–571.

Yimsiri, S. and K. Soga (2001). Effects of soil fabric on undrained behavior of sands. In S. Prakish (Ed.), *Fourth International Conference on Recent Advances in Geotechnical Earthquake Engineering and Soil Dynamics*, San Diego, California.

Yimsiri, S. and K. Soga (2002). Application of micromechanics model to study anisotropy of soils at small strain. *Soils and Foundations 42*(5), 15–26.

Yimsiri, S. and K. Soga (2010). DEM analysis of soil fabric effects on behaviour of sand. *Géotechnique 60*(6), 483–495.

Yoon, J. (2007). Application of experimental design and optimization to PFC model calibration in uniaxial compression simulation. *International Journal of Rock Mechanics and Mining Sciences 44*, 871889.

Yu, A. (2004). Discrete element method an effective way for particle scale research of particulate matter. *Engineering Computations 21*(2/3/4), 205–214.

Yunus, Y., E. Vincens, and B. Cambou (2010). Numerical local analysis of relevant internal variables for constitutive modelling of granular materials. *International Journal for Numerical and Analytical Methods in Geomechanics 34*.

Zdravkovic, L. and J. Carter (2008). Contributions to Géotechnique 1948-2008: Constitutive and numerical modelling. *Géotechnique 58*(5), 405–412.

Zdravkovic, L. and R. Jardine (1997). Some anisotropic stiffness characteristics of a silt under general stress conditions. *Géotechnique 47*(3), 407–438.

Zeghal, M. and U. El Shamy (2004). A continuum-discrete hydromechanical analysis of granular deposit liquefaction. *International Journal for Numerical and Analytical Methods in Geomechanics 28*(14), 1361–1383.

Zeghal, M. and U. El Shamy (2008). Liquefaction of saturated loose and cemented granular soils. *Powder Technology 184*(2), 254–265.

Zhang, L. and C. Thornton (2007). A numerical examination of the direct shear test. *Géotechnique 57*(4), 343–354.

Zhou, Z., H. Zhu, A. Yu, B. Wright, and P. Zulli (2008). Discrete particle simulation of gas solid flow in a blast furnace. *Computers and Chemical Engineering 32*(8), 1760–1772.

Zhu, H. and A. Yu (2008). Preface to special edition on simulation and modeling of particulate systems. *Particuology 6*(6), 389.

Zhu, H., Z. Zhou, R. Yang, and A. Yu (2007). Discrete particle simulation of particulate systems: Theoretical developments. *Chemical Engineering Science 62*(13), 3378–3396.

Zhu, H., Z. Zhou, R. Yang, and A. Yu (2008). Discrete particle simulation of particulate systems: A review of major applications and findings. *Chemical Engineering Science 63*(23), 5728–5770.

Zhuang, X., A. Didwania, and J. Goddard (1995). Simulation of the quasi-static mechanics and scalar transport properties of ideal granular assemblages. *Journal of Computational Physics 121*, 331–346.

Zienkiewicz, O. and R. Taylor (2000a). *The Finite Element Method, Volume 1 The Basis* (Fifth ed.). Oxford: Butterworth Heinemann.

Zienkiewicz, O. and R. Taylor (2000b). *The Finite Element Method, Volume 2 Solid Mechanics* (Fifth ed.). Oxford: Butterworth Heinemann.

Index